# THE WORLD'S HIGHEST MOUNTAINS

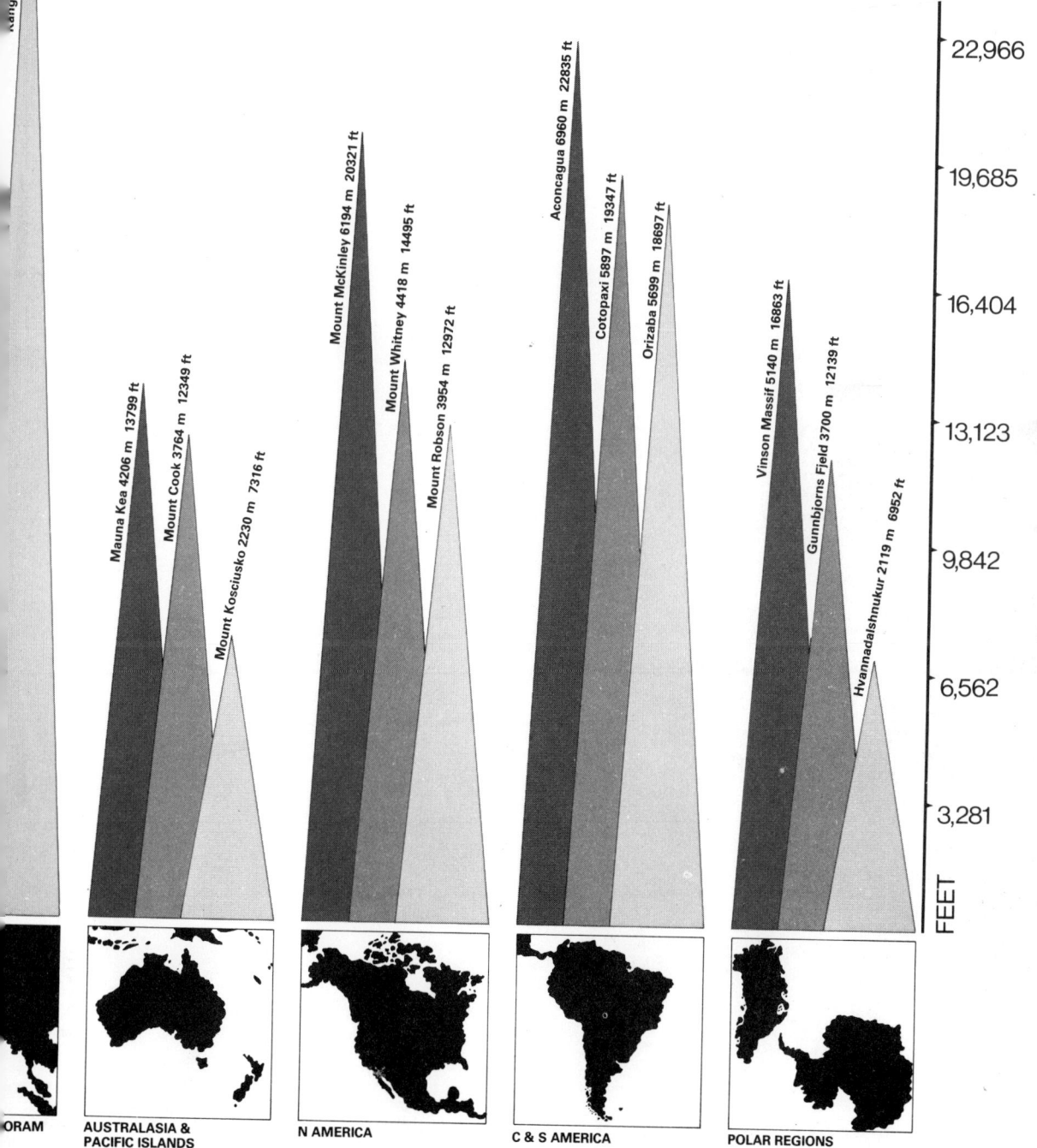

29,527

,247

22,966

19,685

16,404

13,123

9,842

6,562

3,281

FEET

Kangchenjunga 8597 m 28205 ft

Mauna Kea 4206 m 13799 ft

Mount Cook 3764 m 12349 ft

Mount Kosciusko 2230 m 7316 ft

Mount McKinley 6194 m 20321 ft

Mount Whitney 4418 m 14495 ft

Mount Robson 3954 m 12972 ft

Aconcagua 6960 m 22835 ft

Cotopaxi 5897 m 19347 ft

Orizaba 5699 m 18697 ft

Vinson Massif 5140 m 16863 ft

Gunnbjorns Field 3700 m 12139 ft

Hvannadalshnukur 2119 m 6952 ft

ORAM

AUSTRALASIA &
PACIFIC ISLANDS

N AMERICA

C & S AMERICA

POLAR REGIONS

# The Guinness book of
# MOUNTAINS
## & MOUNTAINEERING
## Facts & Feats

## by Edward Pyatt

**GUINNESS SUPERLATIVES LIMITED**
2 CECIL COURT, LONDON ROAD, ENFIELD, MIDDLESEX

**Editor: Anne Smith**

**Layout: David Roberts**

© **Edward Pyatt and Guinness Superlatives Ltd, 1980**

Published in Great Britain by Guinness Superlatives Ltd,
2 Cecil Court, London Road, Enfield, Middlesex

**British Library Cataloguing in Publication Data**
Pyatt, Edward Charles
  The Guinness book of mountains & mountaineering.
  1. Mountaineering      2. Mountains
  I. Title
  796.5′22        GV200

  ISBN 0 900424 49 4


Guinness is a registered trademark of Guinness Superlatives Ltd

Filmset, printed and bound by Hazell Watson & Viney Ltd, Aylesbury, Bucks
Colour separation: Newsele Litho Ltd, Milan

# Acknowledgements

Few books can be prepared in isolation. Usually, as here, the author must benefit from the work of others who have gone before; I acknowledge my indebtedness to many such.

In detail, I wish to thank:

Paul Sharp for the maps;
Ron Treble for the sketches;
Pat Johnson (of the Alpine Club Library) for assistance with sources;
Ann Babbage for photographic work;
Friends in the Alpine Club, and elsewhere, for rallying round and offering large selections of pictures;
Staff and managers of tourist offices, air-line offices, picture agencies, etc, whose unfailing courtesy smoothed the acquiring of still more pictures.

Finally, and most important of all, I thank my wife for her continuous encouragement throughout the task and for rounding it off for me by producing the index.

Edward Pyatt
Hampton, 1980

# Contents

# Preface

This is a book about many aspects of mountains. A series of essays on mountain topics introduce the book, followed by facts and feats relevant to individual mountains or ranges. There is a gazeteer of world mountains, with particulars and dates of first ascents of the major mountaineering problem peaks, and appendices of national mountaineering information and a calendar.

The text is divided into areas, countries or major cordilleras and these are further divided into ranges or groups. Within a range or group the mountains are treated in topographical order wherever possible (ie in the case of reasonably linear arrangement) and the direction of listing is indicated; otherwise they are arranged in descending order of height. In each case the list is cut at an appropriate proportion of the height of the highest point, but augmented by a few outstanding lower peaks.

In Alpine first ascents, and occasionally elsewhere, the names of the party are separated from those of their guides by a semicolon. In most early Alpine ascents the guides provided the technical 'know-how' for the climbing; in the Himalaya the case is not quite the same – originally the Sherpas were no more than porters, but they are increasingly assuming the role of guides as time goes on. In other ranges the exact relationship between members of the party is not always clear and is in any case usually irrelevant.

Metric measurements are used throughout (where the height immediately follows the name of a mountain, 'm' is excluded) – many famous mountaineering nations have always used them. The rest of the world is still engaged in reluctant change. True heights are a subject of endless controversy. In many cases it is difficult, or even impossible, to find out who is the real authority and what degree of uncertainty can be assigned to his measurements. Issues, such as changes in snow cover, refraction in optical methods, determination of and agreement concerning sea-level, difficulties of access and so on, complicate the situation almost indefinitely. Does your authority know better than someone else's and which is right anyway?

There are of course vastly more summits than are mentioned here. The selection, entirely due to the author, is based usually on height relative to that of the highest point of the country, area or range. Lower points of special merit are also sometimes included. In such an exercise no two people can possibly agree on the criteria and there could be endless arguments as to what should or should not be included.

The author and publishers would welcome corrections and updatings of text, different figures for heights and first ascents, and other peaks which possibly ought to be included. Only a co-operative process of this sort can produce a text entirely free from blemish. I trust that this version goes quite some way along the road to the ideal.

# An Introduction to Mountains and Mountaineering

Mountaineering parties crossing a snow-field (M Millet)

## WHAT IS A MOUNTAIN? WHAT IS A HILL?

Here is how the *Encyclopaedia Britannica* answers the question:

Landform that rises prominently above adjacent land, exhibiting a confined or narrow summit area and considerable local relief. Mountains are generally understood to be larger than hills, but the term has no standardized geographic meaning . . .

The *Oxford English Dictionary* on the other hand attempts some sort of distinction:

Hill – A natural elevation of the Earth's surface rising more or less steeply above the level of the surrounding land. Formerly the general term, including what are now called mountains; after the introduction of the latter word, gradually restricted to heights of less elevation; but the discrimination is largely a matter of local usage, and of the more or less mountainous character of the district, heights which in one locality are called mountains in another are reckoned merely as hills. A more rounded and less rugged outline is also usually connoted by the name . . .

The word first appeared about 1000 AD and figures of course in the opening passage of Langland's *Piers the Plowman* (1362) – 'In a Mayes Morwnynge on Malverne hulles Me bi-fel a ferly . . .'

Mountain – A natural elevation of the Earth's surface rising more or less abruptly from the surrounding level, and attaining an altitude which, relative to adjacent elevations, is impressive or notable . . .

The word first appeared in 1205.

The climber will be satisfied with the all-embracing definition of that famous mountaineer, Geoffrey Winthrop Young, 'earth set on earth a little higher', which brings all elevations of the earth's surface within his purview. Another mountaineer/writer, Frank Smythe, included both hills and mountains within one integrated philosophy – 'Those who love hills need go no higher than the summit of Holmbury Hill (North Downs, England, 261 m). They will discover there that height counts for little and it is the hill that matters. Low hills teach us that height, be it a mere two or three hundred feet, is something precious.'

Specialised scientific studies of the last few decades have refined the definitions somewhat. One geographer (W. G. Moore) adds formation to the definition – 'Mountains may be formed by earth movements, by erosion (the more resistant rocks being left while the softer rocks surrounding them are worn away), or by volcanic action'. Another (R. Peattie) says that a mountain must have 'conspicuity', and we know what he means, but he ventures on to dangerous literary terrain in claiming also 'individuality' for them. (Any apparent individuality resides in the observer, mountains have no human personalities.) For him a mountain 'is a conspicuous elevation of small summit area'. A plateau, he adds,' is a similar elevation of larger summit area with at least one sheer side'.

Finally a geologist (B. Booth) has suggested an even more restricted definition – 'a mountain is not just an elevated area, it is also underlain by strongly folded and faulted rocks, or indeed by large igneous masses such as granite batholiths or piles of ancient tuffs and lavas. If on the other hand the area is underlain by sedimentary rocks which are more or less horizontal or gently dipping, they should not be called mountains'.

Our definition in this book will be as all-embracing as possible – the mountaineer's definition.

## Plate Tectonics and the Origins of Mountain Ranges

The revolutionary theory of Continental Drift was first put forward in 1912 by Alfred Wegener. Based on similarities in outline between the various land masses, on coordination between geological and biological features of lands now far apart, and on evidence of marked climatic differences over the ages, he proposed that all had once been one large land mass (Pangea), which had subsequently broken up and the pieces drifted apart to their present positions. While this explained many hitherto puzzling features of the present-day world, the problems remained of why the primeval land mass should have divided and what then drove the pieces over the surface. For some decades the theory foundered on these objections. Wegener died during the course of an expedition to Greenland in 1930.

During the 1950s two new areas of research threw fresh light on these problems. The detection of the direction of magnetisation in various rocks indicated an apparent wandering in the position of the earth's magnetic pole; since this in fact must more or less coincide with the rotational axis, the observed phenomena clearly indicated a one-time movement of the continents. The second area lay in undersea exploration, which charted the existence, hitherto completely unsuspected, of huge mountain systems beneath the oceans – 65000 km long, 1000 km wide and 2 to 3 km high, covering one-third of the total surface area of the world. A central rift valley with a depth of up to 3 km was found running along the centre of the chain. It was subsequently discovered that material (driven by some internal mechanism, possibly by

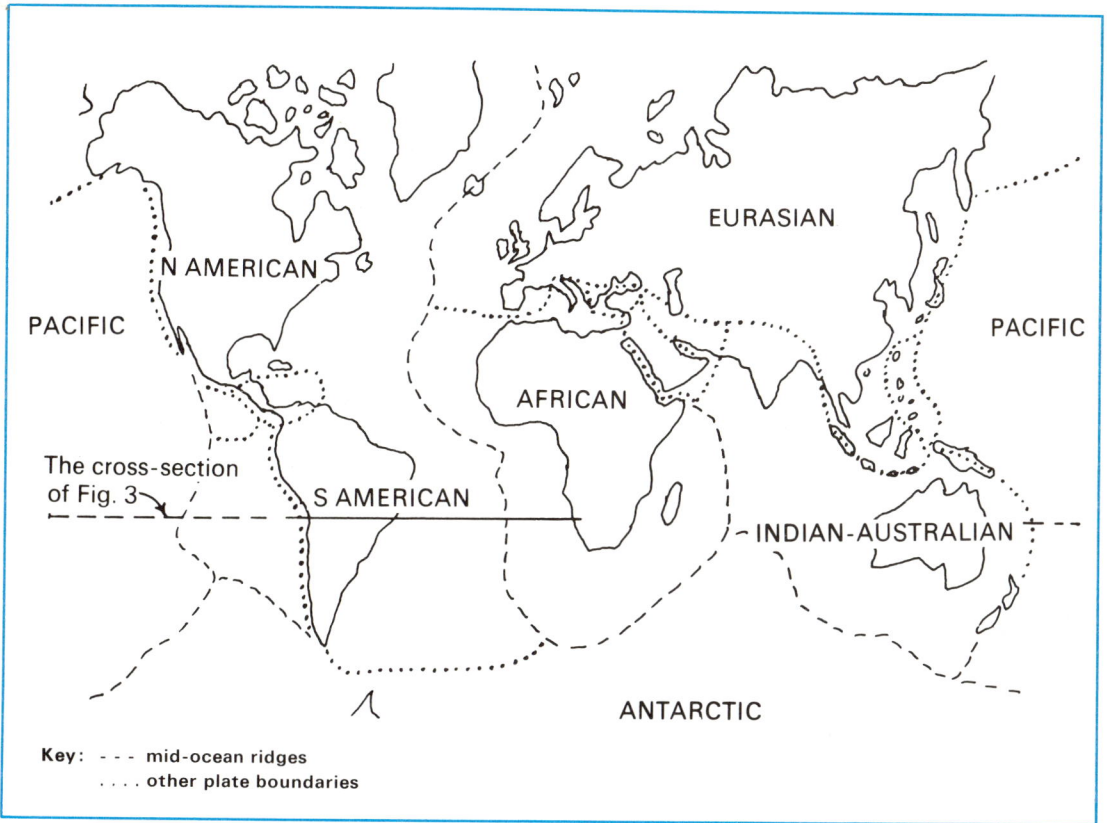

**Fig. 1 The major tectonic plates.**

convection cells in the interior) was welling up at these rift valleys and spreading the sea-bed away on either side at a rate of a few centimetres a year. The theory of Plate Tectonics was born.

The surface of the earth is now thought to be covered by a series of plates (see Fig. 1), 50 to 100 km thick, which carry on their backs the continental masses, consisting of less dense rocks, and which move by virtue of the sea floor spreading, as outlined above. All short-term activity of the earth, in the form of earthquakes and volcanoes, takes place at plate boundaries, while long-term effects, like mountain building, indicate the positions of former plate boundaries.

Three different types of plate boundary are recognised: (a) *ridge axes*, as outlined above, where plates are diverging and new material is welling up from the interior of the earth; (b) *transform faults*, where plates, or portions of plates, are sliding past one another and material is neither evolved nor destroyed; and (c) *subduc-*

*tion zones*, where one plate rides over the top of another, forcing much of the plate material at the junction into the molten interior of the earth, where it is consumed (see Fig. 2).

Three different types of subduction zone are distinguished:

(a) *Continent-Continent boundary* (see Fig. 2(a)).

The lighter continental masses carried on the plate do not sink appreciably with it into the interior but rather are raised at the junction to form mountain ranges. Some plate material is carried up and appears in the final composition of the new range alongside the continental material; this consists largely of sedimentary rocks formed by the weathering of the continental mass and deposited at the continent's edge. Movement ceases when the forces driving the plate are no longer able to sustain the uplift, which subsequently starts elsewhere on the plate. In this way a mountain range is formed in the centre of a land mass.

(a)

(b)

(c)

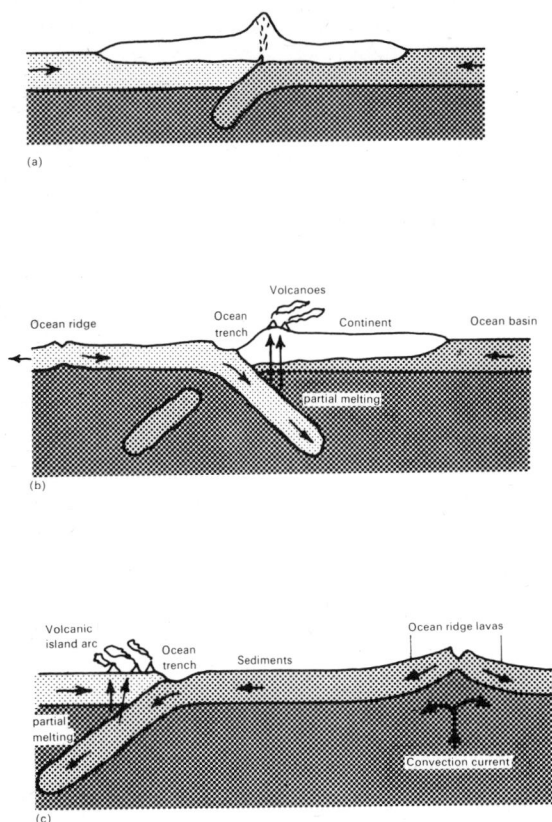

Fig. 2   (a) Continent-Continent boundary (formation of the Himalaya by collision of the Indian and Eurasian plates). (b) Ocean-Continent boundary (formation of the Andes by the Pacific plate dipping beneath the S American plate). (c) Ocean-Ocean boundary (formation of Japan, the Philippines, etc, by Pacific plate dipping beneath Eurasian plate)

(b) *Ocean-Continent boundary* (see Fig. 2(b)).

The ocean plate subsides beneath the continent-bearing plate forming an ocean trench close to the edge of the land mass. Meanwhile the land mass is uplifted into a mountain range parallel to the trench. Partial melting of the subsiding plate produces volcanic materials which are ejected at points of weakness along the range, forming volcanoes (eg the Andes).

(c) *Ocean-Ocean boundary* (see Fig. 2(c)).

Once again a trench is formed, but here the volcanoes rise from the sea-bed to release the materials produced from the down-plunging plate. These and any squeezed-up plate material combine to form an island-arc (eg Aleutian Islands, Japan, the Philippines, etc – Challenger Deep in the Marianas Trench is $-11\,033\,\mathrm{m}$).

The ridge axes, which are invariably beneath the seas, have numerous transform faults associated with them. These appear at right-angles to, and produce discontinuities in, the ridge lines. Other transform faults which are actually plate boundaries, appear on the land surface. Best known is the San Andreas fault, which menaces the cities of California, where the Pacific plate is sliding past the North American plate. Earthquake shocks invariably originate at plate boundaries – shallow at mid-ocean ridges and transform faults, deep-seated at ocean trenches; in fact the plate boundaries were determined in the first place largely from earthquake studies.

Figure 3, a cross-section along the Tropic of Capricorn, illustrates most of the major tectonic features.

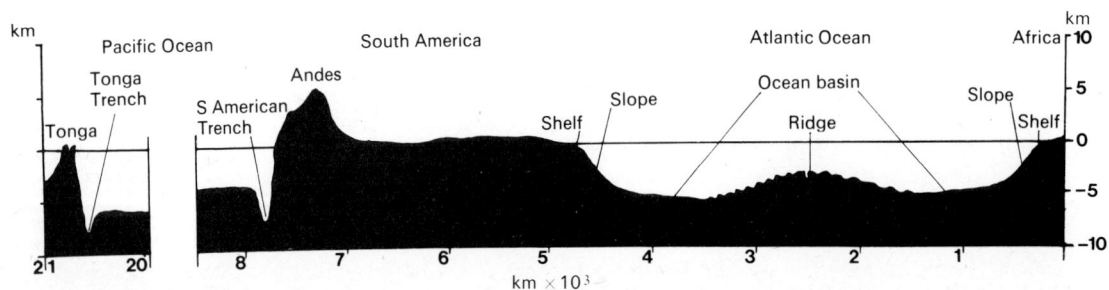

Fig. 3   Cross-section along the Tropic of Capricorn (illustrating all the main tectonic features)

# The Origins of Mountain Forms

The new science of Plate Tectonics provides in broad terms an explanation of the origins of the major mountain ranges (cordilleras). The mountains thus elevated are sculpted by ice, snow, water, wind and sun, and by chemical and biological processes into the tremendous variety of forms found on the earth, so that every country and every range has its own characteristic mountainsides and summits.

There are three broad classifications of rocks, based on their mode of formation – igneous, sedimentary and metamorphic.

**Igneous rocks** are formed from molten material from the interior of the earth, which solidifies on cooling, either (a) beneath the surface (in a form characterised by large crystals, eg granite), whence it is later exposed by erosion of the covering layers, or (b) on the surface (in forms characterised by very small crystals, eg basalt).

**Sedimentary rocks** are formed from the deposition of material broken down by the weathering of other rock types and transported to new sites by water or wind. Sedimentary rocks are invariably found in layers which mark successive changes in the conditions of deposition during the formative stages.

**Metamorphic rocks** are produced when earth movements, involving heat or pressure, act on igneous or sedimentary rocks, thus shales become slates, sandstones become quartzite, etc.

The continuous erosion process tends all the time to make more and more sedimentary rocks at continental margins, so that the mountain building processes envisaged by the theory of Plate Tectonics usually operate on this class of material. The sedimentary strata are crushed, folded and metamorphosed by the mountain building forces; they are penetrated by the welling up of igneous materials; they are faulted when folding can no longer be sustained, and break into huge blocks, often as large as a whole mountain range.

Figure 4 (*a*) shows simple fold mountains. The Jura on the Franco-Swiss frontier is classical fold country – the inter-plate forces, which produced the complex folding of the Alpine chain further south, were here reduced to producing relatively simple longitudinal folds. The range is limestone and reaches 1718 m at the

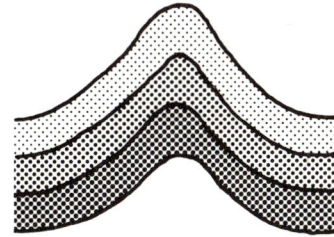

(a)

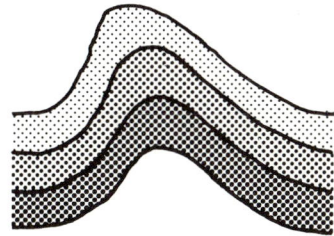

(b)

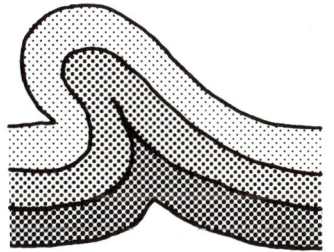

(c)

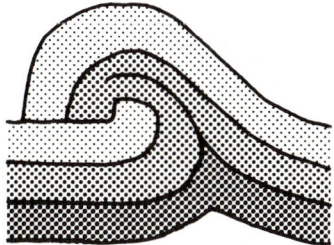

(d)

Fig. 4   (a) Simple fold. (b) Asymmetric fold.
(c) Recumbent fold. (d) Nappe

Crête de la Neige; it is 250 km long, 50 km broad at the centre, tapering to a few kilometres at either end. The main rivers run in the folds along the length of the chain, sometimes breaking through the intervening ridges using valleys cut by streams which were there before the folding took place.

Figure 4 (*b*) shows asymmetric folding, which might easily arise in a more complex situation. In Fig. 4 (*c*) the fold is bent over until it reclines on the surface away from the producing force. Figure 4 (*d*) shows the situation where the rock is no longer able to conform to the folding forces; fracture occurs and the upper limb of the recumbent fold rides along past the lower, forming a feature known as a nappe (table cloth). Nappes are common in the Alps, the complex origin of which has been slowly unravelled by generations of earth scientists.

Faults occurring in strata subject to folding also produce mountain land-forms. Figure 5 (*a*) illustrates the rift valley, or graben, the classic example of which is the valley of the middle Rhine between Basle and Strasbourg. The surface was upwarped by gentle folding to a point where the rock could no longer sustain the forces. Two considerable parallel faults developed, sloping in towards one another, and the block between them slipped down relative to the land outside the faults. Now the Rhine flows in the valley formed by the slipped block, while on either hand rise the Vosges (west – Ballon de Guebwiller 1423) and the Black Forest (east – Feldberg 1493) with steep slopes facing the river and gentler slopes away from it. Other examples are the Rift Valleys of Africa, the Jordan-Dead Sea depression and Death Valley, California.

When a block of country rises between faults, either by an uplift between them or by a sinking outside them, the resulting land-form is known as a block mountain, or horst (Fig. 5(*b*)). Graben and horst sometimes occur alongside one another, eg the Vosges and the Black Forest, delimited by further faults on the sides remote from the Rhine graben, can be looked on as horsts. The Sierra Nevada of California and the Tetons of Wyoming are similarly block mountains.

Combined folding and a single fault can produce a land-form as shown in Fig. 5(*c*). The steep face thus exposed, known as a fault scarp, often takes the form of a crag.

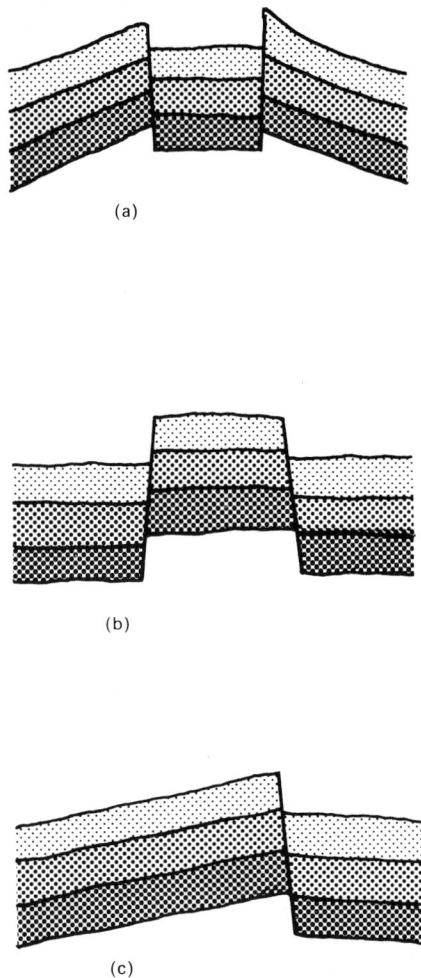

(a)

(b)

(c)

Fig. 5  (a) Rift valley or graben. (b) Blockmountain or horst. (c) Combined tilting and faulting

An essential feature of every volcanic mountain is a vent through which molten material is, or has been, ejected from the interior of the earth. This may take the form of ashes, rocks and boulders or liquid lava, and the slope of the volcano will be determined by the proportions of these and by the viscosity of the lava (see Fig. 6(*a*)). Alternative layers of ash and lava produce a strato-volcano, which is in fact the commonest type. If volcanic activity ceases, say by transferring to another outlet, leaving a plug of lava to solidify in the vent, subsequent erosion of the cone leaves the plug upstanding as a crag (see Fig. 6(*b*)) – as at Shiprock (New

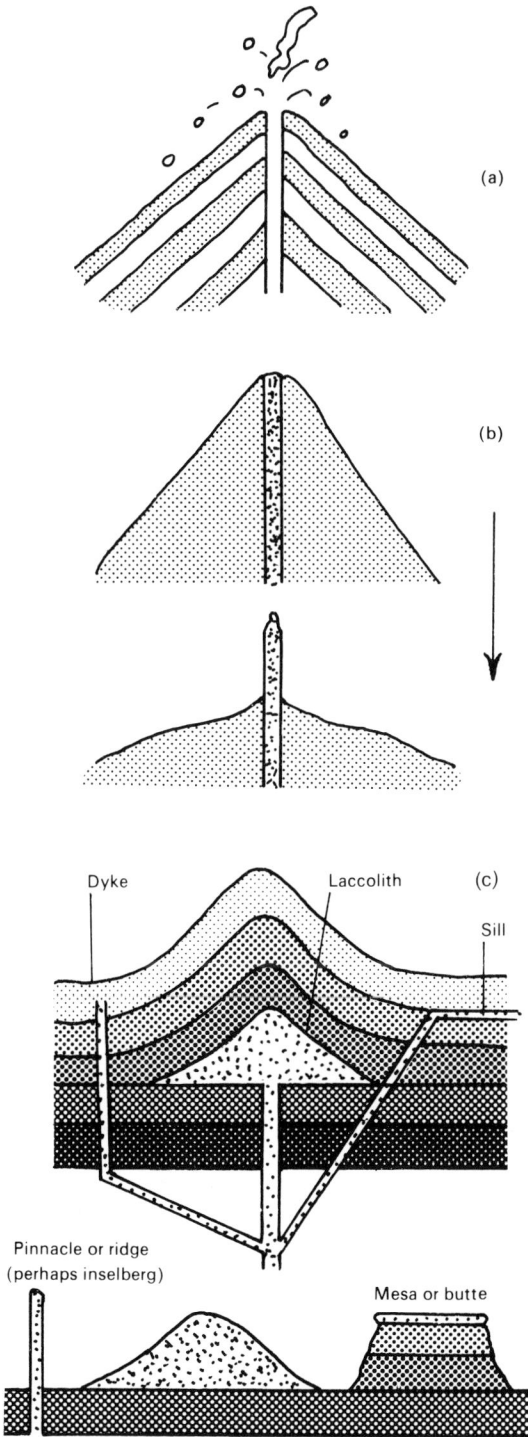

Fig. 6 (a) Strato-volcano (ash/lava cone).
(b) Volcanic plug and possible landform after
erosion. (c) The origins of some volcanic forms

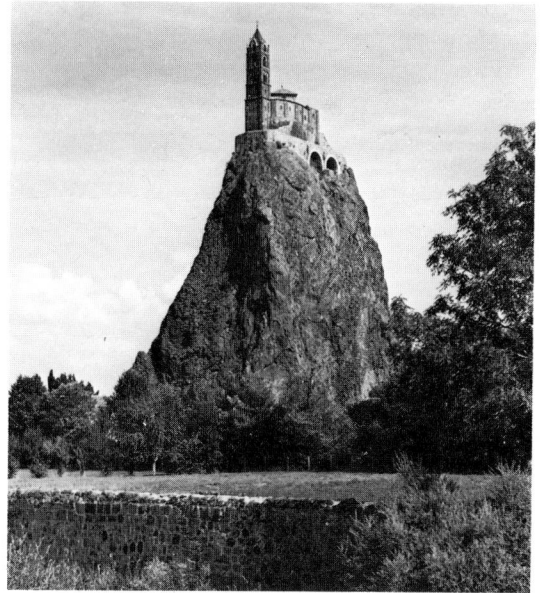

Volcanic plug, le Puy, France (French Government
Tourist Office)

Mexico), Devil's Tower (Wyoming) and the
Rocher Saint Michel (Le Puy, Auvergne).

Molten material which has solidified beneath
the surface is called a plutonic rock. Some of the
resultant land-forms are shown in Fig. 6(c). If
the molten material, welling up, raises the strata
above and 'blisters' the land surface, the resultant
dome-shaped mass is known as a **loccolith**. Flow
of material horizontally between strata produces
a sill, upward into vertical faults or fissures, a
**dyke**. The types of land-form produced by
subsequent erosion of the overlaying rocks are
shown also in Fig. 6(c). The dyke has become a
wall or a pinnacle, the laccolith a dome-shaped
hill; the sill has protected the rocks beneath it
from further erosion and produced a **mesa** or a
**butte** (see Fig. 6(c)); a sill may also become a line
of crag. Many examples of these are found in
the desert states of western USA; there are dyke
ridges in the Spanish Peaks of south Colorado
and the Crazy Mountains of Montana.

The **inselberg** (common in Mozambique and
Rhodesia, also found in central Australia and
around Rio de Janeiro) is a specialised outcome
of the erosion process. An isolated mass of hard
rock, such as a granite dyke, is isolated by
erosion, the steep sides are maintained by
exfoliation (cracking away of outer layers by
weather action), while the flat base is maintained

and lowered by sheet flooding. Successive elevations of the land surface keeping pace with the erosion of the base have enabled some examples to attain heights of 300 to 450m.

**Relict mountains** are formed by considerable erosion of other mountain land-forms, so that their final form depends entirely on the erosion process and bears little or no relation to the original. Thus the continued erosion of simple domed strata can proceed so far that the synclines of the original folding form the new summits, while the anticlines have entirely disappeared

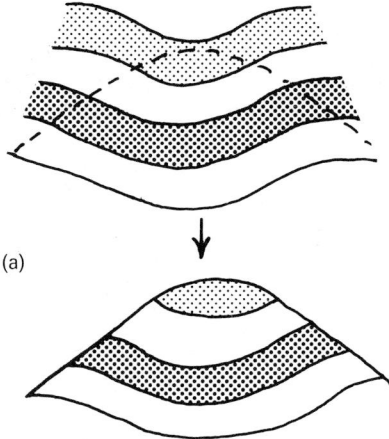

(a)

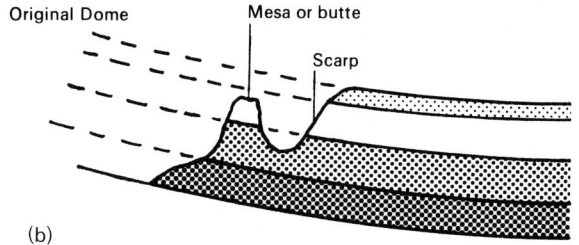

Original Dome          Mesa or butte

Scarp

(b)

**Fig. 7   (a) Folded rocks after denudation. (b) Cuesta**

Earth pyramids at Euseigne, Switzerland

(see Fig. 7(*a*)). The process is assisted by the fact that in the folding the synclinal portions were strengthened by compression while the anticlinal portions were weakened by tension. Or the synclines can be protected by lava flow, as at Montagne de la Serre in the Auvergne, where such a flow, originally in a valley, now occupies the top of a hill ridge.

The mountains of Snowdonia in North Wales and of the English Lake District are the remains of domes which once rose thousands of metres above the present level and which still show something of the stream pattern originally initiated on the dome. Figure 7(*b*) demonstrates another outcome of the erosion of a dome, known as a **cuesta**. Harder strata are left forming steep scarp slopes facing towards the original centre and, if these are sufficiently hard, the outcrop may take the form of a crag. Small-scale local erosion processes can isolate a portion of the strata in front of the scarp to give a mesa or a butte.

An interesting possibility is an apparent horst formation which was in fact originally a graben. The rocks outside the graben are eroded away more rapidly than the graben itself, which is left higher than the surrounding countryside.

Almost horizontal strata subjected to long-period erosion by water are carved into blocks and gorges; the Grand Canyon of the Colorado River is the supreme example. The maximum depth is 1900m, the rims vary from 8 to 25km apart, erosion has taken place at a rate of $14km^3$ per kilometre of river. Isolated blocks and pinnacles in the Canyon and its branches have the dimensions of mountains and present the same type of problems of ascent.

The so-called **'Badlands' topography** results when an arid region of comparatively soft rocks is violently attacked by torrential rain storms and worn into a pattern of gullies and ravines separated by columns, spurs and platforms. A normal protective covering of vegetation is unable to grow. Notable examples are found in western South Dakota and, best known of all, Bryce Canyon in Utah. A similar process on a smaller scale, where selective erosion results from the existence of a large capping boulder, has produced the famous earth pillars at, for example, Euseigne in the Val d'Hérens, Switzerland, and the Demoiselles Coiffées of Ubaye, France.

## Mountains beneath the Sea

During the last 50 years refined techniques of echo-sounding have revealed relief features on the ocean floor comparable with, or even more striking than, those of the land masses. Previously the ocean bed had been thought to be of low relief, consisting of gentle slopes and flat plains loaded with an extensive covering of sediments derived from the rivers and coastlines of the land. Indeed a proportion of the ocean bed is of this form, but there are also huge ranges and isolated peaks.

Figure 8 shows the average form of the surface of the globe if the surface of the sea is considered to be level. Twenty-nine per cent of the surface lies above sea level, divided as follows: 0 to 180m, 8 per cent; 180 to 900m, 13 per cent; 900

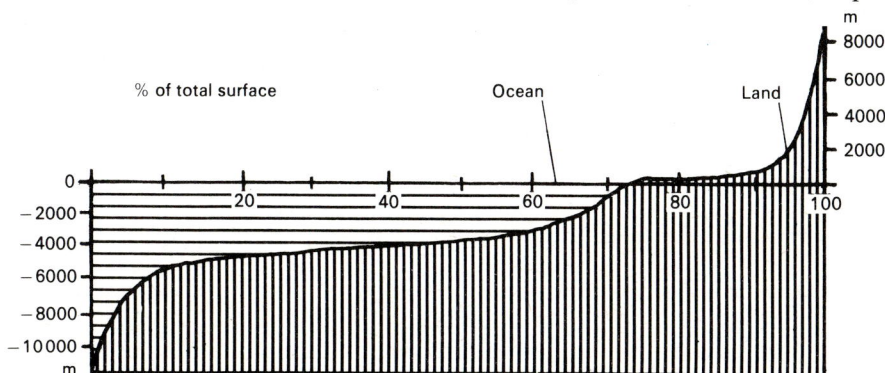

Fig. 8   The hypsographic curve (mean elevation of land 840m; mean depth of sea 3800m)

to 1800m, 5 per cent; 1800 to 3600m, 2 per cent and above 3600m, 1 per cent, with a mean elevation of 840m. Seventy-one per cent of the surface falls below sea level, divided as follows: 0 to −180m, 5 per cent; −180 to −900m, 3 per cent; −900 to −1800m, 2 per cent; −1800 to −3600m, 15 per cent; −3600 to −5400m, 41 per cent and over −5400m, 5 per cent, with a mean depth of −3800m.

The portion down to −2000m, which may perhaps be considered as part of the continental masses, comprises the continental shelf down to −200m and the continental slope connecting this with the deep sea plains. Striking features of the continental slope are a number of sub-marine canyons associated with the mouths of some great rivers. The Congo Canyon can be traced for 230km to a depth of −2300m, the Hudson Canyon for 290km to a depth of −4875m. The Monterey Canyon, off California, extends for 80km offshore to a depth of −2750m, with a cross profile similar to that of the Grand Canyon.

The area between −2000 and −5000m deep is called Abyssal Plain and Abyssal Hills; these features are widespread across the globe. Slopes are mostly gentle though sometimes sub-marine peaks protrude through the sediments, and there are occasional scarps.

The deep sea trenches, which have a notable role in mountain building processes, are a feature of the deepest 10 per cent of the ocean. A prominent line of them pass round the edges of the Pacific Ocean; paralleling the west coast of the American continent are the Peru-Chile (−8066) and Middle American (−6662) Trenches; then follow the Aleutian (−7822), Kuril (−10542) and Japan (−10374) Trenches on the north and north-west sides of the Ocean. The western Pacific has many others – the Ryu Kyu (Nansei Shoto) (−7507), Marianas (where Challenger Deep −11033m is the deepest known), Yap (−8597), Palau (−8138), Philippines (−10497) and Sunda (Java) (−7450) Trenches; north-east of New Zealand are the Tonga (−10882) and Kermadec (−10047) Trenches, while minor deeps between Papua and the New Hebrides reach down to −9165m. The dimensions of the trenches are impressive, with depths considerably in excess of the heights of the highest mountains. That part of the Marianas Trench which is deeper than −5900m is 800m wide and 32km long (the bathyscaphe *Trieste* descended to the floor of it in 1960). The depth of the Middle American Trench exceeds 4400m over a distance of 2000km and exceeds 5500m over 600km.

Huge mountain systems, with a total length of 65000km lie beneath the oceans. The mid-Atlantic Ridge runs from Iceland in the north to Bouvet Island in the south, roughly bisecting the Ocean, of which it occupies the middle third. Its width varies from 800 to 1400km and its height from 2500 to 4000m. Along the crest is a deep rift valley, 1000 to 4000m deep and 25 to 45km wide. Features which rise above sea level are Iceland, the Azores, Ascension and Tristan da Cunha – all sites of volcanic activity associated with the accepted theory of ocean-floor spreading.

Detailed sea-bed surveys have revealed the existence of isolated sub-marine hills of various types – particularly numerous below the Pacific Ocean. A **sea-mount**, which should have more than 1000m of local relief, rises to a sharp crest; a similar feature having a flat top is called a **guyot**. There are thought to be 10000 of them in the Pacific alone, many rising 3000m or more above the ocean floor. The biggest sea-mount so far discovered is close to Tonga and rises 8690m to a summit of −365m. Both these features probably originated in volcanoes, but the guyots owe their flat-topped form to erosion consequent upon their exposure above sea-level at some time in their history.

These mountains beneath the sea fall well within the general definition of mountains propounded above. They are substantial topographical features, the discovery of which has thrown considerable light on the changing forms of land surfaces above sea-level. Unfortunately their tremendous scenery is for ever hidden from us.

## Volcanoes

Volcanicity is defined as including all processes by which solid, liquid and gaseous substances are ejected from the core of the earth, either into the crust or on to the surface. The resulting mountain forms are discussed under 'The Origins of Mountain Forms'.

**Eruption of Merapi, Indonesia, 1954 (Indonesian Embassy)**

The type of eruption depends on:

(a) the nature of the outlet orifice, whether it be a crack or fissure or a localised vent (this last producing the characteristic conical volcano form);
(b) the composition of the ejected material – acidic materials are viscous, welling up slowly and building steep cones, while basic materials are more fluid and give rise to flatter cones spread over much greater areas.

The most recent fissure eruptions were at Tarawera in New Zealand in 1866 and at Laki, Iceland, in 1783. Previous outflows have on occasions been on a considerably larger scale, having sometimes been built up by a succession of outpourings. For example, the Deccan Plateau of India covers a million km², with thickness varying from 1200 to 1500m and up to 3000m near Bombay. The Columbia-Snake Plateau of the western USA has an area of 650000km² and a thickness of 1500m. There are extensive areas also in the Stormberg lavas of South Africa and the Parana plateau of Brazil.

Eruptions from a central vent can take the form of a single explosion (acting like a safety valve) after which activity ceases, as in the Eifel district of West Germany. Frequently, however, a particular eruption is just one phase of a continuing vigorous activity.

The ejected material can be gaseous, liquid or solid:

**Gases** are mainly superheated steam, carbon dioxide, sulphur dioxide, hydrogen sulphide, chlorine and fluorine. Indeed, even now, the pollutant sulphur dioxide in the atmosphere still derives more from volcanoes than from the industrial activities of man. Sulphur dioxide and steam produce sulphuric acid which kills off the vegetation around the outlet.

**Water.** The steam condenses to form heavy rain, which, mixed with fine dust simultaneously ejected, produces destructive mud-flows (such as overwhelmed Herculaneum). Disastrous floods also result, incidentally, from the melting of summit snow and ice by ejected lava.

**Lava** is the general name of molten rock; acid rocks are viscous, basic rocks more fluid.

**Solid debris,** of which a number of states are distinguished: (a) solidified lava – angular fragments; (b) pumice – solidified froth and scum; (c) scoria – cindery masses; (d) *lapilli* – small fragments; (e) volcanic bombs – rounded masses of lava, ejected presumably in a half-molten state and shaped by passing through the air; (f) tuff – finely divided material or dust which eventually forms partially or completely consolidated masses.

Volcanoes can be classified in a number of
ways:

*According to degree of activity*
1   *Active volcanoes*, which may be constantly
erupting, eg Stromboli, or erupt periodically,
eg Vesuvius.
2   *Extinct volcanoes*, which have not erupted in
historical times or are believed to be incapable
of further activity, eg Mount Egmont,
Aconcagua, etc.
3   *Dormant volcanoes*, which are apparently
dead but in fact are only sleeping, eg Katmai.

*According to the nature of the eruption*
Seven types can be distinguished (see Fig. 9):
1   *Fissure or Icelandic* type, where the outlet
was extended, taking the form of a crack or a
line of vents, eg Laki.
2   *Hawaiian* type. There was no explosion; the
very fluid lava spread out widely, eg Mauna Loa.
3   *Strombolian* type. The lava is less volatile;
trapped gases escape spasmodically, with red-
hot clots of lava, eg Stromboli.
4   *Vulcanian* type. More violent than the last,
with explosions irregularly spaced. Dark ash-
laden volcanic clouds of cauliflower shape are
thrown out, eg Vulcano.
5   *Vesuvian* type. A paroxysmal extension of
Strombolian and Vulcanian types in which long

**Vesuvius in eruption (Aerofilms)**

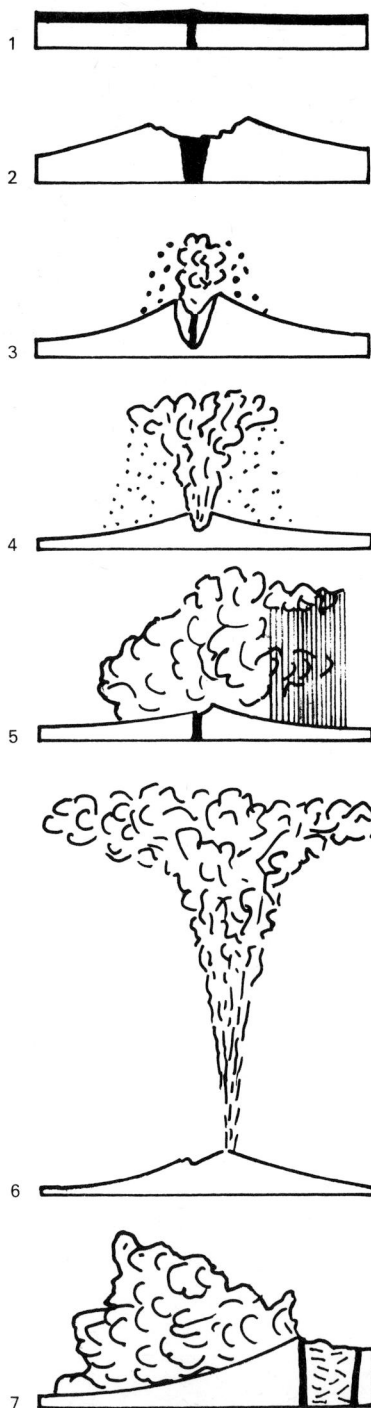

Fig. 9   (1) Fissure or Icelandic type. (2) Hawaiian
type. (3) Strombolian type. (4) Vulcanian type.
(5) Vesuvian type. (6) Plinian type. (7) Peléan
type

Volcanic cone, Japan (Japan Information Centre)

periods of quiescence are followed by violent expulsion of solids and gases. Dark clouds rise to considerable heights and shower ashes over a wide area, eg Vesuvius.

6 *Plinian* type. The most violent of the Vesuvian type erupting columns of gas to a height of several miles.

7 *Peléan* type. The ultimate in high viscosity and delayed explosiveness. Upwards escape of material is prevented by a plug in the vent, so that a mixture of hot gases and incandescent ash (known as a *nuée ardente*) escaping through lateral cracks rolls down the slopes annihilating everything in its path.

*According to composition*

1 *Shield volcanoes*, are formed from fluid basic lavas in the absence of explosive activity. A large diameter, low angle cone results.

2 *Dome volcanoes*, are formed from viscous acidic lavas, giving a dome-shaped mountain with steep convex sides, sometimes without a crater. Similar shapes are formed when there is no ejection on to the surface, the strata being blistered but not actually fractured.

3 *Cinder cones*, are formed when explosive activity gives rise to showers of fragmented material, which piles up round the vent, eg Volcan de Fuego in Guatemala is a cinder cone of 3300m.

4 *Composite cones*, are formed from alternate layers of ash and lava flow. Most of the highest volcanoes in the world are of this type.

Volcanoes are invariably situated along plate boundaries. There are several hundred active at the present time and the major volcanic mountains are distributed as shown in the Table below.

| DISTRIBUTION OF MAJOR ACTIVE VOLCANIC MOUNTAINS | |
|---|---|
| Japan and Formosa | 64 |
| The Indonesian arc | 60 |
| Central America and the Antilles | 41 |
| Kurile Islands | 39 |
| Sarighe Island to New Britain | 37 |
| South America | 34 |
| Solomon Islands to New Zealand | 33 |
| Kamchatka | 25 |
| The Aleutians | 18 |
| Alaska | 15 |
| Mexico | 12 |
| The Philippines | 12 |
| Western USA | 3 |

Extinct volcano crater, Diamond Head, Hawaii, now houses Federal Aviation Agency installation (Barnaby's)

Some other terms in volcanicity:

## Caldera
After the ejection of molten or other material the chamber in which it was formerly contained is left substantially unsupported. Sometimes a wholesale collapse of the walls and roof has resulted, leaving behind a considerable basin-shaped crater bounded by steep cliffs. If the volcano remains active, new cones form inside the caldera, eg there is an active volcano in the caldera of Aso (23 × 17 km) in Japan. If the volcano is extinct a large lake may form inside the caldera; the best known of these is Crater Lake in Oregon, where a more recent cone, itself also extinct, remains as an island in the lake.

## Fumarole
This is a hole in the ground ejecting steam and gases, a side effect of more spectacular volcanic activity. The Valley of Ten Thousand Smokes in Alaska, close to the catastrophic Katmai eruption of 1912, now a National Park, has many square kilometres riddled with fumaroles.

## Solfatara
This is a volcano close to extinction which only emits steam and certain sulphurous gases.

## Geyser
Another phenomenon of volcanic regions, this ejects a jet of hot water or steam at regular or irregular intervals to heights of up to 100 m. The highest recorded was at Tarawera in New Zealand in 1901 – 450 m. A narrow tubular fissure penetrating the earth fills with percolating water; the water at the lower end, coming in contact with hot rocks, is converted to steam at high pressure; this expands and ejects the whole contents of the fissure. The best known geyser areas are New Zealand, Iceland and western USA (particularly in Yellowstone National Park, Wyoming, where the geyser known as Old Faithful used to function regularly at intervals of $66\frac{1}{2}$ min; however, recent changes have disturbed this regularity). The word 'geyser' comes from the name 'Geysir' of a particular example in Iceland.

## Hot Springs
These result from a similar phenomenon without the explosive action. Water at an elevated temperature flows continuously from the ground in places where there are no signs whatsoever of volcanic activity.

## Mud Volcanoes
If the erupted water is muddy, a conical mound of mud may be formed with a crater on the top. This is usually a very late stage in volcanic activity; but similar results may be found in non-volcanic regions, where the ejection is due to the generation of gases beneath the earth's surface, eg in the Crimea, at Baku on the Caspian Sea, and in South Baluchistan.

## Geothermal Power

In those regions notable for their geyser and hot spring activity, in particular western USA, Iceland and New Zealand, these phenomena are utilised for power generation. The first geothermal plant was constructed at Larderello in Tuscany (Italy) in 1897, when the natural steam was led straight to boilers. In 1904 the steam was used directly in a piston engine driving a dynamo. The first steam turbine was introduced in 1913. The output of the plant was steadily increased and, though it was wrecked in World War II and rebuilt, by 1950 it was producing 6 per cent of Italy's power needs (1·8 GWh per annum).

Geothermal heat has also been harnessed in the West Indies, El Salvador, Nicaragua, Japan and the USSR.

# Snow, Ice and Glaciers

Snow is the stable crystalline state of water at low temperatures and is formed either by growth direct from water vapour, or by the crystallisation of supercooled water droplets around nucleii of ice or dust. The structures of snow flakes show an almost infinite variety – needles, plates, stars, etc – in all manner of spatial arrangements; in fact it has been suggested that there have never been two identical snowflakes in the history of the earth. As soon as the flake is sufficiently heavy it begins to fall, aggregating with others on the way down. The spiky nature of the flakes means that on landing they can cling to steep, even vertical, faces and grow into all manner of bulges and overhangs. In windless, low-temperature conditions snow accumulates in a layer as light as down. The wind piles the layers into drifts in hollows and forms cornices on the lee sides of ridges. Subsequently, abrasion between crystals rounds them to form a more compact mass.

At high altitudes (4500 to 5000m), where the temperature is always low, the intense radiation from the sun is concentrated into hollows protected from the wind, which deepen while hummocks gradually build between. Snow columns are produced which subsequently turn to ice; they can reach 10m high in the mountains of Central Asia and in the Andes, where they are called *nieves penitentes*. Above 6500m the snow seldom melts except close to rocks; holes thus form on ridges which, combined with the cornices, make progress along them very difficult.

The snow which falls on the steeper parts of high mountains is either blown by the wind, or falls in the form of avalanches on to lower angle slopes or hollows further down. At these higher temperatures the crystals begin to melt, the snow layer settles and becomes more compact.

The crystals change shape continuously; fine points disappear as the ice sublimes to water vapour, which condenses on the central more massive parts of the crystals forming them into larger and larger grains. These processes slowly convert the snow to a compact form known as névé. The melting and freezing shatters and abraids the underlying rock, so that the size of the hollow is gradually increased forming a cirque (circular recess – see Fig. 10) beneath the névé. Several cirques cutting back into a mountain give rise, after thousands of years have elapsed, to pyramid-shaped peaks joined by narrow walls (ridges or arêtes) around the heads of the cirques. These features are conspicuous in mountain areas formerly glaciated, now well below the snow-line. Between the névé and the

**Névé slopes and bergschrund (C D Milner)**

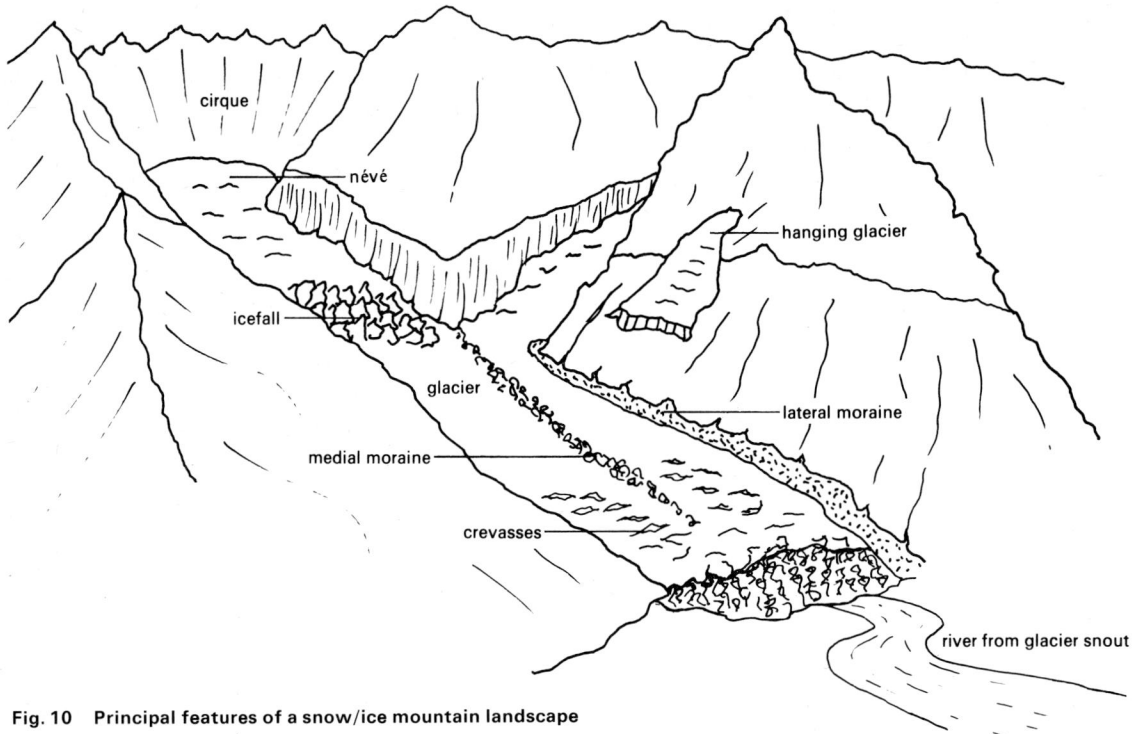

Fig. 10   Principal features of a snow/ice mountain landscape

back wall of the cirque a huge horizontal crack (bergschrund, rimaye) is often found.

As snow continues to pile up in the cirque, the lower layers of the névé convert by stages to ice which is driven towards the valley by the pressure above. This moving ice is a glacier, which flows on downwards to a point where the flow of ice from above is balanced by melting and vaporisation (ablation). In the Alps, glaciers flow at up to $1\frac{1}{2}$ m a day; in the polar regions, rates as high as 25 m a day have been recorded. Glaciers flow faster in summer than in winter, since ice formation and final ablation both take place more quickly.

Glaciers can be divided into three groups:

1   *Ice sheets* which cover the whole land, as in Greenland or the Antarctic Continent. As snow continues to fall in the interior the ice sheet expands towards the surrounding coastlines, where mountain ranges split it into separate tongues. These end at the sea, calving off into huge masses called icebergs, which float off to become a menace to shipping, before finally melting in more temperate waters. There are smaller ice sheets in many high latitude countries

– Iceland, Norway, Alaska, Patagonia, etc, usually where a single block of mountains is covered with glaciers flowing down on all sides.
2   *Mountain or valley glaciers*, the type found above the snow-line in high mountain areas. Fed by mountain snowfall, they flow down to the valleys (see Fig. 10).
3   *Piedmont glaciers*, formed where several glaciers debouch on to the plains below the mountains and there unite to form an extensive ice sheet. The Malaspina glacier in Alaska, which has an area of 3800 km, has now gouged out a basin 180 to 250 m below sea-level at the centre. It is so slow moving that trees grow on the surface.

The height of the snow-line in various mountain areas is a pointer to the extent of the glaciation. This is sea-level in the polar regions; 600 m in south Greenland and south Chile; 1500 m in south Norway and south Alaska; 2750 m in the Alps; 4400 m in Assam; 5300 m in Africa and the central Andes; 5800 m in the Punjab Himalaya. The height of the snow-line depends also on aspect, being higher on slopes facing the sun; on the steepness, lying longer on

The Gorner and Grenz glaciers with Monte Rosa and Lyskamm, Switzerland

gentle slopes; and on humidity, higher in a wet region than in a dry.

Boulders, which have fallen from the slopes on either side, forming what are called lateral moraines, are transported by glaciers and deposited at the snout in banks called terminal moraines. When one glacier flows into another, the joint ice-stream continues valleywards with the former lateral moraines united to form a medial moraine. Small boulders warmed by the sun tend to sink into the ice, but large boulders protect the ice beneath them from the rays of the sun and sometimes remain supported by a pillar of ice – a glacier table.

The rocky material, abraided at the glacier bed and then carried along by the ice, scours out the floor and the side walls of the glacier valley until it becomes U-shaped in section. Such valley shapes indicate the one-time existence of glaciation in mountain areas now below the snow-line. Smaller branch glaciers, which wear away their valley floors less quickly, tend to get left behind high on the side of the main valley as hanging glaciers. The ice may fall from them periodically on to the main glacier below or it may melt before reaching the edge, but sooner or later a cirque is formed here too.

A glacier moves by plastic flow of the ice (ie it constantly changes shape as it moves). When, due to irregularities in the bed, the internal forces become too great, cracking, sliding and freezing occur to assist the movement. Due to friction against the containing walls the centre flows faster than the sides and

Crevasse (M Millet)

Mountaineers on the Grindelwald glacier, 1905, Switzerland

the surface faster than the base layer. The resulting strains often produce cracks (crevasses) too large to be closed by the sliding and freezing of normal flow. Crevasses are also formed where the valley narrows or widens or bends, or where there is a step in the valley floor. In the last case an unstable jumble of ice towers (seracs) is formed and is known as an ice-fall, a great obstacle to travel upon a glacier, which otherwise can provide a passage into the heart of the mountains. Crevasses when snow covered also demand considerable care on the part of the glacier traveller.

Towards the end of a glacier melt water flows over the surface, plunging sooner or later into one of the crevasses to join a considerable river beneath the ice. This pours out at the snout, loaded with debris abraided by the ice. Many of the world's greatest rivers are thus glacier fed.

Ninety-nine per cent of the available fresh water in the world is in the form of ice, comprising a total volume of around 30 million km$^3$. Ninety-nine per cent of the ice is in the Antarctic and Greenland ice sheets, the other one per cent is in smaller glaciers and snow-fields. The remaining one per cent of fresh water is in rivers, lakes, soil moisture and vapour in the atmosphere. More than 1/10th of the land surface (three per cent of the total surface of the globe) is covered by ice; during ice ages the area covered is almost three times as large. One quarter of the land is covered by snow for more than four months each year. During an average year snow for some time covers half the land surface and, for short periods, up to 85 per cent. Sea ice has a large seasonal variation with an average maximum coverage of six per cent of the world. Snow and hail account for five per cent of the total annual precipitation.

The largest glaciers are in the polar regions. Then come the glaciers of Alaska (Hubbard 120 km; Logan only a little shorter), the Pamirs (Fedtchenko 77 km; Inylchek 70 km) and the Karakoram (Siachen 72 km; Hispar 61 km). The Tasman glacier in New Zealand is 30 km; the Aletsch, the longest in the Alps, is 26 km (with the ice some 600 m thick).

The sliding or falling of snow or of ice to lower levels on a mountainside involves the process of avalanche – a dramatic and highly dangerous feature of all mountain areas. The earliest mention, in Strabo's *Geography* (64 to 36 BC), reveals a reasonable insight into the mechanism. Subsequently, the more spectacular avalanche accidents usually accompanied the movement of armies through mountains, but there was periodic loss of life and property damage in all alpine areas.

Snow avalanches may spread from a single point or fall from a large area leaving behind a wall. The whole depth of the snow cover may be involved or only some of the upper strata. The constituent snow may be wet or dry. The slide may take place on an open slope or be confined to a gully or valley. The flow may take place along the ground or through the air. A striking feature of air-borne powder snow avalanches is their high velocity, for example the Glärnisch avalanche of 6 March 1898, which fell 1750 m, was estimated to travel at 360 km/hr. Modern observations (175 to 290 km/hr, with larger ones reaching 350 km/hr) tend to confirm the accuracy of this early judgement. The accompanying air blast can devastate hectares of mature forest in seconds, for example on 18

February 1962, in the Inn Valley of the Upper Engadine, snow which had travelled around 4 km through a vertical height of 2000 m snapped off or uprooted a square kilometre of larch and pine (120 to 150 years old) on the Swiss side, also producing considerable further damage on the Austrian side. On 11 January 1954 near Dalaas in Austria a 120 tonne locomotive was lifted from the rails and slammed against the station. The power involved in a large-scale wet snow slide has been estimated assuming 200 000 m³ of snow with a 2000 m vertical drop at a speed of 36 km/hr; it comes to twenty million horse-power.

There is a wealth of statistics from the Alpine countries because the range is relatively thickly populated and because of the large numbers of visitors who come for the winter sports. Relevant research is carried out at the Swiss Federal Institute for Snow and Avalanche Research at the Weissfluhjoch, above Davos. In the greater ranges the avalanches must inevitably be on a larger and more impressive scale.

Ice avalanches are invariably the result of glacier movement; the ice moves to the edge of a drop and falls over. On 30 August 1965 one million m³ of ice broke off from the Allalin glacier in the Saas Valley, Switzerland, killing 88 workers on a hydro-electric site. The largest

avalanche accident on record took place in Peru on 10 January 1962. Between two and a half and three million m³ of the ice-cap of Huascaran Norte broke off and plunged from 6400 to 2400 m in the Santa Valley. The concussion started a secondary avalanche of 300 000 m³ which followed the first. As this mass flowed over the lower slopes more ice and boulders were torn away making the mass up to five million m³. Six villages were completely destroyed and three more partially; 4000 people and 10 000 animals lost their lives.

In very bad seasons the toll of avalanches is considerable. In 1950–51 234 people in Switzerland were buried in avalanches (probably the total for Alpine countries was 650 to 700). Some 1400 head of livestock were killed, while in some areas half the wild life was destroyed. A total of 2500 buildings were ruined as well as 60 km² of timber. In Switzerland total damage expenditure amounted to £2·2 million, while another £1·25 million had to be earmarked for the building of defences.

These defences take the form of sheds or tunnels for railway and road. For habitations woodland is planted, and deflecting walls and wedges of masonry are built, while higher up in the breakaway zone fences are erected to prevent the initiation of a fall.

## Minerals and Single Crystals

Metallic ore deposits are usually found in igneous or in nearby metamorphosed rocks, which means in mountainous or formerly mountainous districts. These materials are most accessible where considerable weathering has taken place, that is in older rather than in younger mountains. That metallic ores have not yet been found to any extent in the Alps or in the Himalaya may well be because they have not yet been laid bare by the weathering process. On the other hand, deposits are plentiful in the American cordillera, particularly at the plate boundary of the Andes, and in the Urals, an ancient plate boundary. Mining in the Andes, which is carried out at altitudes up to 5200 m or more, has produced a wide range of metals, as well as borax and sulphur. The miners live permanently at a height of up to 4500 m, a zone which would otherwise not be inhabited; the

outstandingly high-level railways in the area were largely built to serve the needs of the industry. The history of North America during the 19th century was punctuated by a series of 'rushes' to various parts of the western mountains when gold or silver had been discovered at some new location. Mining communities sprang up, thrived after their fashion and then faded away, leaving behind the 'ghost' towns which still dot the landscape.

Mineral substances when cooled from a molten state solidify to a characteristic crystalline form. If the cooling was fairly rapid, growth would be initiated more or less simultaneously at many sites throughout the mass, resulting, in the case, for example, of an igneous rock-like basalt ejected on to the surface, in the formation of large numbers of very small crystals. On the other hand, very slow cooling, which takes place

Large single crystal in S Dakota quarry (By courtesy of
the trustees of the British Museum (Natural History))

when ejected material is trapped below the surface, has sometimes led to the initiation of
growth in only a few isolated sites and produced
large gem or gem-type crystals, so-called single
crystals. These are valuable in some cases as
precious stones and in recent years in basic
scientific studies of the properties of materials.

Nature, having almost unlimited time at her
disposal, sometimes produces the ideal conditions for the growth of very large crystals. A
specimen of calcite discovered in Iceland was
$6 \times 2$ m; a crystal of beryl from Albany, Maine
weighed 18 tonnes and measured $5 \cdot 5 \times 1 \cdot 2$ m. At
Lacy Mine, Ontario, a mica crystal $4 \cdot 2 \times 9$ m
and weighing more than 90 tonnes was found. A
crystal of quartz from Hintze in Switzerland
weighed 640 kg. A crystal of spodumene 12 m
long has been reported. Only in recent years

have the processes been well enough understood
for them to be reproduced in the laboratory.
Quartz, ruby, sapphire and a host of others have
now been created artificially.

When a molten material consisting of several
constituent substances is cooled, the one with
the highest melting point crystallises out first.
It has unlimited space in which to grow and thus
fairly large crystals may be formed. Materials of
lower melting point crystallising later do not
find the same unlimited space. When the cooling
process is a result of ejection towards the surface, constituents having the higher melting
points, solidifying earliest, are deposited at
lower levels. Constituents with lower melting
points are deposited progressively higher. Thus
it is possible that a mine dug for one metal can,
by sinking to a lower level, tap an ore of a
different metal (eg in Cornwall, England, where
mines around the granite bosses, dug originally
for copper, were later deepened to find tin).

## Optical Phenomena Among Mountains

In connection with the sun (and to a lesser
extent with the moon) two classes of optical
phenomena can be distinguished:

1 Parhelia – concentric with the sun (or
moon); and
2 Anthelia – opposite the sun (or moon).

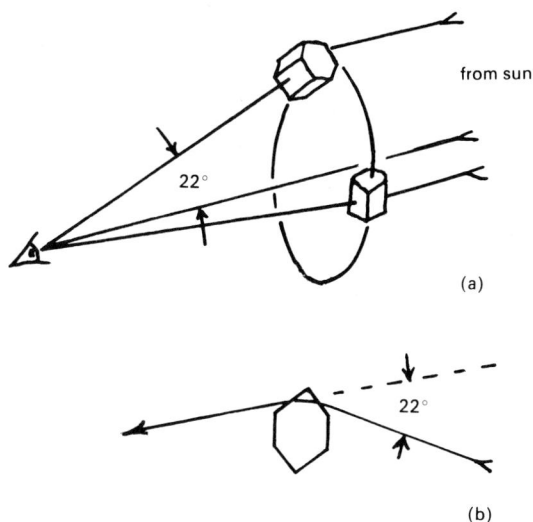

Fig. 11   Formation of haloes

## PARHELIA

### Haloes (see Fig. 11)

The name for general phenomena embracing large circles, bows, crosses and points of light, which result from refraction in hexagonal ice-crystals in the high cirrus clouds of the upper atmosphere. The 22° halo round the sun is the most common; it is often obscured by the brightness of the sun which has to be blocked off before the circle can be observed. A halo is coloured from red on the inside to blue on the outside, since the refraction is colour selective. 'Mock suns', which can appear on either side of a ring halo, are concentrations of light arising from a preponderance of ice crystals having their axes vertical. If there are many plate-like crystals, simple reflection effects produce light pillars, or even crosses, centred on the 'mock suns'.

During the descent of the Matterhorn, after the accident of 14 July 1865 (see p. 104), Whymper reported seeing '. . . a mighty arch, rising above the Lyskamm, high into the sky. Pale, colourless and noiseless, but perfectly sharp and defined, except where it was lost in the clouds, this unearthly apparition seemed like a vision from another world; and, almost appalled, we watched the development of two vast crosses, one on either side.' There does not appear to be a ready explanation for this, since he states clearly in a footnote that the sun was at their backs, so that it could not have been one of the straight-forward halo phenomena. We are left with a) the possibility that halo phenomena may also in rare cases be associated with 'mock suns' or reflected circles of light at 180° or even at 120°, b) some peculiar geometry of ice crystals, or c) a poor third – the supernatural.

### Coronae

These are diffuse rings of light, sometimes coloured, round the sun (or the moon) produced by diffraction in layers of fine water droplets. The radius is much less than that of a halo and the appearance less well-defined.

## ANTHELIA

### The Spectre of the Brocken (see Fig. 12)

This well-known optical phenomenon was first observed on the Brocken (Harz Mountains) and thus was named for it. The explanation is simple reflection. When the sun is low in the sky, at sunrise or sunset, and fog or mist is present in the opposite direction, an observer sees a shadow of himself, apparently huge, cast on the fog or mist. Since the sun's rays are effectively parallel, the shadow is in fact man-size and close at hand in the nearer portions of the mist, yet the illusion is strong, particularly when there is some visibility in depth, that it lies at the limit of visibility and is therefore much bigger. When observers are some distance apart each can only see his own 'Spectre', though the other shadow may appear formless like a tunnel in the mist. Observation must take place more or less along the line of the producing rays.

### Glories

A glory is a series of concentric coloured rings (with red on the outside) often seen in association with the Brocken Spectre around the shadow

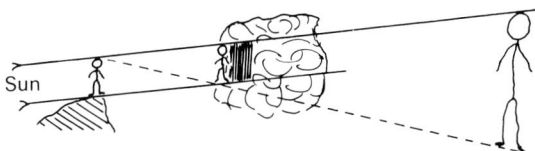

Fig. 12   The Spectre of the Brocken

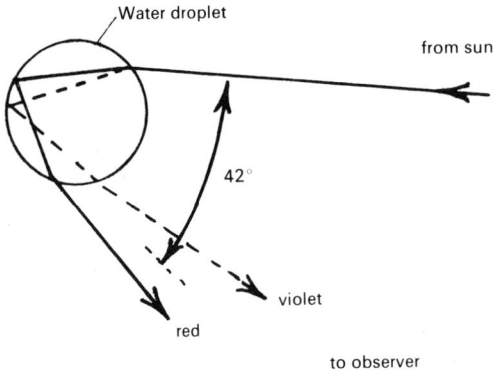

Fig. 13   Formation of a rainbow

of an observer's head cast on a mist. The explanation appears to be complicated – probably light is back-scattered from droplets in the mist and then refracted in further droplets to project a corona-type phenomenon towards the observer. The maximum diameter is around 12°. A fog-bow (here called a Brocken Bow) is sometimes produced under similar circumstances – the exact nature of the phenomenon depending on the size of the mist droplets.

*Rainbows and Fog-bows* (see Fig. 13)
These are formed by refraction and internal reflection in droplets of water suspended in the air. The angle between incident and emergent rays is about 42°, and the bow is formed by light from all points in the sky making this angle with the eye of the observer. Wavelength selectivity of the refraction process produces violet at the inner edge and red at the outer. A fainter concentric bow with colours reversed, which can sometimes be seen at an angle of 51°, is formed by two internal reflections in each droplet.

If the droplets are very small the colours mix and a grey fog-bow results in place of the multicoloured rainbow; this may be due to scattering rather than refraction and reflection and is much smaller in diameter, only 15°.

## LIGHT AND COLOUR OF THE SKY, THE GREEN RAY

Viewed from low levels the sky appears blue due to selective scattering by the molecules of the air, which tends to leave this colour preponderant. Sunlight itself, lacking blue, appears to have a pale-yellow hue. At higher levels, above the major part of the scattering layer, the sky is found to be blue-black and the sun and the moon bright white. There is much more radiation at the violet end of the spectrum and the skin (face cream) and the eyes (goggles) have to be protected from burning.

This change in scattering properties with altitude causes the mountaineer to err in estimates of distance, for at high levels objects are farther away than they appear to the eye conditioned at low levels.

When the sun is low in the sky, its light is relatively refracted by the long atmospheric path length. Yellow and orange light is substantially absorbed in the atmosphere, while blue and violet light is scattered, leaving red as the predominant colour in sunrise and sunset. The *alpengluh*, seen any fine morning among snow mountains, which for a short time are bathed in pink light, is a striking example of this effect.

However, there is also some green light present, and at the extreme position of the sun (at the point of disappearing below, or appearing above, the horizon) a small amount of it, originating between the absorption and scatter bands mentioned above, can appear momentarily as a coloured flash. Since the processes of absorption, scattering and refraction have all to be operating, and since the observer's view has to be clear to the horizon, this phenomenon, known as the Green Ray, is only occasionally seen.

## INTERVISIBILITY

Intervisibility between points on the earth's surface depends on the curvature and on the complex variability of the optical properties of the air along the line of sight.

Over distances of a few kilometres, simple geometry on a map enables the intervisibility of two points to be determined, taking account of the heights of intermediate points. For greater distances the curvature of the earth has to be taken into account and the effect of intervening heights can no longer be simply determined.

However, this is not too serious, since the air path travelled by light from the object to the viewing eye rarely has uniform properties throughout its length. Vertical temperature gradients are invariably to be found in the

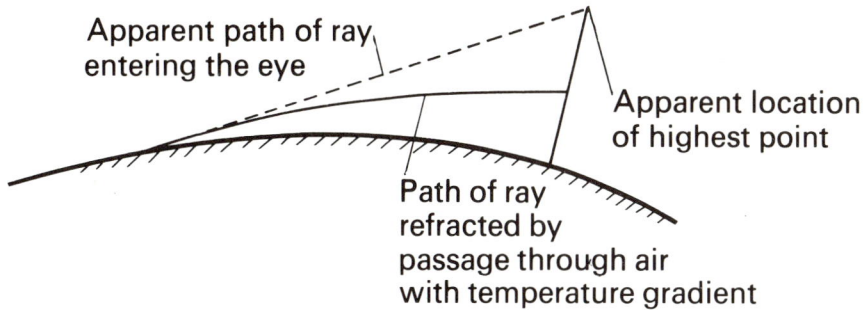

Fig. 14

atmosphere close to the surface of the earth and, since the optical properties depend on temperature, the path of such a ray can be refracted, or curved. The most usual, the most striking effect is to bend the path of the ray towards the earth, so that distant objects appear apparently at a higher level.

At high latitudes the temperature gradients vary widely. When the gradient reaches a certain critical value the refractive capability of the air will be sufficient to equal the curvature of the earth, which will then appear flat; above this value it will appear concave. Such temperature gradients, relatively commonplace in higher latitudes, particularly for lines of sight over the sea, have led to some interesting observations. From July 1939, there is a well authenticated report of a sighting of Snaefells Jokul (1446) from a ship in the Atlantic 500 km away. Gunnbjorns Fjeld (3700) has been seen from north-west Iceland (300 km away) and Vatna-

jokull (2000) from the Faeroe Islands (nearly 400 km away). It has been suggested that these phenomena may explain why Celtic and Norse exploration centuries ago spread towards the north-west, and in general for the ancient and medieval concept of a flat earth.

On land, the possibilities are limited since lines of sight, except in big areas of plains, are likely to be interrupted by intervening heights. Mount McKinley has been sighted from Mount Sanford (370 km apart); Mount Kilimanjaro from Mount Kenya (320 km apart); lines of comparable length may well exist in parts of Asia. In Europe the longest lines are likely to be across the north Italian plain from one part to another of the Alpine chain. ·

It is worthwhile noting that in measuring the height of Mount Everest by observations from the plains of India, corrections to the figures to take account of refraction can amount to as much as 420 m.

## Atmospheric Electricity Among Mountains

Thunderstorms and lightning take place, of course, all over the world wherever the requisite electrical conditions become established. The mountaineer, however, is more likely than most to be a close spectator. His peaks are very effective producers of storm conditions, while his viewpoint will often be close to, or even above, the thunderclouds.

When warm moist air rises above mountains and its temperature falls, the water vapour con-

denses, first as water droplets then higher up as ice crystals forming visible clouds, in which very turbulent circulation is taking place. As the vapour, the droplets and the crystals coalesce into rain drops or hail, there is a separation of electrical charges, as yet not completely understood, which results in a negative charge on the cloud-base and a positive charge on the cloud-top. The negative charge on the base induces an equal positive charge on the ground giving a

Lightning on the Matterhorn, drawing by E Whymper
(AC Collection)

Now a faintly luminous discharge begins to move down from the cloud in zig-zags of 50 to 100 m, the so-called stepped leader. When this reaches to within 50 m of the ground, an upward streamer of ions starting from some point on the ground reaches up towards it. The electric circuit is complete and thousands of amperes pass for a few microseconds in a flash which is intensely luminous since the air in the strike channel reaches a temperature of 30000°C. If there is sufficient charge on the cloud there may be several discharges along the same path taking place one after the other. The total energy released can amount to $10^9$ Joules. Inside a minute the cloud can recharge itself and the process be repeated.

In high mountains the summit may directly discharge a cloud, the stepped leader seeming to branch upwards from a prominent feature such as a rock peak or gendarme.

The almost instantaneous release of such a large amount of energy can cause serious damage to anything in its path. Small buildings may be wrecked, trees and rocks exploded, trenches blasted in the ground, fires started and electrical and communications equipment damaged. Holes, several centimetres across, have been punched in metal objects on mountain tops (eg the Madonna on the Petit Dru and the cross on the Géant) – the metal being vaporised at the point of contact. A strike on rock produces vitrification forming glass-like substances called fulgurites; in very steep places a stroke may lead to a rock fall (as on Il Gallo in the Bregaglia Alps during the 1920s).

A human body can be subjected to a direct strike or to a side flash when the body provides a path of lower electrical resistance alongside an object such as a tree that has been directly struck. The effects are failure of the respiratory system or disturbance of heart rhythm, with possible side effects of loss of consciousness, loss of memory, temporary paralysis lasting for up to a few hours, burns, temporary blindness and ruptured eardrums. On the exposed upper parts of a mountain pointed rock features like pinnacles, gendarmes or high steps on ridges will probably offer some protection; the lightning will tend to be initiated close to the top, leaving a lightning shadow for some distance down on either side. The mountaineer should still belay himself to the rock to avoid being

potential difference of up to $10^9$ volts between cloud and ground. Lightning is the transient high-current electrical discharge which cancels the potential difference within a cloud, between two clouds or between cloud and ground. The last is the spectacular lightning stroke most usually seen.

In open country the lightning process may operate as follows. As the charge separation continues in the cloud the potential between cloud-base and ground increases to a point where the air between starts to ionise. The first indication of this, denoting that a lightning stroke is imminent, is the appearance of luminous brush discharge (streams of ionisation which occur close to sharp points in the field where the gradient is the greatest) accompanied by a hissing noise. Such signs are a warning that evasive action is necessary.

shaken off by the mechanical shock of the strike. In order not to provide a path for a side flash he should avoid sheltering in vertical cracks or chimneys running with water, underneath dripping overhangs or close to any flowing water and he should crouch rather than stand.

Thunder is the acoustic signal generated by the rapid expansion of the air heated to a high temperature in the strike channel. The noise comes from the whole of the path length so that observers some distance apart will record quite different thunder patterns. It appears that under storm conditions sound waves are often refracted upwards, so that the thunder corresponding to distant flashes may never reach the ears of the observer.

St Elmo's Fire was the name given to the luminous electrical brush discharge when it was observed originally on the masts and rigging of ships, and later on aircraft wing tips.

Andes Glow (or Andes Lights) are terms used to describe illumination seen at night near certain Andean peaks. Though most frequently reported from South America, the phenomenon has also been observed in the European Alps, Mexico and Lappland. It has been seen in a cloudless sky and therefore is probably not conventional thunderstorm lightning as described above. It may well be large scale St Elmo's Fire arising from points on the mountain top; or, alternatively, the flash to earth of a charge which can build up on the upper surface of a haze layer, which nearly coincides with the mountain summit.

# Mountain Winds

On top of a mountain it is almost invariably colder than at the foot. In spite of the fact that the sun's rays are more powerful on the mountain top, the short wavelength radiation passes through the air without heating it. In fact the air receives its warmth from below – first, by reradiation in the infra-red of the energy received from the sun at lower levels and second, by contact with the earth and convection. The temperature gradient on a mountainside, known as the lapse rate, is typically 1°C for 165m of ascent. This is an average value; the actual value depends on the humidity and on the temperature.

In daytime the air close to a mountain, and thus heated by contact, is warmer than the air at the same level over the plains. The air near the mountain will rise and cooler air from above the plains will flow in to take its place, producing a wind up the mountain. At night the conditions are reversed, cold air near the mountain sinks and is displaced by warmer air from above the plains, producing a wind blowing down the mountain. This is known as a katabatic wind. The cold air descending at nights can be trapped in hollows in the mountainsides producing temperatures far below those observed on higher ground nearby. A case has been recorded of a hollow 170m deep in a plateau with no outlet below 50m. In the base of the hollow noon temperature in January was 0°C, falling to −25° to −27°C each morning; meanwhile on an adjacent hill top at the same altitude as the base of the hollow the temperature scarcely ever sank below 0°C and reached 8° to 10°C each day. A winter minimum of −51°C has been recorded in the hollow, while the minimum at the other site was only −19°C. This sort of situation can prove a menace to plant growers.

Examples of large scale katabatic winds include: the Mistral, blowing from the Cevennes down the Rhone Valley to the Gulf of Genoa, which is dry and cold and can gust at over 100 km/hr; the Bora, a similar wind in the Adriatic originating in the cold Hungarian basin; the Pampero, a cold blast off the Andes which sweeps the Argentinian pampas; the fall winds in Norwegian fiords; tremendous flows from the ice-caps of Antarctica, Greenland and Alaska.

Draughts of cold air are sometimes found blowing from cave-like openings at the ends of glaciers or from cave mouths on mountainsides. Such a cavity is open at both ends; the air within is cooled by contact below the temperature of the air outside and flows downwards and out at the lower end.

Another phenomenon of mountain regions produces warm winds which have distinctive names in various parts of the world – Föhn, Chinook, Zonda, etc. These are produced when a wind drives moisture-laden air up and over a mountain. As it rises, the temperature falls and some of the water vapour condenses as mist or rain; this process, however, generates some heat, so that the fall in temperature is not as great as if the air had been dry initially. (The lapse rate for moist air varies from 130 to 270 m/°C depending on the initial temperature, in contrast with a rate of 100 m/°C for dry air.) Now, when the air begins to flow downwards on the far side of the mountain, condensation ceases and the temperature rise comes to depend on the dry air lapse rate. The result is that the temperature rise on the down slope is greater than the temperature fall on the up slope; the wind reaches the plains much warmer than it was at the start while, because of the condensation that has taken place, it will also be drier. By this process a mountain range such as the Alps can raise the temperature of an air-stream by 10 to 15°C with a relative humidity of 30 per cent or less on the lee side.

These warm dry winds can produce some startling effects, not only melting snow cover with great rapidity with resulting avalanche danger, but also on the physiology and psychology of humans exposed to them. The Föhn blows from south to north across the Alps for some 40 days a year; in summer it dries timber and increases the danger of forest fires; it speeds up the arrival of spring, and helps to ripen crops in the autumn. The Chinook, which blows from west to east across the Rocky Mountains of Canada and the USA, has been known to raise the temperature from −25°C to above freezing point. The rapid melting and evaporation of snow cover exposes the grass and enables cattle to be kept out of doors in Montana all the year round. The Santa Ana is a similar north or east wind in lower California.

The sport of gliding relies substantially on upward-flowing air currents generated in hill and mountain areas. Air warmed by contact with the ground rises in currents called thermals which are utilised for soaring by the glider pilot, who seeks them by flying above the appropriate parts of a range.

He also makes use of a second phenomenon – the so-called lee waves; these have enabled gliders to reach quite astounding heights. When water flows over an obstacle the surface does not rise at the obstacle, but falls slightly while the rate of flow speeds up; a series of crests and

troughs of decreasing height is formed downstream as the water regains its original level. A similar phenomenon can be observed when air flows over a mountain range. Currents of air, rising and falling rhythmically, are set up within the flow; there are considerable areas of upcurrent and a path can be chosen for a glider following an ever rising line. A characteristic pattern of stationary clouds sometimes forms parallel to the range, at right angles to the wind direction, in the crests of the lee waves. These conditions can be found in many hill ranges whatever the altitude. They are often specially well developed in the Sierra Nevada of California, the summits of which rise 2700 m above the Owens Valley to the east, so that gliding altitude records have been made there (14 100 m) (see Fig. 15).

The most spectacular feature of these lee waves is the occasional formation of a giant air current called a rotor, in which a mass of air rotates about a horizontal axis directly beneath the crest of the lee wave. It can sweep up a wall of dust from the valley to a height of 5000 m and is very dangerous to any glider or aeroplane happening to fly into its path. Lee waves are one of the possible sources of that general menace to aviation known as clear air turbulence. Upward and downward motions of 600 m/min have been found at 7600 m in winds of 80 to 200 km/hr, but now that the reasons for their formation are better understood steps can be taken to avoid them.

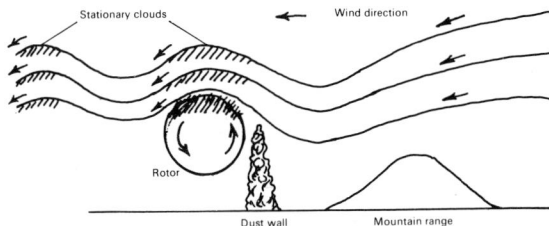

Fig. 15  Lee waves

# Mountain Medicine

Some ten million of the world's population live above 3650m, mostly in South America, about 1/5th of them in Central Asia. On the Central Asian plateau there is a pastoral/agricultural economy, while in summer yaks are driven up to pastures between 4875 and 5175m. Similarly in South America the pastoral/agricultural communities live between 3650 and 4250m, above which freezing temperatures occur at some time during every month of the year. Here, however, mining villages are found up to 5350m, the highest permanent dwellings in the world. Attempts to establish communities at even higher levels, where some of the mines are sited, failed due to the onset of mountain sickness, so this would appear to be the greatest altitude to which man can become permanently adjusted.

Above this height acclimatisation to altitude can take place for a limited period, but sooner or later the opposite process – high altitude deterioration – sets in and it becomes necessary to return to lower levels. Airline passengers take their low level environment with them in the form of a pressurised cabin; travellers on the trans-Andean railways are provided with oxygen to sustain them on their journey. Otherwise, mountaineers are the only people who spend much time above 5500m or so. The longest period spent at 5800m has been about 90 days; above this the periods shorten until two nights is the maximum at 8400m. However, in 1978 R. Messner and T. Habeler demonstrated that after appropriate acclimatisation it is possible to reach even the summit of Mount Everest (8848m) without the use of oxygen.

Adaptation to the high mountain environment involves adaptation to cold (shared with polar exploration) and adaptation to altitude. The first is achieved in the short term by provision of suitable shelter and by correct selection of clothing; considerable advances have taken place since World War II in wind proofing and insulation and in man-made materials having specially desirable properties. The effect of wind – the so-called wind chill factor – is of extreme importance. The chilling effect of a 65 km/hr wind at −12°C blowing on to bare skin produces an equivalent still-air temperature of −37°C. (The Americans have a 30-30-30 Rule – when exposed to a wind of 30mph (48km/hr) with a temperature of −30°F (−35°C), human flesh freezes solid in 30 sec.)

For long term adaptation the body seems to adjust to increase heat production (eg native porters walk bare-footed in snow; there is a case of a Sherpa who slept outside tents with bare feet at temperatures of −15°C at 5175m, yet his skin temperature never fell below 10°C). It appears too that ability to expose hands to cold is somehow increased by constant exposure.

For sea-level dwellers adaptation to altitude in the short term requires slow progress to higher and higher levels. Rapid transfer, say by mountain railway or téléphérique, from the valley to a point even of less than 3500m leads to distressing symptoms of shortage of breath – soon alleviated by return to lower levels. Adaptation to higher levels can be achieved within a few days by methodical upward progress, though some individuals have a ceiling above which they cannot reasonably go. The higher the ascent the shorter the period before deterioration sets in and a descent becomes imperative.

In the long term the body can readjust rate and depth of respiration, while the capacity to diffuse oxygen through the lung walls is increased; the red blood cell count increases enabling more oxygen to be held in the blood; changes take place that enable oxygen to be transferred more easily to the cells; the size of the right-hand side of the heart increases. Complete acclimatisation, such as attained by the mountain dwellers of Central Asia and South America, takes years to come about. Certainly people who have lived all their lives at these altitudes are on average shorter than their compatriots at sea-level, yet their lungs and hearts are larger.

The following are the principal disorders of altitude:

**Hypothermia** (exposure, general cold injury) denotes a reduction of heat content of the body due to the conditions to which it is exposed. The inner core temperature is reduced. Loss of consciousness accelerates the process and death may occur within a few hours. Cold/wet conditions, such as often met in Britain, are a major cause. Women are less susceptible than men, older men less than adolescents, short thick-set adolescents less than tall thin ones.

**Frostbite** denotes the formation of ice crystals between the cells of the affected part of the body. It attacks extremities (hands, feet, ears, nose) which the body's defence mechanism tends to sacrifice in order to save the inner core. Superficial frostbite affects only the skin and the tissues immediately below it. The afflicted parts blacken, but this eventually peels off to reveal replacement tissue below. Deep frostbite affects also bone, muscle and tendon and permanent loss of tissue is usually inevitable. However, after months the blackened portions slough off at the surface between dead and healthy tissue and amputation has only to be carried out as a last resort if infection occurs.

**Hypoxia** is the name given to early manifestations of oxygen lack, which vary with speed of ascent, duration of exposure and individual susceptibility. The symptoms are breathlessness, headache and impairment of mental function. If a man goes from sea level to over 7000m he will become unconscious after 15 min, to 9000m after only 2 to 3 min. South American Indians living at up to 5350m can withstand exposure at 7000m for long periods, as can mountain climbers who have acclimatised by slow upward progress.

**High Altitude Deterioration** describes the eventual effects of prolonged sojourn over 5350m. The higher the climber goes the more rapid and more severe the eventual deterioration, even to one who appeared to be fully acclimatised. Loss of weight and decrease in work capacity are the usual symptoms, though outside factors, such as poor living conditions and inadequate food and fluids, are likely to have similar effects and contribute to the condition.

**Acute Mountain Sickness** (seroche, puna) was first described by Fr Joseph de Acosta when crossing an Andean pass of 4275m in 1750. It is a result of too rapid a transfer over a considerable elevation and produces nausea, headache, vomiting, lassitude, weakness and loss of work capacity. The remedy is rest or descent, the prevention a slower rate of ascent. Because of the relatively rapid ascent involved, passengers on the trans-Andean railways are given oxygen to mitigate the distressing symptoms; the local Indians achieve the same results by chewing coco leaves, which contain cocaine.

**Chronic Mountain Sickness** (Monge's Disease), only reported so far from South America, implies an inability to acclimatise at all over 3650m.

**Pulmonary Oedema** is due to fluid accumulating in the lungs and can rapidly prove fatal. The onset is relatively sudden even to people who otherwise appear to be acclimatising well. The sufferer becomes breathless, cyanosed and weak with a rapid pulse. The first recorded case was the death of Dr Jacottet in 1891 at 4235m after climbing Mont Blanc (4807).

**Cerebral Oedema**, a similar condition affecting the skull, is likely to leave permanent damage.

**Thrombosis,** where blood clots in the veins, recorded on some high altitude expeditions, is precipitated by dehydration and inactivity. Though occurring relatively infrequently, it may well prove fatal.

**Mental Symptoms** are produced in some mountaineers at high altitudes. Their resulting irrational behaviour may be dangerous to themselves and their fellows.

Since World War II there has been extensive study of many of the aspects of mountain medicine with the result that the causes of the various disorders are much more clearly understood and treatment methods have been developed to minimise their impact. Mountaineers who have spent long periods at high altitudes without experiencing the more serious of the disorders outlined above do not appear to suffer any lasting or long-term effects.

## Living in Mountains

Hills and mountains, abounding as they do in natural defensive sites, have always served as a refuge for the beaten or the oppressed and often, too, as a hideout for those outside the law. A raised site, having steep or precipitous slopes on either hand, giving a good view of the surrounding country and approached only by narrow and difficult ways, would well serve their various purposes.

Thousands of years ago on the chalk hills or cliff tops of England primitive man built forts defended by rows of ditches and banks, unas-

*Above:* Arch, South-western desert, USA (G Flanagan). *Above right:* Bryce Canyon, Utah, USA (G Flanagan)

*Below:* Bernina-Scerscen, Switzerland (W Kirstein)

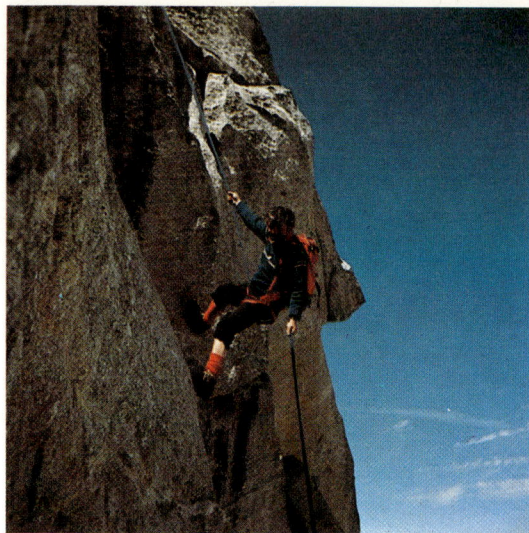

*Above left:* Mountaineering party on the Mettenberg, with the Eiger behind, Switzerland

*Above:* Abseiling (W Kirstein)

*Left:* Ballooning in the Bernese Oberland, Switzerland

*Below:* Surveying in Britannia Range, Antarctica (C Swithinbank)

sailable with the weapons of the time, which he
may have occupied continuously or, maybe,
only when his enemies came. For similar reasons
the Pueblo Indians of Colorado built their
homes on buttes and mesas, which could easily
be defended against their enemies. Among limestone hills early man used the natural caves for
shelter, leaving his artefacts and the rubbish of
his living to be covered over on the floors and
sometimes making primitive paintings on the
walls. In softer rocks elsewhere similar shelters
were hollowed out for occupation.

Throughout the ages unassailable summits
have continued to attract as easily defended
sites. Before the days of gunpowder the medieval
castles of Europe and the Middle East, often
built on isolated hills or on the ends of spurs,
were certainly more or less impregnable and
usually succumbed only to starvation. Religious
buildings were often similarly placed, eg in
Greece, monastries crown the hills of the
peninsula of Mount Athos. Further south at
Meteora others are sited on steep rock pinnacles; visitors to Aghios Stefanos, built on a
rock separated from the mountain behind by a
ravine where there is now a bridge, had formerly
to be drawn up in a net; Metamorfosis was only
accessible by rope ladder or basket haulage,
now there is a rock staircase. There are temples
on Tai-Shan and Omei Shan in China and on
Fujisan and Ontakesan in Japan. A chapel
crowns Mount Sinai and there is a chapel and a
cathedral on volcanic plug pinnacles at Le Puy
in France. Religious houses for the succour of
travellers were built from the earliest times on
the summits of the principal Alpine passes, that
on the Grand St Bernard, for example, dates
from the 9th century.

Sometimes whole towns were perched on hill
or plateau tops. An example is Machu Picchu in
Peru, an Inca stronghold on top of a steep hill,
600m above the Urabama River. Three sides are
precipitous and encircled by the river, the fourth
is a steep ridge guarded by a fort.

Some world countries are at high level – the
average level of the Tibetan plateau is nearly as
high as the summit of Mount Blanc. Lhasa, the
capital, stands at 3000 + m, but there are monastic buildings, eg at Rongbuk below Mount
Everest, as high as 4900m. Cuzco in Peru
(population over 120000), which was the capital
of the Incas, stands at 3350m, while there are

**Hut on the Matterhorn (AC Collection)**

mining villages as high as 5350m. Of Bolivia,
220000km² are above 3650m; La Paz, the
capital, which is at 3625m, has a population of
over 660000.

In the early days of mountaineering, peaks
were climbed up and down in one continuous
trip from the habitations of the valley. As
increasingly difficult problems came to be
tackled it was realised that a starting point near
the snow-line would enable more daylight hours
to be spent on the mountaineering proper, so
tents, equipment and food were carried up by
extra porters to the highest rocks and returned
by them to the valley the next morning after the
climbing party had set off upwards. However,

The Mountet Hut with the Obergabelhorn,
Switzerland

Mountain camp, Alaska (H Adams Carter)

some of this complication could be eliminated if a large overhanging rock could be found for shelter, improved perhaps by the building of a rough wall on the open side. Some of these bivouac sites, or gîtes, such as the Couvercle on the Mer de Glace on Mont Blanc, became famous and were frequently used in the last century.

At the best of times gîtes were draughty and damp and the obvious answer to the overall problem was to build huts on sites which were safe from avalanche and completely convenient for the mountaineering, rather than use fortuitously occurring rocks. At first they were primitive shelters with straw for bedding and a few cooking utensils, and climbers had to bring their own fuel. Later a local man, probably an ex-guide, was enrolled as warden in the summer – the structure was thus better preserved and fuel and meals could be provided. Nowadays the main huts in the Alpine countries, some accommodating over 100 persons overnight, are sited on the highest rocks immediately below the summer snow-line and they provide an excellent service. There is a dining room, common room and kitchen, quarters for warden and guides and dormitories for men and women. Climbers' huts covering the whole range from primitive to palatial are found now in mountains all over the world. They contribute substantially

to mountaineering achievement and to the social life of the sport; they provide shelter and rescue facilities in the event of accident or bad weather. Some can be reached by road, others are supplied and provisioned by helicopter. Overcrowded huts are a sad feature of the popularity of mountaineering, so that camps and bivouacs must once again be sought by the climber who wishes to be apart from his fellows.

More remote ranges have always demanded the use of tents, and special models for mountains have gradually evolved. Some used ice-axes or ski-sticks for poles and were held down by snow or ice piled on flaps at the sides. Double tents, originally developed for polar work, have also been used at high altitudes.

Later, when difficulties increased to the point where climbs took several days to complete, the bivouac sack came into use. This was a water-proof pole-less tent supported by the climbers' bodies; it certainly served to pool the party's warmth, even if not much sleep was possible. Around the same time some very small huts were constructed here and there in various parts of the Alps to serve as a basic shelter, accommodating, at the most, half-a-dozen climbers. Relative inaccessibility confined their use to experts.

Bivouacs above the snow-line are taken in snow caves – a trench excavated in a snow field

A mountain bivouac, the Alps (M Millet)

roofed by a ground-sheet covered with snow or a hole in a snow bank walled in at the end. A partially choked crevasse can also serve a similar purpose in an emergency. In less desperate circumstances in igloo, a dome-shaped structure of snow blocks, can be constructed.

During the last two decades the bivouac box has largely replaced the tent for high mountain work. It is a framework of angle-iron covered by stout canvas and is indeed more secure and weather resistant in fierce snow and wind conditions. On hard and lengthy rock climbs individual hammocks hanging from pitons are nowadays used for bivouac.

Most mountain areas have their distinctive styles of architecture, based on the materials locally available for the construction of dwellings, on national individuality and on the methods devised for protection from the elements. The subject is a vast one. A distressing feature of the last two or three decades is the proliferation of ski resorts in all mountain countries. Huge complexes of bedrooms and public rooms, with shops, supermarkets, casinos, dance-halls and all manner of recreational facilities have sprung up on mountainsides previously deserted. The ramifications of the ski industry are likely to increase in the coming years.

## Mountaineering

Originally (since 1610) a mountaineer was a dweller among mountains. Only slightly more than a century ago, coinciding with the rise of mountain climbing as a sport in the European Alps, did the word come to signify a climber of mountains, and the sport itself to be called mountaineering.*

At the beginning the aim of the mountain climber, whom henceforward we may call mountaineer, was the ascent to the tops of Alpine mountains by the easiest and most

* Note the recent coining of the words 'coasteering' for climbing activities carried out at coastlines, 'oceaneering' for underwater sporting activities at sea, 'orienteering' for a sport of cross-country running combined with route-finding, 'canyoneering' for climbing on the walls of the Grand Canyon.

straightforward route. When all the major summits had been climbed (the last in the Alps was the Meije in 1877), mountaineers extended their activities to include ascents of mountains by routes other than the easiest or most straightforward, and ascents of minor summits on mountainsides and on ridges. In due course the achievement of any selected line anywhere on the mountain, that is the process of climbing itself, replaced the attaining of mountain tops as the true objective of the sport.

These developments have gone farthest in the Alps, where the multiplication of climbers' routes has been considerable, so that in many cases there are numbers of different lines on the same area of mountainside. Most routes have also been climbed in winter months (between November and March) when extreme cold and short days aggravate the climbing problems. Many routes have been climbed on skis and there is a new vogue for their use on extremely steep descents. Then there is the *direttissima*, which tries to avoid any deviation right or left of a straight line from the base of a face to the summit, and the lengthy traverse passing over several moutains with possibly some bivouacs on the way.

Developments elsewhere are mostly less advanced. While many summits in the Andes and the great ranges of Asia have fallen to mountaineers during the last three decades, there still remain many lower points as yet untried, while major peaks are coming to be tackled by alternative lines. In more remote parts of the world large numbers of summits are still unclimbed

An early boot (AC Collection)

because of physical, or even political, inaccessibility.

Starting from the basic items of a rope and nailed boots, mountaineering equipment has developed over the years to a sophisticated collection of paraphenalia, which now supports a sizeable industry. In the earliest days the purpose of the rope was unclear for, while it enabled stronger climbers to help the weaker to ascend, it united the party in such a way that, if one fell, all fell. In fact, in the beginning, guides held the rope but did not tie on to it. However, in due course it was realised that temporary attachment of the rope to the mountain enabled most of the party to be safeguarded most of the time, and certainly while one of them was making a potentially dangerous move. This came to be called belaying.

In the first place boots were nailed for the protection of the soles and thus for prolonging the wear. But nails of suitable shape could also help to adhere to small projections and rugosities. Metal nails of a wide range of types were in use for almost a hundred years, only being replaced as late as the 1940s by rubber soles with a nail-like pattern moulded on as an integral part.

The mountaineers' first 'tool of the trade' was the alpenstock – a pole some 2 m long, spiked at one end so that it could be driven into snow, and reasonably soon fitted with a cutting blade at the other to enable ice to be fashioned. Over the years the latter function came to predominate; the head with an adze blade for snow and a pick for ice functioned best on the end of a shaft of 90 cm – the ice-axe. Guides, mostly, but also some amateurs, carved fantastic ladders of steps for hundreds of feet up tremendous ice slopes with their aid.

The first belays were convenient pieces of the mountain – pinnacles, prongs or projections of rock, knobs or bosses of ice – many just serving very temporarily as the climbers passed on their way. The axe could be used to fashion a belay; the axe itself could serve as a belay with the shaft driven into snow or wedged across a fissure in ice. It also served as a third leg or arm in many diverse situations.

The turn of the century saw the first application of crampons in mountaineering. These are frameworks of spikes strapped to the boots which enable the wearer, after suitable confi-

**Modern ice climbing (J Tasker)**

Nowadays the adze blade is often replaced by a hammer head, which is used for the insertion of pitons, metal spikes which enable the mountaineer to make belays where the mountain does not provide them.

World War II marked the introduction of new materials for mountaineering ropes, which till then had been made from natural fibres and easily became unmanageable when wet or frozen. Man-made fibre ropes were markedly more resilient, their tremendous extension under loading making them much more suitable for absorbing the energy of a falling climber without breaking. First developed under the stimulus of military need, the last two decades have seen great advances in manufacturing methods, in specification and in testing under conditions which simulate those of the mountain.

Clothing has been similarly revolutionised in the same period. A century ago, climbers wore heavy hard-wearing tweeds; then as the sport expanded between the wars the range of clothing types increased to embrace all manner of garments transferred from more mundane applications. Close-woven windproof materials (such as Grenfell Cloth) made their appearance, as

**Mountaineering party on a crevassed glacier in the Stubai Alps (By courtesy of the Austrian National Tourist Office)**

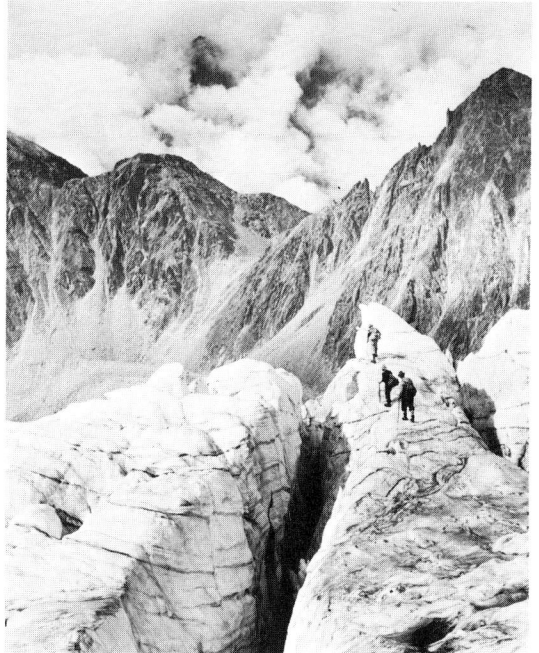

dence has been attained, to walk up ice slopes of quite considerable steepness, thus making obsolete the prolonged cutting of steps. Crampons had been in use for centuries by travellers in various mountain countries to assist them in going about their ordinary business, being more appropriate under some conditions than the more widely used snow-shoe. Now since World War II the addition of two further spikes pointing forwards at the front of the crampons has enabled them to be used on almost vertical ice faces.

For such steep ice the conventional ice-axe was no longer suitable; it was too heavy and unwieldy and could not be swung in the constricted space. Shafts have become shorter, while picks have assumed a variety of special shapes now that the tool serves as a movable handhold rather than as a fashioner of steps.

**Mountaineering in Alaska (H Adams Carter)**

well as heavy canvas, corduroy, moleskin and so on. Again the big changes came with World War II. Windproof and weatherproof garments, produced originally for the Armed Forces, were disposed of on the surplus market and clothed the new generation of climbers. Then followed specialisation as ingenious designs of fabric, in many man-made materials, and of clothing itself offered increased protection, combined with lightness and strength. Nowadays, too, the crash helmet, to protect the wearer against falling stones or in the event of turning over in a slip, has entirely displaced the felt hats, balaclava helmets and so forth of yester-year.

In the Alps climbs are almost invariably made from one of the very large numbers of huts now available, though even today the tent or bivouac is occasionally preferred. Climbs of the highest standards may involve one or more bivouacs on the way up.

Tents, or bivouac boxes, continue to provide shelter in all the more remote mountains. Very high mountains, needing more than a day to climb, are overcome by placing a series of camps higher and higher until it is possible for the ascent to be completed from the topmost. This is the standard technique of expedition mountaineering. Many climbers are involved in placing and provisioning the camps, but only occasionally do more than a few reach the summit.

Alternatively, if a small party tackles a big mountain by carrying all their equipment to a first camp, strikes that and continues to a second, then to a third and so on, they are said to make the ascent 'Alpine style'. They are cut off from the foot of the mountain – in contra-distinction to expedition style mountaineers who have a supply line running down through their chain of camps, by which they can be helped if in difficulties.

## Rock Climbing

In mountains with substantial snow and ice cover rock climbing plays only a comparatively minor role in mountaineering. On the other hand, below the snow-line when mountains are too steep to be climbed by walking, rock climbing becomes the whole of mountaineering. The supply of inaccessible summits is limited, so to exercise his skills the rock climber has had to

**Modern Alpine rock climbing (M Millet)**

seek lines on the steeper sides of mountains or even on crag faces which may be only an incidental feature of the mountainside. Thus in many cases he has had to abandon the established goal of the mountain climber – the attaining of summits. There certainly were for him a number of minor elevations in the form of isolated rocks and pinnacles conforming to the traditional pattern – but many of these had been climbed anyway in earliest times, either for utilitarian purposes (eg the sea-stacks of St Kilda for sea-bird's eggs) or for the challenge of inaccessibility they presented to the adventurous (Toverkop, South Africa; Mont Aiguille; Devil's Tower, Wyoming and so on).

Nowadays his activities are mostly directed towards climbing for its own sake, wherever it may be found, however insignificant the rocks against the background of the landscape as a whole. Thus he also climbs on crags in flat country away from mountains, on quarries where nature has neglected to expose climbable rocks at the surface, and on sea cliffs. However, all these climbing sports share common techniques and equipment, and in fact the whole pageant from sea cliffs and rock outcrops to the Andes and the Himalaya is regarded as one integrated whole by the true enthusiast.

It is interesting to note the interplay in a non-mountainous country such as Britain between the mountaineers, inspired to tackle the problems of home mountains as practice for mountaineering in foreign snow and ice ranges, and the local schools whose interest stemmed from the obvious challenge presented by their local rocks. These groups operated in parallel at first, but soon came together; finally specialisation in the home activity produced a sport no longer a sideline of snow and ice mountaineering, but one having its own traditions, methods and ethics, and these may well differ from country to country.

Shorter expeditions, carried out under much more predictable conditions, enabled rock climbers among low mountains to carry out their activities more deliberately. At first they carried ice-axes but these were soon left behind; they used ropes much the same as did mountaineers in the Alps but before long the technique of belaying became formalised – only one of the party moving at a time with the rest securely belayed to the mountain. At first the belays were large spikes, bollards of rock, boulders and stones wedged naturally in cracks; as standards advanced much ingenuity was displayed in the utilisation of smaller and smaller features –

thinner ropes were used for climbing and belays were made with very thin loops indeed and even with loops of wire; while, where nature did not provide anything suitable, pitons (metal spikes) were hammered into cracks and mechanical wedges of a variety of forms inserted wherever they would grip. Thenceforward it was only a short step to using these same belaying devices as actual holds for hands and feet. The introduction in the 1940s of man-made fibre ropes of considerably enhanced elasticity, in place of the natural fibre ropes hitherto employed, led to great advances over the whole range of climbing techniques.

In the early days soft iron nails were placed all round the edges of the boot soles with wings at the sides to protect the welts. Hard steel nails from the Continent were in vogue between the wars, while World War II saw the advent of soles with moulded rubber nail patterns. In the first decade of the century, tennis shoes, which conformed to the minor contours of each foothold, were introduced for the hardest climbs; these gave way later to the very cheap rubber-soled plimsoll; nowadays there are a host of special designs of tight-fitting footwear.

Rapid descents are accomplished by a technique, known as abseiling, involving controlled descent of a doubled rope. The rope, passed behind a projection or through a rope loop, is pulled down and the manoeuvre repeated. The rope alone can be used but certain modern gadgetry can serve to facilitate the process.

The climber's route – the line he chooses on the crag – is an interesting concept. Geoffrey Winthrop Young has described them as 'pre-existing, but concealed lines of possibility contrived by nature and by time up steep and unknown cliffs'. Above all else the rock face must present an acceptable mix of vertical lines of weakness (clefts and cracks providing an upward way which is continuous, yet bordered on either hand by rock which cannot be climbed except at a markedly higher standard) and horizontal lines of weakness (providing holds, and ledges suitable to divide the climb into acceptable lengths, or pitches, for the leader). Where vertical lines predominate, the climbing is likely to be very hard; where horizontal lines predominate the route becomes capable of so much divergence on either hand as to lose a separate identity. Advanced safeguarding tactics and the development of climbing skills have in some places multiplied the 'concealed lines of possibility contrived by nature', and, where nature has failed to provide, man has supplied the deficiency with pitons and other devices. The unclimbable areas in between have shrunk until the lines are hardly independent.

The Great Walls of the world are the ultimate expression of rock climbing. Variously tackled in the course of the last half century have been: the three great North Faces of the Alps (the Eiger, the Matterhorn and the Grandes Jorasses); the rock walls and pinnacles of the deserts and canyons of western USA; the walls of the Cumberland Peninsula on Baffin Island; the rock peaks of Patagonia and of the Dolomites; the Trolltind Wall in the Romsdal in Norway. Some of these have their difficult rock problems augmented by the presence of ice and snow and by atrocious weather.

# Skiing

Stone Age skis, dated to 2500 BC, have been found preserved in peat bogs in Scandinavia and Finland. Throughout the ages their use for travel, for hunting, and for purposes of war is authenticated in numerous historical documents. Scandinavia remained the home of skiing which, by the beginning of the 19th century, began to emerge as a sport as well as a utility. By the middle of the century Norwegians had introduced ski to Australia (where a ski club was founded at Kiandra in 1878), Japan and California (where an enterprising immigrant inaugurated a ski-carried postal service across the Sierra Nevada). An exhibition of ski-jumping was given in Christiania in 1868, while annual ski races were being run in various Scandinavian countries by 1870.

Within the next decade the sport began to be practised in Switzerland, Russia and the Black Forest. The appearance of skis in Scotland in 1892 was reported by a mountaineer, who added – 'In the Alps it is not unlikely that the sport may eventually become popular' – he cannot have imagined the scale of his second sight! However, it was the crossing of Greenland by Fridtjof Nansen in 1888 which really fired men's imaginations. Systematic ski schools started in the 1890s and books on technique began to appear.

Ski-mountaineering, wrote Arnold Lunn, is 'the result of the marriage of two great sports'. The ski exploration of the Alps began with passes – the first outstanding expedition being that led by W. Paulcke in 1897 from Meiringen by the Oberaarjoch and the Grünhornlücke to Belalp and Brigue. Paulcke attempted Monte Rosa in 1898, while Hoek (another ski pioneer) climbed the Finsteraarhorn and the Mönch in 1901, and the Wetterhorn in 1903. The first ski ascent of Mont Blanc came in 1904, by which time seven other 4000m peaks had also been climbed. New expeditions in the Alps by natives of the Alpine countries and by the British followed fast in succeeding years, including long-distance trips from hut to hut along mountain ranges, like the famous High Level Route from Chamonix to Zermatt.

In due course skis appeared in more distant ranges. In 1930 Frank Smythe took ski almost to the summit of Ramthang Peak (6700) and the next year R. Holdsworth reached Meade's Col (7138). In 1934 Norman Watson's ski party crossed the Coast Range of British Columbia in an early approach to Mount Waddington.

The last decade has witnessed a new and startling development – the descent on skis of climbing routes of high standard and of quite unusual steepness (up to 60°). These feats are of course accomplished solo and as yet there are few practitioners. Skis have been taken to the South Col of Everest (by an expedition employing 800 porters and 27 tons of equipment), from which Y Miura descended into the Western Cwm with the aid of parachute braking. A skier, Sylvain Saudan, has descended from the top of Mount McKinley in Alaska and from the top of Nun in the Punjab Himalaya – descent from an 8000m peak will certainly be attempted within the next few years.

Other aspects of this popular sport can be briefly noted. Langlauf is long-distance cross-country racing on skis, of which the Scandinavians are the leading exponents. Ski-jumping from the end of a specially prepared ramp also originated in the same part of the world, where the jump at Holmenkollen was one of the first. The world record is 179m. Both these are the subject of Olympic contest.

Ski mountaineers on Cevedale, Ortler Alps (W Kirstein)

Ski-tracks in the German Alps (Photo: Beckert/ Kölbl – Garmisch-Partenkirchen)

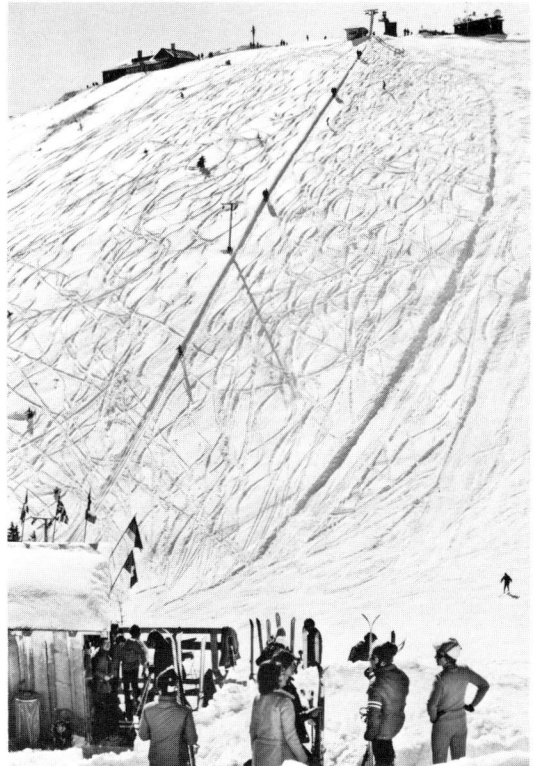

Downhill ski-ing is now tremendously popular. A recent estimate suggests that £10 000 million per year is now spent on equipment and holidays in all parts of the world; the industry provides one million jobs. In Alpine countries one and a half million beds in Winter Sports resorts register 100 million bed/nights each winter; there are 2000 aerial cable-cars and 8500 ski-tows. In Switzerland alone the Winter Sports industry contributes £900 million to the gross national product, while half the passenger fares paid on the Federal Railways derive from cable-cars and ski-tows. To cater for the growing number of enthusiasts there is a vast programme of resort building in every world country that has mountains of the requisite height to provide accommodation, transport, access, snow-making machines, rescue services, après-ski facilities and so on. These constitute perhaps the major threat to the mountain environment.

## Caving

Caves form a significant part of the world mountain scene. Nearly all caves are in older limestones (newer limestones like chalk are not strong enough to support chambers), but there are caves also in lavas, gypsum and ice. The caves in limestone result from the slight solubility of the rock in acid rain-water, which acts continuously therefore to enlarge horizontal and

**In Mammoth Cave, Kentucky (Barnaby's)**

vertical fissures in the mass. The depth of caves is set by the thickness and tilt of the strata, the length by the absence of serious faulting. Surface water running on to the limestone disappears into swallets, sinks or pot-holes and burrows through these fissures, penetrating through the years to lower and lower levels. The result is a honeycomb of caverns with dry passages higher up and water-courses at a lower level; finally on reaching the valley water-table the water emerges from the limestone once again, often flowing as a wide and fast river from an archway – the so-called resurgence. As erosion proceeds, a hill may finish up pierced with caves from top to bottom, a complex system which may be entered at many levels, though in the lower reaches the passages may be waterlogged and impenetrable.

Mountain Limestone is perhaps the commonest rock type in the world; almost all countries have hills and mountains of it, providing considerable climbing opportunities above ground. Thus there are caves to explore everywhere. As in mountaineering, exploration has proceeded further in Europe, where cavers have been in action for a century or more. Further afield innumerable huge caverns await discovery and exploration.

**The deepest known caves are in France** – the Gouffre de la Pierre St Martin ($-1332$), the Gouffre Jean Bernard ($-1208$), and the Gouffre Berger ($-1141$). **The cave with the longest passages** is the Flint-Mammoth Cave System in USA (290km). Holloch in Switzerland is 120km, and at the same time over 800m deep; Peschtscheva Optimititscheskaya in the USSR is 104km. However, the exploration of a cave system of this magnitude is a gigantic task, comparable with a mountaineering assault on a major Himalayan peak, involving successive camps progressing into the depths, so that development must, of necessity, take place only slowly. A proportion of known caves are maintained as show places, illuminated and open to the public on conducted tours. Some, too, are important archaeological sites.

The caver has much in common with the mountaineer. His also is a sport of exploration, of penetration into the unknown, but while advances in mountaineering can nowadays be made only by the most expert, there is still plenty of scope below ground for the determined enthusiast. The mountain climber descends at the end of his day, but the caver must climb at the end of his – a much more arduous procedure. Common techniques and equipment are used, although the ethics of caving permit things which the ethics of climbing forbid and vice-versa. A few individuals have made their marks on both sports.

# Walking

'Mountaineering', writes Francis Keenlyside, 'is an elaboration of the simple experience of walking up or down hill'. Indeed most climbing of mountains is basically walking.

On many snow mountains a step can be fashioned where required by a smart kick with the climbing boot, while the ice-axe serves only as a walking-stick. Among lower mountains only a few peaks require the use of the hands. Uphill walking is mainly a rhythmic plod, heels down at every step so as to employ the strongest leg muscles for lifting, coordinated with rhythmic breathing. Suitable occupation for the mind while this process is in train is one of the chief problems of the hill climber. Another is the 'false summit', which once attained reveals another a short distance further on; several such may have to be surmounted before the true top is underfoot.

**Walkers' route in Switzerland**

On gentler hill terrain any line of ascent will serve, but steeper hillsides, rock strewn slopes, gorges and thick vegetation cover demand paths if satisfactory progress is to be made. In these cases some form of waymarking has to be introduced to ensure maintenance of the line where from below many alternatives seem possible. This takes the form of signposts, cairns of stones or splashes of coloured paint on rocks and trees; the last method is prevalent in continental Europe. The walker has to supplement such indications by understanding and using a map and compass and by having a 'feel for the country'. To be able to move freely on very steep ground, and to be able to use the hands for simple climbing and scrambling, are important assets on really rough terrain.

Some countries have catered for the walker by setting up a network of long-distance footpaths, suitably mapped, guidebooked and way-marked; many of them traverse hill and mountain country. The USA led the way with the Appalachian Trail (Katahdin, Maine to Oglethorpe, Georgia – 3300 km), the Pacific Crest Trail and others. After World War II a number of long-distance footpaths were designated in England (totalling around 2500 km), while France is developing a complex network all over the country (more than 10000 km), among them many which take the traveller to vantage points in the Alps and in all their lesser ranges. Switzerland has a series of long-distance routes running from east to west across the country, totalling 1600 km, again in some cases sampling the scenery of the highest peaks (in addition there are over 30000 km of shorter way-marked paths). Long-distance routes have been set up in West Germany (Flensburg on the Baltic to Genoa on the Mediterranean is some 2000 km) and in Sweden (the 350 km Kings' Way through the northern mountains).

## Mountain Railways and Cableways

These can be divided into three categories depending on the means used to support the passenger cars and on the motive power, as follows:

### 1 Rail support, rail drive

(a) Simple adhesion (ie the weight of the engine on the rails is sufficient to make the wheels turn in contact with the rail) as used on conventional railway systems. The maximum gradient is around 1 in 15 on metre gauge (eg Montreux – Bernese Oberland Ry) and around 1 in 35 on standard gauge (eg Lötschberg Ry). Ingenious methods of climbing by simple adhesion have been used in various parts of the world, eg by using spiral tunnels to increase the length of the line between two points, or by zig-zagging, where either the engine has to run round the train each time there is a change of direction or else it alternately pushes and pulls. In Switzerland the summits reached by adhesion running are 2257 m at the Bernina Hospice on metre gauge and 1240 m at the Lötschberg Tunnel on

High viaduct, Swiss railways, Switzerland

The Furka-Oberalp railway and zig-zags of the road near Gletsch, Switzerland

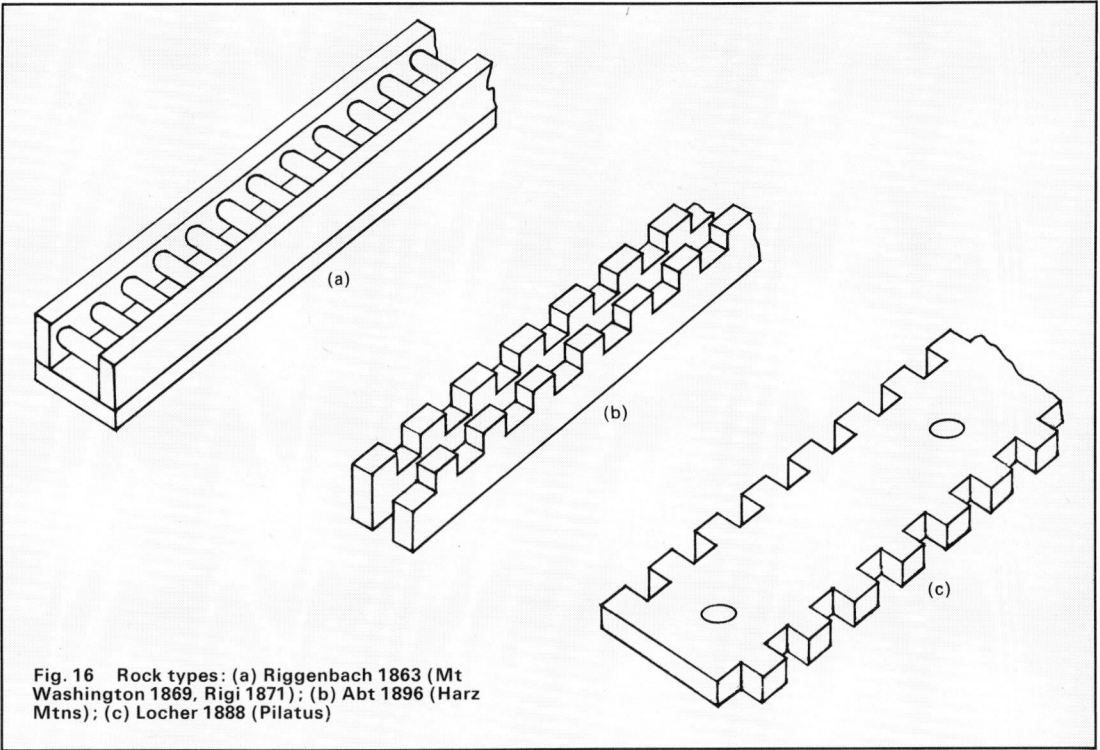

Fig. 16 Rock types: (a) Riggenbach 1863 (Mt Washington 1869, Rigi 1871); (b) Abt 1896 (Harz Mtns); (c) Locher 1888 (Pilatus)

**Pilatus railway, Switzerland**

standard gauge. Metre gauges reach considerable altitudes in South America (qv).

(b) *Augumented adhesion*. An additional horizontal rail provides extra adhesion for horizontal wheels on the engine. Known as the Fell System, this was used on the railway which crossed the Mont Cenis before the boring of the tunnel, and in the crossing of the Rangitata Range in the North Island of New Zealand.

(c) *Rack and pinion* (cog railways; Zahnradbahn; chemin de fer à cremaillère). Here additional toothed rails of various types (see Fig. 16) mesh with toothed wheels on the engine and thus provide a non-slip driving means. The first railway of this type was the Mount Washington Cog Ry of 1869, engineered by Sylvester Marsh. The Rigi Ry (Riggenbach System) followed in 1871. The maximum gradient in Switzerland is on the Pilatus Ry (Locher System) – 1 in $2\frac{1}{2}$; the Jungfraujoch holds the height record at 3454 m.

The Abt System, introduced in the Harz Mountains in 1896, is now used by a high proportion of systems.

**Early engine on the Rigi railway, Switzerland**

**Cable-cars on Mount Norquay, Alberta (Travel Alberta)**

## 2 Rail support, cable drive

*Funicular* (drahtseilbahn; funiculaire). Here the passenger cars run on rails but are raised and lowered by wire cables driven from engine houses at the top or the bottom; the descending car usually forms a counterbalancing load. Such systems are popular throughout the world of tourism. Sometimes water, taken on at the top

**Téléphérique in the Austrian Alps (By courtesy of the Austrian National Tourist Office)**

and discharged at the bottom, is used as the driving means.

## 3 Cable support, cable drive

In all cases the passenger carrying means are supported by cables and hauled either by the same or by other cables, powered by engine houses at the top or the bottom. The cables are supported wherever possible on pylons, but sometimes there are very considerable spans where the exposure beneath the passengers' feet is measured in hundreds of metres. Gradients up to 1 in 1 are commonplace. There are four groups, distinguished by differences in the passenger carrying means as follows:

(*a*) *Téléphérique* (luftseilbahn). These employ large cabins holding up to 100 people. Invariably one car is rising while the other is descending and the hauling cable only moves when the cars are in motion. The Sunrise Peak Aerial Ry in Colorado, installed in the mid 1900s, was of this type; the earliest example in Switzerland was erected in 1908 from the foot of the Upper Grindelwald glacier to a point at the west corner of the north face of the Wetterhorn. **The highest point reached in Switzerland by téléphérique** is 3413m on the Stockhorn above Zermatt, but the Aiguille du Midi cableway at Chamonix reaches to 3843m, while the **height record is held by** the Piz Ospejo line in Venezuela (4764). Many such lines are in use all over the world, not only for tourists but for goods also.

(*b*) *Chair lift* (télésiège; seselbahn). The passengers sit on single or double seats suspended from the driven cable, which runs continuously; the seats are attached and detached as required

Seselbahn, Grindelwald, Switzerland

at the terminal stations. The **first cableway of this type in Switzerland** was installed in 1944 from Trubsee to the Joch Pass. **The highest point reached** is La Chenalette (2772) above the Grand St Bernard. In this style of cableway there is usually no protection for the passengers other than a small sheltering roof.

(c) *Télécabin* (gondelbahn). Similar to (b) but the passengers' seats are enclosed in a small cabin or capsule to provide shelter from the elements. These **first appeared in Switzerland in 1950–51** from Crans to Bellalui; this seems the most likely form for further expansion. At present **the highest point in Switzerland reached this way** is the Col de Menouve (2764) above the Grand St Bernard.

(d) *Skilift* (téléski). The skiers attach themselves to bars moving upwards on a cable and are dragged uphill on their skis. These are simple to construct and are widely used at ski resorts everywhere.

A comprehensive selection of the various types outlined above can be examined in the Grindelwald/Lauterbrunnen/Mürren district of the Bernese Oberland and are shown in detail in Fig. 17.

Fig. 17    The railways about Grindelwald.

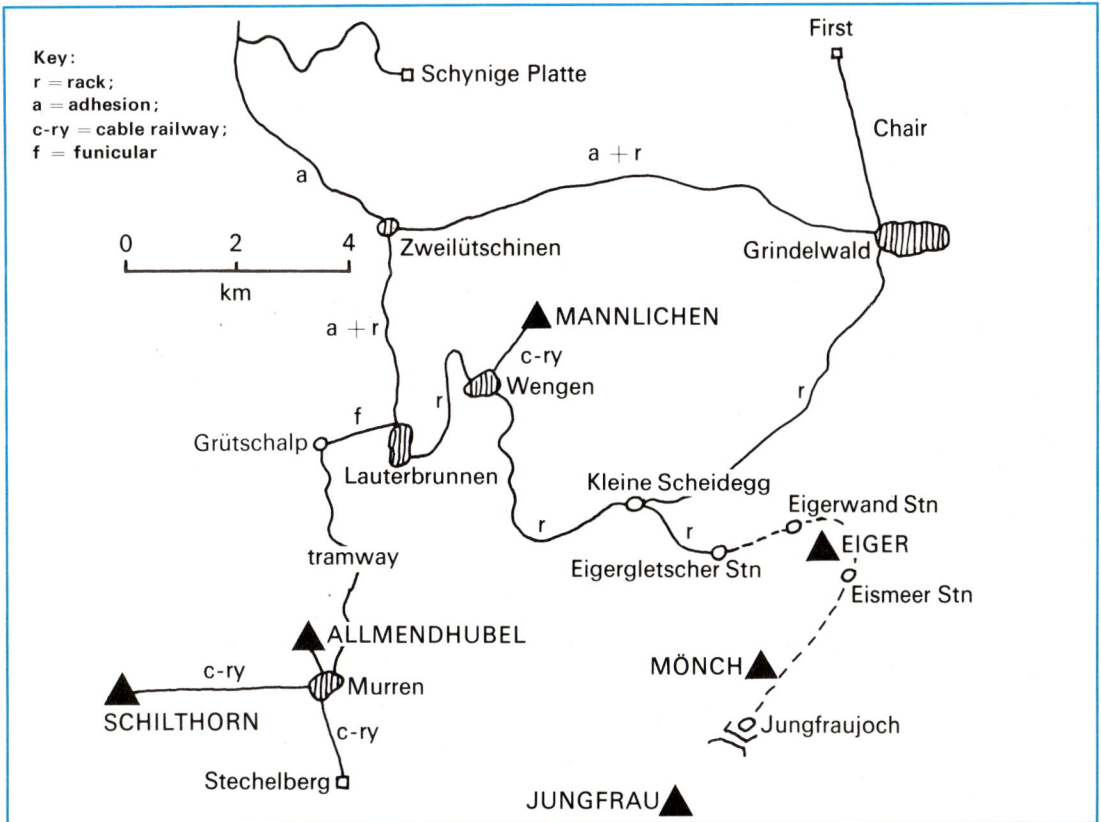

Key:
r = rack;
a = adhesion;
c-ry = cable railway;
f = funicular

# Pastures and Forests

In many parts of the world, mountainsides are cultivated by terracing the slopes to prevent erosion and by extensive systems of irrigation, where ingenious leats (channels) enable every drop of running water to be directed to maximum advantage. In mountains like the Alps, where cultivation of grain fields and vineyards is possible on the lower slopes, higher fertile places are used almost entirely as pasture for cattle. At one time a system known as transhumance was used to utilise these various levels. The farming community occupied as many as five or six different levels at different times of the year. Winter was spent in the valley, while higher up various temporary groups of chalets were occupied only during the summer months, the cattle being removed progressively uphill as the snow melted. Herdsmen and cheese-makers accompanied the herds so that the milk could be converted on the spot to more readily transportable cheese. Later in the year these high levels were abandoned and men and animals returned to the valleys. Now only immature and beef cattle are moved in this way. Milk animals stay at lower levels all the year round; the cheese is made in valley factories, with which small-scale manufacture cannot compete. A recent trend is the introduction of milk pipe-lines from Alp to valley, so that once again the high pastures can be usefully utilised.

A simplified system of transhumance is employed in countries of lower mountains (eg Wales) where sheep roam the hills in the summer and live in winter on the plains beneath. In California the cattle are moved to upper pastures in the summer when the valleys become too dry. In less fertile areas sheep and goats are preferred to cattle since they can thrive on less nutritious foodstuffs. The Indians of Peru, driven from the valleys to the mountain slopes by European invaders, are an example of a pastoral community based on sheep and llamas.

Most hill and mountain areas support extensive forests. The upper limit, known as the tree-

Chalets on an alp looking across to the Täschhorn (EA Shepherd). *Inset:* Hay drying, Julian Alps of Yugoslavia (Yugoslav National Tourist Office)

line, above which trees will not grow, is determined mainly by temperature. In desert areas there is a lower tree-line also, below which conditions are too dry for growth. In tropical and even warm temperate mountains prolific vegetation provides one of the chief barriers to travel and access. It is costly to work mountain forests for timber, but this is nevertheless done in places where there is nothing more accessible. Wood is widely used in all mountain areas for heating and as a building material.

## Maps and Survey

After World War II the aeroplane and then the satellite took over the burden of mapping the mountains of the world, so that there are no longer blanks on any maps.

Previously the process was tedious and time-consuming and maps were built up gradually, with frequent revisions, as expeditions probing into the unknown reached more and more viewpoints. The most elementary type of survey was the compass traverse, carried out by the explorer as he progressed through unknown country, which then enabled others to follow the same route. A compass was used to measure the directions of the various legs of the journey, while the lengths of the legs were estimated from the speed of travel. Compass bearings were also taken to outstanding objects on either hand, so that it might be possible on returning home to tie the whole thing into existing maps.

Expeditions having mapping as a prime objective used conventional triangulation techniques, measuring large numbers of angles with a theodolite after having set up a base line of length accurately determined by a calibrated tape. Some compass bearings enabled the survey to be coordinated with other maps. Cairns on shoulders and summits of mountains provided prominent survey points and served later, where reasonably accessible, as viewpoints to many more distant and less accessible places. Even so there constantly arose problems of identification of objects from different viewpoints. An explorer's alternative to the theodolite was the plane-table on which the map was actually drawn on site using an alidade – a ruler with a viewing telescope attached.

Heights, and hence contours, were determined in both cases by tilting the viewing system and measuring vertical angles. Ground based photography was a useful tool for surveyors – vertical and horizontal angles could be measured from panoramic photographs, while a pair taken from either end of a known base-line could be used in an automatic plotting machine to draw a map directly.

Aerial surveys produced strips of pictures, taken at a known height, which were also utilised for the automatic plotting of maps. Photographic techniques were applied in different wavelength ranges, contrasting and emphasising many surface features of human or ecological interest, and thus contributing substantially to land-use studies, geographical survey and so on. However, all these methods, operating in the visible part of the spectrum, needed good weather conditions for successful application and moreover could only be employed during daylight hours.

Of recent years these limitations have been overcome to a large extent by the use of radar operating in the microwave (3 mm to 1 m) wavelength range, which is not attenuated by atmosphere and is unaffected by clouds and darkness. The actual technique is known as Side Looking Airborne Radar (SLAR) which uses a fixed antenna directing a beam of radiation laterally from the aircraft and receives reflections from objects along a strip at right angles to the flight path. Distortion due to the sideways view and resolution along the flight path are corrected electronically. Though sensitive to a completely different part of the Electromagnetic Spectrum, radar pictures have many points of similarity to photographs and skilled operators are said to be able to interpret them with equal facility. Using different wavelength ranges enables different features to be detected, eg a vegetation cover can be penetrated and the land surface beneath revealed. As an example of the use of this technique, parts of Panama and Colombia were surveyed in 1967. In six hours of imaging time within a period of six days, 17 000 km$^2$ of terrain were imaged. Previously, 15 years of photographic aerial survey had resulted in only

30 per cent of the area being covered because of almost continuous cloud barrier.

But more impressive still are the results of the Landsat programme carried out by NASA in the USA. Three satellites circle the earth in a polar orbit at an altitude of approximately 920 km and every point is crossed once every six days in daylight hours. Here also the detectors operate in various different spectral regions, each seeing a different aspect of the terrain below. By this means the whole of the land surface of the world, plus some adjacent sea areas have been mapped to a scale of 1 : 250 000. Only 20 per cent of this area had previously been mapped adequately to this, or to larger scale. Pre-war expeditions to the great mountain ranges spent months producing a map which this programme can cover in much greater detail in less than a minute – such is the astounding march of progress.

Determination of the heights of mountains has also presented problems. Climbing scientists in the early days included among their equipment a mercury barometer and also a boiling point apparatus (since boiling point is related to atmospheric pressure is related to height). It was necessary to know the barometer reading in the valley at the same time as it was being read on the summit. At low levels the pressure falls 1 mm for every 10·5 m of ascent, but, since the lower layers of the atmosphere are more dense than the upper, the fall is not proportional to height. One millimetre of fall needs 13·4 m of ascent at 2000 m, 17·2 m at 4000 m and 22·5 m at 6000 m. The method would appear therefore to present wide possibilities of error, since little is known about the exact shape of the atmospheric envelope anyway. The aneroid barometer (1848) provides a portable alternative.

Heights are determined also by angle measurements from base lines in the plains or from adjacent, and perhaps more readily accessible, summits. Here the existence of an accurate map of the terrain leads to an improvement in the measurement accuracy. When completely covered by snow and ice the summit of a mountain is not a very stable entity – its height can certainly vary. The accuracies claimed for some height determinations are therefore open to question. As an example, the highest point in Norway is often said to be Glittertind, a mountain whose snow cover makes it sometimes higher and sometimes lower that its neighbour Galdhopiggen. The latter is rightly named the official national summit.

A considerable problem in the determination of height by optical means lies in the refraction of light along the line of sight as it passes through air layers at different temperatures (*see* 'Intervisibility'). There are always temperature gradients particularly close to the surface of the earth.

## *In the Air above the Mountains*

Man obtained his first view down on to mountains from a hydrogen-filled balloon. The **first crossing of the Alps took place in September 1849** when Francisque Arban floated from Marseille to Turin over Monte Viso. As early as the 1860s J. Glaisher reached the height of Mount Everest in a balloon, nearly perishing in the attempt. Another famous crossing of the Alps was in October 1898, when Prof. A. Heim, Dr J. Maurer and Dr Biederman, in the balloon *Wega* (3268 m³) piloted by Capt. E. Spelterini, floated from Sitten to Langres. In 1931 Prof. A. Piccard made an ascent into the stratosphere above the Alps in the balloon *FNRS* (14 000 m³). After World War II sporting ballooning became increasingly popular and an International Balloon Week was held at Mürren in August 1962, when numerous ascents were made. In 1957 D. Simons in America reached 31 000 m in a balloon of 85 000 m³.

The **first crossing of the Alps by hot-air balloon** had to wait until 1972 when D. Cameron and M. Yarry in the balloon *Cumulo Nimbus* (3970 m³) floated from Zermatt (1600) over the Gorner Glacier and Monte Rosa to Biella (365), 60 km away in Italy. The maximum height attained in the course of the journey was 5425 m.

Early airships began flying among mountains in the first decade of the present century. In 1908 *Graf Zeppelin* flew from Friedrichshafen to Lucerne and in 1910 *Parseval* from Munich around the Zugspitze. However, the likelihood of atmospheric disturbance in the neighbourhood of high mountains would have discouraged commercial flights from actually crossing them regularly. An airship could always go round anyway!

The **first crossing of the Alps by aeroplane** was accomplished in 1910 by a Peruvian, G. Chavez, who flew from Brigue over the Simplon Pass. Unfortunately he crashed on landing at Domodosolla and was killed. Four months later J. Bielovucic became the first to cross and survive. **The first successful landing on a mountain** took place on the Jungfraujoch in 1919, and in 1921 Francis Durafour set down and took off at 4275 m on Mont Blanc.

The ceiling of the aeroplane gradually increased as engines grew in power, though fliers still sought out passes when they could see them. Navigation in cloud was a serious problem in a mountain district with strong winds, powerful up-currents and abrupt changes of level in the land surface. Lost planes sometimes crashed in these circumstances. Even when the weather was fine it was still possible to be trapped in a mountain valley between converging walls, the rising valley floor and the ceiling of the aeroplane.

Soon aeroplanes were being used in polar regions (eg by the Watkins Greenland Expedition in 1930–31, the objective of which was to find landing sites for trans-polar commercial flying), while in 1932 an expedition to Alaska landed men and equipment on the Muldow glacier below Mount McKinley.

By the 1930's the altitude record stood at over 12000 m. The Bristol Pegasus engine appeared to offer a reasonable prospect of flying over Mount Everest, so in 1933 an expedition was organised to overfly and take pictures using Westland aircraft. At that time all climbing expeditions had to approach from the north (Tibetan) side, but as flying on that flank was not allowed, the pictures, although outstanding, proved of little use to mountaineers.

During World War II great developments took place in the design of aeroplanes and in their utilisation to drop or land a wide variety of objects. After the war, therefore, when mountaineers once again began to visit distant ranges, aeroplanes increasingly came into use. Most expeditions to Alaska now flew in both men and supplies to glaciers or flat snow-fields. In 1959 Terris Moore landed a Super-Piper-Cub on the summit of Mount Sanford, two heli-

Ballooning, Switzerland

copters touching down there at the same time. Seaplanes were also used to lift expeditions, landing on the mountain lakes of north Canada.

During the 1950s a Swiss pilot, Hermann Geiger, made a substantial contribution to the techniques of mountain flying. Having found that unpredictable air currents played havoc with accurate parachute dropping of supplies and materials, he set about perfecting a method of landing and taking-off in very constricted spaces. He landed in fact uphill, where the slope considerably reduced the length of runway needed to slow the plane to a standstill; he took off downhill using the slope now to gather speed rapidly. He first landed on the Kander glacier near the Mutthorn Hut; numerous similar feats culminated in landing on a 1 in 2 slope only 25 m wide in front of the Rossier Hut on the Dent Blanche. In the days before the helicopter attained its present capability his contribution to rescue work and to general supply activities was considerable.

An aeroplane was taken to the Himalaya by

the Swiss expedition to Dhaulagiri in 1960, but it crashed and had to be left on the mountain. In the last decade, planes and helicopters have now become commonplace in mountains everywhere for supply and rescue services. Much of the trek from Kathmandu to Mount Everest has been eliminated by an air service to Lukla, but this has created new problems of mountain sickness in the participants due to the too rapid ascent. The Italian Everest expedition of 1973 used helicopters to ferry their supplies into the Western Cwm thus avoiding the major difficulties of the Khumbu Icefall. The question now is how long will such activities continue to be accepted as legitimate aids to mountaineering.

Great air-liners fly every day far higher than the highest mountains. In heated, pressurised cabins and dressed in their everyday clothes, the passengers pass rapidly across rugged mountain lands, which can only be traversed on the ground by great skill and organisation. For them mountains are no longer a barrier.

The flying sport of high lands is gliding. It often happens, especially in hill districts, that cold air from high in the atmosphere sinks down and displaces hot air which has been warmed by contact with the earth's surface. This rises in so-called thermals, up-currents which enable the glider pilot to soar. Another phenomenon he utilises is that of lee waves (see Mountain Winds) and sometimes slope lift formed when the wind blows against a hill ridge. The altitude record of 14100 m for gliders was set up in February 1961 by Bikle using lee waves in the Californian Sierra Nevada alongside Mount Whitney.

Two of the world's longest mountain ranges – the southern Alps of New Zealand and the Allegheny Mountains of eastern USA – have contributed most of the distance records (for a journey out and back to the start). These are very dissimilar in type, but presumably the orientation relative to the prevailing wind is the significant factor. The New Zealand fliers worked almost entirely on lee waves, the Americans with slope lift. The current record was set up in May 1977 when K. Striedieck flew from Lock House, Pennsylvania to Tennessee – an overall distance of 1616 km. By contrast the record for distance in a straight line (1460 km) is held by H. W. Grosse with a flight from Lübeck to Biarritz, certainly not the product of a single mountain range.

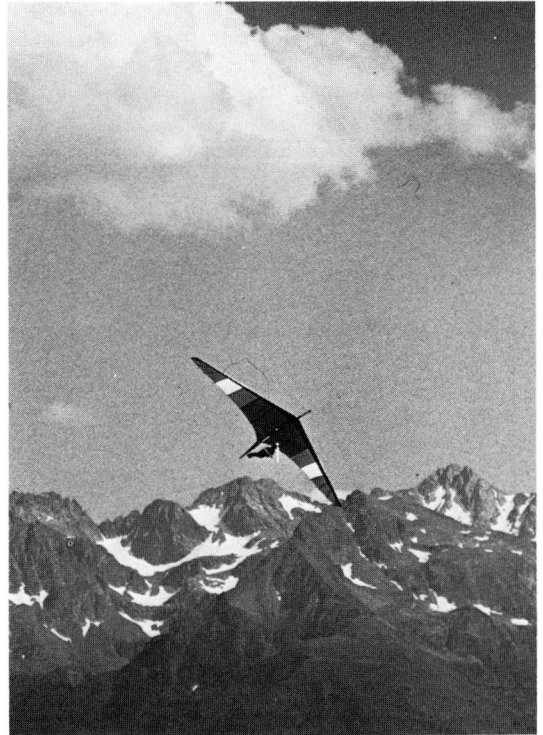

**Hang gliding, Chaîne de Belledonne, France (Alan James)**

Recent years have seen a revival of the hang-glider pioneered originally during the last decade of the 19th century. In contrast to the high performance glider, this is little more than a personal pair of wings; it can take off without assistance, but the pilot needs to project himself from an elevated position in order to have some space for soaring. In 1973 R. Eipper leaped by hang-glider from Mount Haleakala in Hawaii landing 15 km away and 2550 m lower, 19½ min later. The chase for records was on. In 1976, four men floated down from the summit of Mount McKinley, their deeds simultaneously recorded for television; the following year Rene Ghilini floated down 4000 m from the summit of Huascarán in Peru; then in 1978 Kilimanjaro, and in 1979 K2 (from 7600 m).

Considerable effort is required to raise the necessary equipment to the take-off point, yet the extension of hang-gliding to the Himalaya cannot be far away. The height record already is a descent from a balloon at 9631 m in California.

Parachuting is basically a military activity, though more recently a sport. German troops tried free-fall jumping (ie without a parachute) on to deep snow fields, but gave up because of losses from concussion and asphyxiation.

Members of the paraclub at the Centre de Grenoble were early pioneers in the sporting field, and jumping is now a sideline winter sport at places like Mont de Lans, Alpe d'Huez, etc.

In September 1961 three men jumped from 600m above Mont Blanc and landed on the summit. During the following year three others wearing oxygen masks jumped on to the summit of Kilimanjaro (5895), where a reception party carrying more oxygen awaited them; the descent to the first village took 14 hours. In 1968 ten

Soviet parachutists jumped from around 8000m to land at 7000m on Pik Lenin in the Pamirs, where a support party of mountaineers was already installed; each carried eight hours of oxygen, three days' food, as well as mountaineering and bivouac gear. Sadly four lost their lives, being blown over a precipice.

Among near-vertical rock peaks like the Dolomites parachutists have occasionally jumped over the edge of a steep crag and floated down to the foot. At Yosemite in California in 1972 a climber, R. Sylvester, skied down the summit slopes of El Capitan and parachuted to the foot of the face, landing in a tree.

## Mountains and War

Mountain walls or ridges, appearing to provide an effective barrier to the warlike ambitions of a state on the far side, undoubtedly give a feeling of security, but this can be shattered at short notice if a sufficiently determined enemy can find a way through or over. History abounds with such incidents even from areas of comparatively low hills such as the campaigns of Montrose in the Scottish Highlands, or those of Robert E. Lee among the hills of Virginia. However, the higher the mountains, the more spectacular the action.

The most potent weapon in high-mountain warfare has always been the avalanche, seldom under control of the contestants but striking down friend and foe indiscriminately. Probably one of the earliest of these struck the armies of Alexander the Great among the mountains of eastern Asia. Certainly avalanches contributed in full measure to the losses sustained by Hannibal in his crossing of the Alps in 218 BC – 18000 men, 2000 horses and several elephants (from a total of 38000 men, 8000 horsemen and 37 elephants). The forces of Rome often crossed the Alps in their campaigns – Pompey by the Col de Montgenèvre in 77 BC, Caesar by the same col in 58 BC and by the Petit St Bernard in 49 BC.

In the Napoleonic Wars there were a number of Alpine campaigns. The Russian General Suvorov forced the St Gotthard during the autumn of 1799 and drove the French down into Switzerland. He then led his troops over the Kinzig Kulm, Pragel and Panixer passes, losing

hundreds in the latter during October. In May 1800 Napoleon's army on the way to the Battle of Marengo lost men and material in avalanches on the Grand St Bernard, as did Maréchal MacDonald in November the same year when he led an army over the Splügen Pass. Three years later General Moncey's army crossed the St Gotthard Pass.

There were fearful losses on the Austrian/Italian front in World War I. The front line ran through the Ortler group, where the Austrians held the Konigspitze and the Italians the Payerjoch. From the latter a 1000m tunnel was driven towards the Austrian positions to a point overlooking the lines of communication to the Konigspitze. Tunnelling techniques were in fact widely employed – lengthy galleries beneath

Italian troops in the Alps, Ruitor glacier
(AC Collection)

enemy positions were packed with explosives which completely wrecked the landscape above. The forces were based on high mountain huts supplied by téléphérique from the valley. Supply tunnels were dug through glacier ice, with staircases and bridges across internal crevasses, to reach the front line. This sometimes consisted of continuous ice galleries along the crests of mountains with openings in the side for machine guns.

In the Dolomites the summit plateau of Monte Piano was held for three years against superior forces occupying higher mountains on either hand – the Drei Zinnen (Tri Cime di Lavaredo) to the east and Monte Cristallino to the south-west. In this part of the front the famous Austrian guide, Sepp Innerkofler, was killed in an attack on the Paternkofel (by the North-north-west Arête, of which he had made the first ascent in 1896); his small party was discovered and overwhelmed by superior Italian forces on the summit.

At least 40000 (maybe double that number) are believed to have died in avalanches here between 1915 and 1918. A two-week snowstorm in December 1916 produced avalanches which swept away whole barracks – 253 died in one incident on the Marmolada. Avalanches released by firing shells into snow-laden slopes were used as weapons, much more effective than artillery. Once, 3000 troops were killed in a 48-hour period on the Austrian side, while similar losses were also incurred among the Italians.

The subsequent peace treaties moved back the frontier; half a million people in South Tyrol were transferred from Austria to Italy; many local names were Italianised, so that they carry alternatives to this day.

During World War II the Alps saw little serious front line action, but in all Alpine countries tremendous struggles developed between occupying troops and local resistance movements. The fighting was particularly bitter in the Vercors, the Mont Blanc area and the Julian Alps. Meanwhile epic mountain expeditions were undertaken by Allied prisoners-of-war escaping from Italy into neutral Switzerland. Most of these escapees knew nothing about mountaineering, they were poorly clad and had little or no food or equipment; some had hardly ever seen snow before. They were guided to the crest of the range and then left to make

Memorial to the partisans, Julian Alps of Yugoslavia (Yugoslav National Tourist Office)

dangerous descents of steep glaciers towards freedom, eg those leading down to the Gorner Glacier above Zermatt. The Swiss maintained patrols which succoured many of these unfortunates, but many more perished from exposure, or in crevasses.

Countries with substantial snow-cover realised long ago the importance of skis in military applications; troops were trained in their use and skirmishes fought over snow-covered terrain. Before World War I the crack units were the Chasseurs Alpins (French) and the Alpini (Italian); the former were in action in the Battle of St Dié in December 1914. Skis were used on the Austrian/Italian front, while a British ski unit was in action in Russia in 1918–19. 1939 found the various countries in an even better state of preparation – the Finns used ski troops with great effect against the Russians; Norwegian commandos used skis against the Germans; German ski troops fought in the Balkans and in Russia. Later the British set up a Mountain and Snow Warfare Training Centre in Scotland; ironically the special Mountain Warfare unit saw its first action on the Island of Walcheren, below sea-level. The needs of war certainly boosted research into equipment, food and high-altitude living, all of which came to benefit the mountaineer in the long run.

The military monuments remain: General Suvorov's is in the Schollenen Gorge of the St Gotthard Pass; the Yugoslav partisans a 5-metre high piton and karabiner (snap-ring) below Triglav; the defenders of Monte Piano a giant cross, and so on – grim reminders that it could happen again. There are also defence works still much in evidence on some mountain frontiers.

# Hydro-electricity

Perhaps the major practical use of mountains lies in the generation of hydro-electric power. The potential energy of snow and ice precipitated at high altitudes and melting and running off to lower levels is tremendous, and considerable ingenuity has been devoted to its utilisation in all mountain countries. Since the supply tends to be intermittent on both a daily and a seasonal basis it is invariably necessary to collect and store the water, releasing it as needed; for this purpose very large artificial lakes have to be constructed by damming. Though mountains would seem to present a large number of potential sites, examination of the local strata to a substantial depth has to be carried out before a selection can be made. The lake basin must not leak; the stability of the dam and the lake sides must be beyond question.

An outstanding system is that of Grande Dixence in the Canton of Valais, Switzerland. Precipitation is collected over hundreds of square kilometres of mountainsides in the valleys of St Nikolas, d'Hérens and d'Hérémence and stored in a huge reservoir in the latter valley. From here it is fed to power stations at Fionnay

and Nendaz to generate 1600 GWh (1600 million units) of electric power per annum. The dam, completed in 1961, is of the mass-gravity type. It is one of the highest in the world (285 m) – twice the height of the Great Pyramid – and contains six million m$^3$ of material. The capacity of the lake is 400 million m$^3$. The feed tunnels are gently inclined from 2496 m above Randa to 2364 m at the lake surface, but where necessary water from tunnels below the general level is pumped up into the system using some of the generated power. To avoid the boring of over-lengthy tunnels above Zermatt siphon tunnels were constructed – one 400 metres deep below the Gorner valley followed by another 270 metres deep below the Zmutt valley. The whole is too complex to be optimised by human operators. It is controlled centrally by a computer at Sion which is fed with data automatically and continuously recorded and transmitted from all parts – water levels, flow rates, temperatures, evaporation losses and so on. Thus flooding is prevented and flow maximised by appropriate control of the various pumping and overflow facilities that are available.

Fig. 18   Some typical cross-sections and relative volumes of dam structures

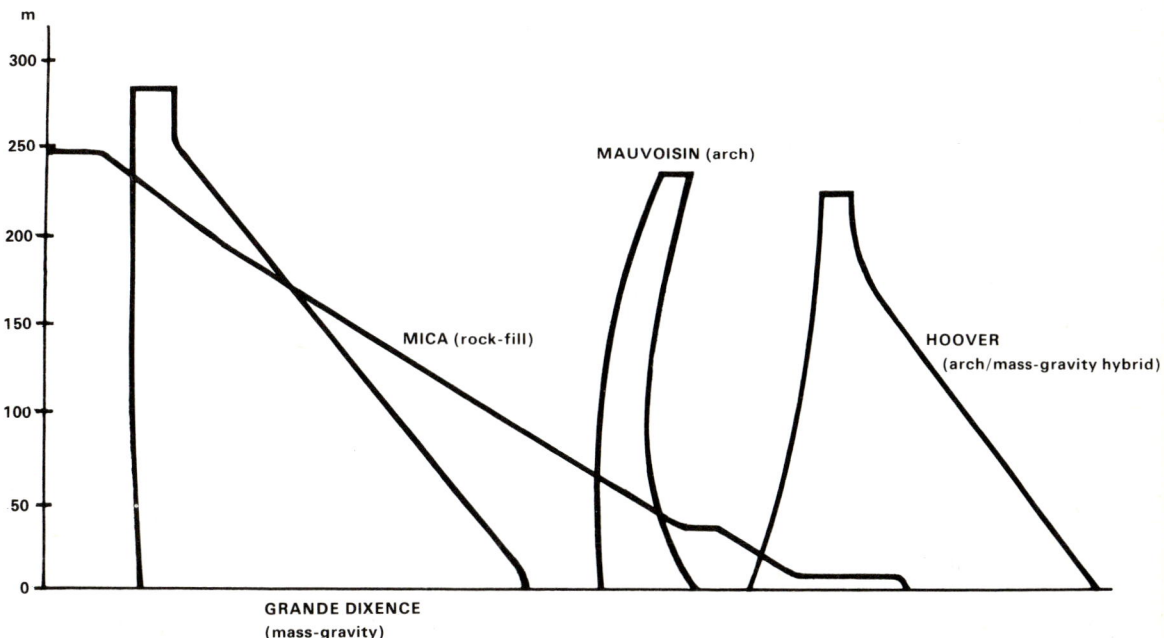

**Grimsel Lake and Dam, Switzerland**

There are three distinct types of dam construction (see Fig. 18):

1   *The mass-gravity type* (as Grande Dixence above) resists water pressure by weight alone, with base and sides keyed to the rock to prevent movement. Fissures in the surrounding rock are filled with concrete for strengthening purposes. Another example in the mountains of Washington, USA, is the Grand Coulee Dam, which is 168 m high and holds back a lake 180 km long.

2   *The rock fill dam* is made by enclosing a valley with a bank of rocks to the required height and then facing the sides with concrete. The Nourek Dam in Tadzhikstan, USSR, when completed, will be 317 m high, enclosing a lake of 10·5 km³ capacity; further upstream the Rogun Dam will be 323 m high. The Mica Dam, close to the confluence of the Canoe and Columbia Rivers in British Columbia, is 244 m high, has a volume of 32 million m³ and holds back a lake 200 km long, between the Rocky Mountains on one hand and the Selkirk and Cariboo Mountains on the other.

3   *The arch type dam* requires a gorge having side walls strong enough to resist the water pressure. A thin curved concrete wall across the gorge, fixed to the base rock, transfers the weight of the water to the side walls by an arching action. Mauvoisin Dam in the Val des Bagnes, Switzerland, which is of this type, is 237 m high; 60 km of boreholes and 900 m of galleries were bored and filled with concrete to strengthen the surrounding rocks. Examples, even higher than this, are under construction in the USSR. The Hoover Dam (221 m) on the Colorado River is a hybrid between arch and mass-gravity types, holding back the 185 km Lake Mead; Sayany Dam (242) on the Yenisei River, USSR, is similarly hybrid.

Some serious problems have arisen in connection with man-made reservoirs of huge capacity. The loading on the earth's surface exerted by such large amounts of water has been found to trigger minor earthquake shocks, landslides and so on. A notable example is the Vajont Dam in north-east Italy, where the action of the water on the strata of the neighbouring hillside produced a considerable landslip (250 million m³ of material). This fell into the lake displacing a very large amount of water, which flowed over the top of the dam. Two thousand lives were lost in the valley below.

The silt carried by a river cannot readily pass a dam and much of it is deposited in the bed of the lake, which must gradually fill (eg it is estimated that Lake Mead above, if no action be taken, will infil inside 300 years).

## A Living from Mountains

Starting in the Alpine countries more than a century ago, tourism grew steadily into a thriving industry. Now it comes close to overwhelming and to destroying for ever many of the essential features which gave it birth. To encourage visitors to a landscape which appeals because of its wildness and remoteness is certain to destroy just those qualities in it; yet often the change arouses little notice except among those who knew the place in its earlier days before the arrival of the road, the airstrip and the hotel. These new amenities soon come to be accepted as part of the scene. For a country such as Switzerland tourism is a major source of foreign exchange and of general prosperity and, as such, is of course exploited to the greatest possible extent.

In earlier and more modest times, apart from hoteliers and providers of transportation, the first to earn some money from tourists were the guides, a body of men recruited from among the local peasants who took their clients up the nearby mountains for a fee dependent on the reputation and difficulty of the route. Many were little more than leaders of walks, but a small number developed into highly-accomplished mountaineers, well adapted to the altitude and toughened by the hard conditions of their lives. Though perhaps not always good judges of weather or of snow conditions, they cut huge ladders of steps up ice slopes and brought the important quality of accumulated experience to mountaineers from flat lands, whose activities had of necessity to be confined to a short holiday period. Guides from the Alps eventually made their mark in mountains everywhere – the Himalaya and Karakoram, Alaska, New Zealand, the Andes, Africa and North America. After World War II the practice of guiding, long established in Alpine and in some other mountain countries, spread into most mountain areas of the world as increasing numbers of mountaineers everywhere sought to make a living from the sport.

Alpine expeditions involving bivouacs or traverses from valley to valley often employed porters to carry surplus baggage. This indeed was the usual introduction for aspirant guides. Later, when expeditions to more distant ranges required days, or even weeks, of approach march, large numbers of porters were required to deliver the climbers to the mountain foot. The now famous Sherpas of Nepal first earned their mountaineering reputation as porters, carrying supplies not only to the mountain but also setting up a line of camps one above the other up

the chosen route. Their competence as mountaineers in their own right gradually emerged, so that now they are equal members in many high-altitude ventures. In some parts of the world the carrying is done by beasts of burden – yaks, camels, llamas and so on, as appropriate; the drivers or tenders are maintaining a porter function and thereby taking a share of their living from mountains. More recently recruited are the pilots of aeroplanes and helicopters which have been increasingly involved in expedition supply service.

Down the years a few photographers have made a certain living from the mountains, and their role has been important since they have produced records of the early history of the subject. Now their numbers have multiplied and our coffee-tables groan under the weight of their picture books.

After World War II there came an upsurge in mountaineering everywhere; the numbers of participants increased dramatically and in due course their needs were being catered for by a greatly expanded equipment industry. Tourism boomed, leading on to organised mountaineering tourism covering all standards of competence in all parts of the world. Many expeditions, sponsored nationally or by industrial or business concerns, or by national clubs, set off for the great ranges in the 1950s and 60s and, in spite of setbacks and tragedies, a great majority of the world's highest peaks fell to the climber. Many more outlets were found for making a living from mountains – guides for all grades of climbing, instructors for ski-ing, instructors and ancilliaries for huge programmes introducing young people to the hills, manufacturers and purveyors of equipment and services, producers of guide-books and specialist magazines and so on.

Now the professional expeditioner has put in an appearance, living between climbing trips on lecture tours, book rights, films and television, endorsement of products, etc. He has to work hard to persuade commercial sponsors to put up money or goods and, while he stays at the top, has a highly successful mountaineering career as a reward for his efforts. Finally, there is the dedicated man who works for a time and then goes climbing until he has no money left – a greater devotion to mountains there cannot be!

# Mountain Flora and Fauna

Plant and animal life has penetrated even the most inhospitable parts of the globe; lichens have been found within 4° of the South Pole and small plants at over 6400m in the Himalaya, while birds fly as high as the summits in South America and central Asia.

Many well-known cultivated plants, now among the staple diets of civilisation, originated in high country. Wheat has been traced to Ethiopia, from where it was transferred to Egypt and the Mediterranean countries and so spread throughout the civilised world. Barley thrives in Tibet at heights above 4300m. Other less familiar grains are teff, a millet produced in Ethiopia, and quinoa, which grows in the high Titicaca region of the Andes. The potato was also developed in the latter area at an altitude of 3600m or so, while the tomato comes from the western slopes of the Peruvian Andes.

Tea, which is extensively cultivated among the foothills of the Himalaya, grows wild in Assam and on Roirama in Venezuela. Coffee came originally from Ethiopia, where it still grows wild. It was transplanted to Arabia and later cultivated world-wide in Java, Brazil and parts of Central Africa.

Cocoa trees grow in profusion on the slopes of São Thomé, off the coast of Africa, where the rainfall is of the order of 10m/annum, so that it has been called Chocolate Island. Saffron grows above 1800m in Kashmir; cinchona, the source of quinine, thrives in the Andes up to 3300m – the natives chew the leaves as a specific against mountain sickness.

Mountains, which have become a survival haven for human and animal minorities, have served similarly for the plant kingdom. The Ape Ape (a plant with 'leaves large enough to hide a man') only survives now in the rain forests around the crater of Haleakala on the island of Maui in Hawaii.

Grasses flourish everywhere at all heights up to the snow-line and form the basis of the pastoral economies found in most mountain countries.

The biggest trees in the world grow on the western slopes of the Sierra Nevada in California; The sequoia, which has survived here, is now preserved in the Sequoia National Park.

Eidelweiss (R D Treble)

General Sherman Tree, Sequoia NP, California (Barnaby's)

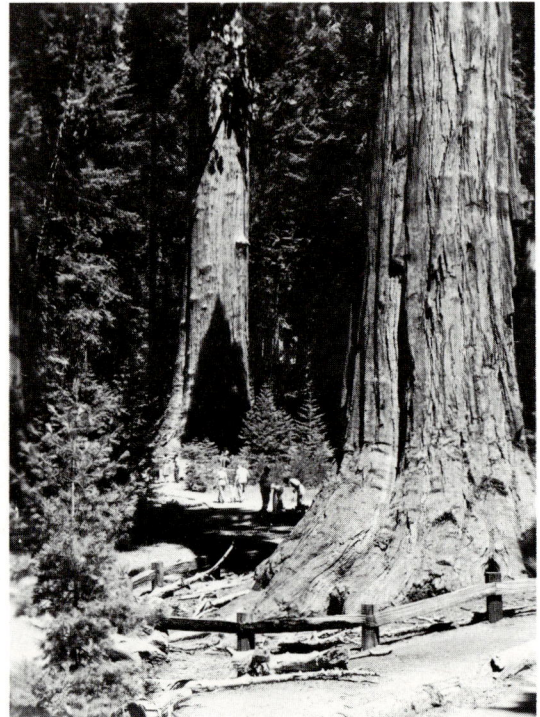

Here the tree called General Sherman is 84 m high, has a girth of 30 m and weighs 2150 tonnes. Bark up to one m thick protects these trees from insect pests, so that their lives can be measured in thousands of years. The redwood in the same area attains even greater heights, but with lesser girths (Founder's Tree is 110 m high and 18 m round).

Some mountain animals are transient visitors from the lowlands adapted in some cases to high altitude living, while others naturally belong there. Some lowland beasts are pastured at high levels during some parts of the year, utilising the grass wherever it may be found. Others have fled to the hills and mountains to escape the depredations of man – notably, for example, in Africa, the elephant, the rhinoceros and the Mountain Gorilla.

Many of the animals which belong truly in the mountain environment are ruminants, such as various goats and sheep. The Rocky Mountain Goat, named for the range, which can be found from Alaska to Montana, is a remarkably sure-footed climber. Other indigenous American species are the Big Horn and the Alaskan White Sheep.

The ibex is a mountain goat found in Ethiopia, the Caucasus and central Asia; in the Alps it is called the bouquetin. The ubiquitous chamois, which is smaller, inhabits many of Europe's ranges. In the mountains of central Asia, too, are found the serow, the ghoral and the takim. From China comes the argali, a larger version of the bighorn sheep, somewhat bigger than a donkey. The burrhel, or blue sheep, is also large and found up to 6000 m. However, the biggest sheep of all is the *Ovis Poli*, with horns two metres from tip to tip and weighing up to 270 kg.

Related to the ox and the buffalo is the yak, used as a beast of burden in some central Asian countries. Incredibly tough and resistant to extreme weather conditions, they carry loads over the highest passes, climbing sure-footed as a goat on slopes which would defeat a horse. Another wild ox, the gaur, living among the hills of India and Malaya, stands over two metres high at the shoulder.

Mountain deer include the sambhur of India, the caribou of North America and the wapiti of the Tien Shan. The musk deer of Tibet and western China, which lives up to 2500 m, is highly prized as a source of perfume.

Fierce wild boars inhabit the mountains of North Korea.

Carnivores are also found among the mountains – the puma, or mountain lion, of western USA, the Siberian tiger of eastern USSR and the snow leopard of central Asia. There are mountain bears, notably the Alaskan brown bear (weighing up to 1000 kg), the grizzly bear of the Rockies (weighing up to 375 kg) and the Himalayan bear. The giant panda, a rare bear-like beast, is confined mainly to the Chinese province of Szechwan, where it lives up to 4275 m. Mountain monkeys include the Gellada Baboons of Ethiopia and the Golden Monkey of Tibet and western China. The huge Mountain Gorilla is confined to the Virunga volcanoes area of central Africa.

Hereabouts, too, we must take note of the 'Abominable Snow-man', probably some member of the Ape family, but still an intriguing mystery in the Asiatic mountain scene. His western American cousin, the 'Big Foot', another mystery monster, is probably a bear.

Four distinctive mountain animals roam the Andes – the llama, a pack animal of great all-round usefulness, the alpaca, the vicuna and the guanaco.

The large birds of the mountains are eagles and vultures. Particularly noteworthy among them are the monkey-eating eagle of the Philippines, the famous Golden Eagle, the lammergeyer, or bearded vulture, which reaches a wing span of three metres and has been seen flying at 6400 m, and the condor of South America, another huge bird which flies up to 6000 m.

The mountain bird, however, is the chough, a member of the crow family, which appears widely in the Alps; some followed Mount Everest climbers up to 8200 m in 1924.

A mouse has been seen in a mountain tent at over 6000 m; a spider was spotted on Mount Everest at 6700 m.

**Mountain dragon, 1660 (E Pyatt Collection)**

# Europe

The Julian Alps of northern Yugoslavia (AC Collection)

## United Kingdom

### 1 SNOWDONIA NP

Snowdon (Y Wyddfa) 1085, **the highest point in Wales** (r-ry to summit).
Crib y Ddysgl 1065
Carnedd Llewelyn 1062
Carnedd Dafydd 1044
Crib Goch 921
Tryfan 917
Aran Mawddwy 905
Cader Idris 892
Cnicht 690, the 'Welsh Matterhorn'.

**Snowdon, North Wales (C D Milner)**

In **England and Wales** there are no high or specially striking mountains, but this is the 'home ground' of a nation which has played a significant part in the development of mountaineering everywhere.

The **Snowdonia National Park** (2163 km²) is the considerably eroded remains of a dome of arched beds, once 6000 m above the present levels. Altogether there are 14 summits over 3000 ft (914 m), the so-called 'Welsh 3000s'. These have been linked in one expedition – 50 km with 5500 m of ascent and descent – in 4 hr 46 min by J. Naylor in 1973. There are numerous climbers' crags.

The summit of **Snowdon** (1085) can be reached by a 800 mm gauge rack railway. Opened in April 1896 and using the Abt system, it climbs to 1064 m in 7·25 km, maximum gradient 1 in 5·5.
    Meanwhile King Arthur and his knights are sleeping in a cave below awaiting the call to arms.

The **outlying mountains of Wales** mostly take the form of a plateau in varied rock types, but there are also some ranges and peaks.

The British Isles

## 2   OUTLYING MOUNTAINS OF WALES

Pen y Fan (Brecon Beacons) 905
Plynlimon 752

## 3   THE HILL GROUPS OF SOUTHERN ENGLAND

a   Bodmin Moor 419, granite.
b   Exmoor NP (678 km²) 520, sand-
    stones and slates.
c   Dartmoor NP (935 km²) 621, granite.
d   Brendon Hills 424
e   Quantock Hills 385
f   Mendip Hills 326, limestone, caves
    and climbing.
g   Cotswold Hills 330, limestone.
h   Chalk hills 297
j   Malvern Hills 425, volcanic rocks.
k   Welsh border hills 546, volcanic
    rocks.

## 4   THE PEAK DISTRICT NP

Kinder Scout 636

## 5   THE PENNINES

Crossfell 893, **the highest point of the
    Pennines.**
Mickle Fell 790
Ingleborough 723
The Cheviot 616

## 6   THE LAKE DISTRICT NP

Scafell Pike 978, **the highest point in
    England.**
Scafell 964
Helvellyn 950
Skiddaw 932
Great Gable 899

There are important sea-cliff climbing areas in western Anglesey, western Pembrokeshire (Dyfed) and on the Gower Peninsula.

The **hills of southern England** are low hills in comparatively young rocks. There are important sea-cliff climbing areas in Dorset, Devon and Cornwall. Mountaineers domiciled in London, far from real mountains, use sandstone rocks in the Weald of Kent and Sussex (50 km), Dorset sea-cliffs (200 km), limestone in the Mendips and the Wye Valley (200 km) and gritstone and lime-stone in Derbyshire (225 to 250 km).

Four thousand years ago when the valleys below the **chalk hills** were thickly afforested these well-drained hill slopes provided homes, religious sites, fortifications and linking trackways for early man. Many relics remain.

The **Malvern Hills** received very early mention by William Langland in *Piers the Plowman* (14th century) – 'Ac on a May morning on Malverne hilles me bifel a ferly'.

The **Peak District National Park** (1387 km²) is an area of dark peat moors with crags of gritstone and of barren light-grey lime-stone with crags and caves.

The **Pennines**, the so-called 'backbone of England', include the National Parks of the Yorkshire Dales (1740 km²) and Northumberland (920 km²). They stretch from the Peak District to the Scottish border. The Pennine Way (430 km) which runs the length of the ridge has been covered in 3 days 1 hr 48 min. Ingleborough is honeycombed by the huge Gaping Ghyll cave system.

The **Lake District National Park** (2217 km²) is the eroded remains of a huge domed structure once high above the present land level; the radial pattern of the original drainage is still retained and there are no roads across the centre of the massif. British rock climbing had its beginnings here in the 1880s; Napes Needle pinnacle on Great Gable was an early conquest. There is a famous long-distance walk involving the maximum number of metres of ascent and descent and as many summits as possible within 24 hours. The current record (J. Naylor) is 72 summits, 11280 m and 170 km.

In **Scotland**, hills above 3000 ft (914 m) are known as Munros after Sir Hugh Munro, who–almost a century ago–produced the first list, of more than 500.

North of the Lowlands, south of the Great Glen, the **Grampian Mountains** include Britain's highest peaks. There are no glaciers, but some almost permanent snow patches. There is notable rock climbing in summer and ice climbing and skiing in winter.

**Ben Nevis** (1343) summit has been reached by motorcycle and in

## 7  THE SOUTHERN UPLANDS

Merrick 842

## 8  THE GRAMPIAN MOUNTAINS

Ben Nevis 1343, **the highest mountain in Scotland and in Great Britain.**
Ben Macdhui 1309
Braeriach 1295
Cairn Toul 1293
Cairn Gorm 1245
Ben Lawers 1214
Ben More 1171
Lochnagar 1154
Bidean nam Bian 1148
Ben Lui 1130
Ben Cruachan 1124
Schiehallion 1081
Ben Lomond 973

**Ben Nevis, Scotland (EA Shepherd)**

## 9  THE NORTHERN AND WESTERN HIGHLANDS

Carn Eige 1182
Mam Soul 1178
An Teallach 1062
Beinn Eighe 1009
Ben More Coigach 998
Slioch 994
Suilven 731, the 'Matterhorn of Scotland'.
Stac Polly 612

## 10  THE ISLANDS OF SCOTLAND

Skye – Sgurr Alasdair 991
       Sgurr Dearg 977
       Sgurr nan Gillean 965
Mull – Ben More 966
Arran – Goat Fell 974
Rhum – Askival 810
Lewis/Harris – Clisham 799
Orkney – Ward Hill 477
Shetland – Ronas Hill 450
St Kilda – Conachair 426

## 11  NORTHERN IRELAND

Mourne Mtns – Slieve Donard 852

1911 by car. There is an annual race to the top and back from Fort William.

In the summers of 1881 and 1882 Clement Wragge climbed the mountain almost daily to make meteorological observations in a stone hut. This was replaced in 1883–4 by an observatory of two rooms and a tower and observations were made continuously henceforward until 1904. There was an average of snow cover for 215 days in a year; snow or sleet fell on 169 days with a total annual rainfall of four metres; the sun shone only 750 hr in the year, and the average temperature in the warmest month was 5°C (41°F).

**Ben Macdhui** (1309), **Braeriach** (1295), **Cairn Toul** (1293), **and Cairn Gorm** (1245) are the major peaks of the Cairngorm Mountains, once the wildest in Britain. However some commercialisation has recently crept in with the development of Aviemore as a ski resort. The traverse of these four peaks was done in 4 hr 41 min by E. Beard in 1964.

Ben Macdhui was climbed by Queen Victoria in 1859. It is also noted as the home of Ferla Mor ('the Grey Man') a notable mountain monster, who chases solitary travellers off the mountain.

In 1772, **Schiehallion** (1081) was selected by Nevil Maskelyne, the Astronomer Royal, as the site of his enterprising experiment to determine the mass of the earth. He argued that an isolated mountain of mass readily calculable from its dimensions would deflect a plumb-line out of the vertical and that the deflection could be measured by observation of the angle between selected star transits and the apparent vertical on the north and south sides of the mass. Part of this angle is accounted for by the curvature of the earth, the remainder is due to the mass of the mountain.

Between 30 June and 24 October 1776, whenever weather permitted, Maskelyne made transit observations, totalling 337 sightings on 43 different stars. Meanwhile, two colleagues were making a careful survey of the mountain, from which another computed its bulk. The observations gave the earth nine-fifths times the density of the mountain; the latter was measured on local rock samples as 2·75 g/cc, thus making the density of the earth 4·95 g/cc (the present accepted figure is 5·517 g/cc). Thus the earth's interior was shown to consist of materials somewhat heavier than those on the surface, and was neither hollow nor filled with water as previously surmised.

The **Highlands**, north and west of the Great Glen, are remote areas of wild mountain country in widely varying rock types.

The highest point of **Skye**, Sgurr Alasdair (991), was not climbed until 1873, and the minor peak of Sgurr Coire an Lochain only in 1896. The summit of Sgurr Dearg (977) is a pinnacle of trap rock known as the Inaccessible Pinnacle. These are the main summits of the Cuillin Ridge, the most impressive peaks in Great

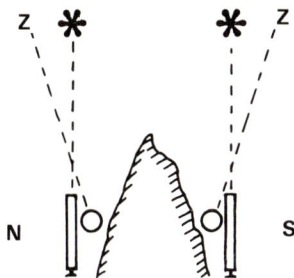

Fig. 1   Apparent difference in the position of a star when viewed from north and south sides of a mountain

Britain. Over a distance of 40 km the level never drops below 750 m (fastest traverse time from end to end 4 hr 5 min).

On **Orkney**, St John's Head (348) is **the highest vertical sea-cliff in Britain**. The Old Man of Hoy (137) is a famous sea stack, the ascent of which was featured on TV.

In the **St Kilda** group the island stacks of Stac an Armin (191) and Stac Lee (166) are impressive. St Kilda islanders climbed difficult rocks in search of sea-birds' eggs long before the advent of climbing as a sport; in consequence rock climbing ability was a male asset much sought after by local maidens.

**Northern Ireland** is not notably mountainous. Lava which has flowed out on top of chalk (the Antrim Lava Plateau) produces startling contrast effects in places, eg Rathlin Island. The columnar lava scenery of the Giants' Causeway is world famous.

# Republic of Ireland

**1   ULSTER**

Donegal – Errigal 752

**2   LEINSTER**

Wicklow Mtns – Lugnaquilla 944

**3   MUNSTER**

Macgillycuddy Reeks – Carrauntual 1041, **the highest point in the Republic of Ireland.**
Mt Brandon 953
Galtee Mtns – Galtymore 920

**4   CONNACHT**

Mweelrea 819
Nephin 807
Croagh Patrick 765
Slieve League 601

The **Republic of Ireland** is not notably mountainous, but there is a great variety of rock type and an impressive range of sea-cliffs.

The Guinness family estate in the **Wicklow Mountains** includes the great crag of Luggala.

The summit of **Carrauntual** (1041) was reached in 1937 by a party which had left the top of Ben Nevis 26 hr previously and had travelled in between by car and aeroplane via the summits of Scafell Pike and Snowdon.

Brendan the Navigator, who may have sailed to Iceland or even America, once built an oratory on the summit of **Mount Brandon.** It is now a famous place of pilgrimage (20 000 people once made the ascent in a day in 1868).

Ever since St Patrick spent 41 days on the top of **Croagh Patrick** (765) in 441, it has been a place of pilgrimage. The special day is the last Sunday in July (in 1951 65 000 people made the climb).

# France (excluding the Alps and the Pyrenees)

**1   MASSIF ARMORICAIN**

**a   Alpes Mancelles,** incorporated in PNR Normandie-Maine (1350 km²).
Mont des Avaloirs 417

The **Massif Armoricain** comprises a number of small blocks of low hills in ancient rocks in Brittany and Normandy; there are a few inland crags (notably at Clecy and Mortain) and sea-cliffs for climbers.

**b   Suisse Normande**
Mont Pincon 365

**c   Monts d'Arrée,** incorporated in
PNR d'Armorique (650 km²).
Mont St Michel et Tuchenn Gador 384

**d   Montagne Noire**
Roc'h Toullaeron 326

**2   MASSIF CENTRAL**

**a   Monts Dômes**
Puy de Dôme 1464, rd to summit,
   television mast, restaurant, etc.
Puy Grand Sarcoui
Puy de Pariou 1210
Puy du Come 1246
Puy Chopine 1181
Puy de Louchadière 1200

**b   Monts Dore**
Puy de Sancy 1885, c-wy to summit,
   **the highest point in France outside the
   Alps and the Pyrenees.**

**c   Monts du Cézallier**
Signal du Luguet 1551

**d   Monts du Cantal**
Plomb du Cantal 1855
Puy Mary 1787
Puy Griou 1694

**e   Aubrac**
Signal de Mailhebiau 1469

**f   Monts de Lacaune,** Montagne Noire,
   Monts de l'Espinouse
Signal de Montgrand 1267
P de Nore 1210
Le Caroux 1091

---

**Pinnacle in the Auvergne
(AC Collection)**

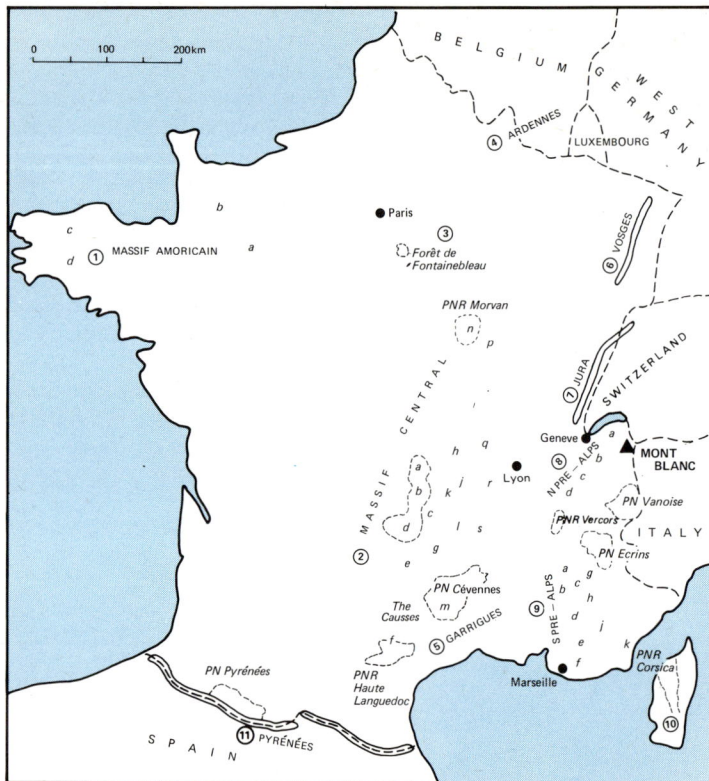

France

The **Massif Central** is a wide area of ancient rocks occupying
around one-sixth of the country, and reaching in places above
1700 m. There are massive limestone beds with the usual spectacular scenery of caves, crags and gorges, as well as relics of some
comparatively recent volcanic activity.

In the **Monts Dômes** area, volcanic activity only ceased around
8000 years ago, so that the countryside exhibits a wide range of
volcanic phenomena almost unaltered.
   The final death blow to the Neptunist theory of the history of
geology (that all rocks crystallised from water solution) was dealt
here in the mid-18th century. Their opponents, the Vulcanists,
were able to demonstrate from direct observation that some of the
local rocks had indeed solidified from the molten state.

The **Puy de Dôme** (1464) played an early role in science, since in
1648, five years after the invention of the barometer, Blaise Pascal
persuaded his brother to climb the mountain in order that they
might make simultaneous observations at base and summit.

---

*Right :* **Skiing poster (French Government Tourist Office).**

HIER ... **CHAMONIX** MONT-BLANC

SPORTS D'HIVER CONCOURS SKI-LUGE-PATIN-BOBSLEIGH

TRAINS EXPRESS AVEC VOITURES DIRECTES DE PARIS ET GENÈVE

DOCUMENT Cie PARIS-LYON-MÉDITERRANÉE   IMPRIMÉ PAR E. PICAUD - PARIS – COLLECTION PAUL PAYOT – PHOTO H. GAILLARD – PHOTOGRAVURE EDIMONTAGNE

*Above:* Scafell Pike, the highest point of the English Lake District (Aerofilms). *Below left:* Stedtind, a coastal peak of northern Norway (B Parker). *Below right:* Ordesa from Torla, Spanish Pyrenees (K Reynolds).

The following volcanic phenomena are worthy of notice: Puy Grand Sarcoui – injection of molten material from below without breaking the surface; Puy de Pariou – a crater partially destroyed; Puy du Come – a double crater with a cone on the rim; Puy Chopine – a plug exposed by denudation; Puy de Louchadière – a caldera.

The **Monts Dore** are also of volcanic origin. This is the nearest ski area to Paris.

The **Monts du Cantal** were once a single volcano 3000m high, ie comparable with the present-day Mount Etna.

The **Monts de Lacaune** are contained within the PNR Haut Languedoc. The **Caroux** is a fine little climbing area.

The **Cévennes** are really the steep southern scarp edge of the Massif Central looming above the Mediterranean coast plains.

The **Côte d'Or** consists of a series of east-facing scarps between Dijon and Lyon. It is a classic wine-growing area and there is also rock climbing.

The **Causses** are extensive limestone plateaux deposited from an ancient sea between the Cévennes and the Rouergue to the west. These were subsequently eroded into the familiar limestone pattern of gorges (such as those of the Rivers Tarn and Jonte), caves, crags (such as the Cirque de Navacelles) and areas of grotesque rock exposure (Montpellier-le-Vieux, Cirque de Mourèze). The Grands Causses are those of Sévérac, Sauveterre, Méjean, Noir and Larzac. Further north-west are the lesser Causses of Quercy, Martel, Gramat and Limogne.

There are many show caves in the area, the best known of which are the Gouffre de Padirac (**one of the finest in the world**, with lift and underground river trips); Aven Armand (descent by cableway) and Grotte des Demoiselles (also with cableway); there are also historical caves with primitive wall paintings such as Lascaux.

The **Forêt de Fontainebleau** is a forest and heath area (170km²) only 50km from Paris. There are 400km of footpaths and many sandstone outcrops and boulders suitable for climbing practice. This is where Paris mountaineers train between mountain trips.

The other crags which they use are all much more distant: Le Saussois, Yonne (200km); Surgy, Nievre (225km); Saffres, Côte d'Or (270km); Cormot, Côte d'Or (315km); Vatteville-Connelles, Eure (110km); Clecy, Orne (250km); Freyr, Belgium (270km) and Berdorf, Luxembourg (340km).

The **Ardennes** is a region of wooded hills with heaths on the summits, the denuded remains of very ancient mountains. They stretch from northern France, through Belgium and Luxembourg

Climbing at Saussois, France
(M Millet)

**g Monts de la Margeride**
Signal de Randon 1551

**h Monts de la Madeleine**
Pierre du Jour 1165

**j Monts du Forez**
Pierre sur Haute 1634

**k Monts du Livradois**
Nôtre Dame de Mons (Bois noirs) 1205

**l Monts du Devès** (Monts du Velay) 1421

**m Cévennes** (PN Cévennes)
Mont Lozère 1699
Mont Aigoual 1565

**n PNR Morvan**
Haut Folin 902

**p Côte d'Or 636**

**q Monts du Beaujolais and Maconnais**
Mont St Rigaud 1009

**r Monts du Lyonnais**
Signal de Saint-André 934

**s Monts du Vivarais**
Mont Mézenc 1753
Mont Gerbier de Jonc 1551, the source
  of the River Loire.
Mont Meygal 1436
Mont Pilat 1432

**t The Causses,** a series of plateaux
  without prominent summits.

**3 FORÊT DE FONTAINEBLEAU**

A forest and heath area without prominent summits.

**4 THE ARDENNES** (extending into
  Belgium and Luxembourg)

French – La Croix Scaille 504
Belgian – Botrange 694, **the highest
  point in Belgium.**
Luxembourgian – Burgplatz 602, **the
  highest point in Luxembourg.**

**5 THE GARRIGUES**

Pic Saint Loup 658

**6 THE VOSGES**

**a South (Grandes) Vosges,** (Vosges
  Cristallines) – Belfort Gap to C de
  Saales.
Ballon de Guebwiller 1423, **highest
  point in the Vosges.**
Hohneck 1362
Ballon d'Alsace 1247

**b Central Vosges** (Vosges Gréseuses)
  – C de Saales to C de Saverne.
Mont Donan 1008

**c North (Lower) Vosges** – N of C de
  Saverne, here is PNR Vosges du
  Nord (825 km²).
Wintersberg 580

**7 THE JURA**

**a French Jura**
Crêt de la Neige 1718, **the highest point
  of the Jura.**
Mt Reculet 1717
Crêt d'Eau 1621
Grand Colombier 1531

**b Swiss Jura**
Mt Tendre 1682
Mt Dôle 1680
Chasseron 1607

**8 THE NORTHERN PRE-ALPS**

**a N of the R Arve,** bounded in the N
  by Lac Leman and the Swiss frontier.

to the German frontier. At many places, particularly alongside the River Meuse at Freyr, there are magnificent limestone crags. There are also caves.

The **Garrigues** is limestone country of modest hills, but with many climbers' crags, between the Cévennes and the Mediterranean coast.

The **Vosges** rise on the west bank of the River Rhine for some 250 km between the Belfort Gap and the German frontier. With the facing Black Forest this is a classical rift valley (or graben) landscape. The geomorphology was unravelled by Hans Cloos in the 1930s after a rail tunnel bored through the Lorettoberg revealed the direction of slope of the boundary faults. The block occupied by the river slipped down between the facing hill blocks as earth forces drove them together, and the strata arched. There are two scenic roads: the Route des Crêtes, from which all the summits are readily accessible, was constructed to serve the front lines during World War I, when the range saw a great deal of fighting; at the foot of the scarp the Route du Vin samples the length of the east-facing vineyard slopes. There is some skiing and a little rock climbing. Attila left behind a gold waggon in the Lac du Ballon, which is still unclaimed.

The **Jura** line the Swiss frontier between Basle and Geneva. This limestone range, 250 km long, is 50 km wide at the centre, tapering to only a few kilometres at the ends. These are classical fold mountains produced by the same forces as those that produced the Alps. The drainage pattern appears odd since the rivers tend to run the length of the folds and have only occasionally broken through from one to another; thus the River Doubs flows 400 km to reach its confluence with the Saône, only 80 km (as the crow flies) from its source. The slopes, steep to the east, gentler to the west, are mostly poor pasture and forest. There is ski-ing, climb-ing and caving.

There are many fine viewpoints towards the Alps, looking across the Swiss plain to the Bernese Oberland, with Mont Blanc looming unbelievably high, further west.

The **northern Pre-Alps**, thrown up by the same earth movements which produced the Jura, are again predominantly of limestone, and exhibit in full measure the characteristic rock scenery – obelisks, spires, pyramids, molars, crests, combs and palisades; there are isolated summits difficult of access, climbers' crags and vast caverns. Deep and impressive gorges mark some of the river courses.

Proximity to big towns has led to development as a ski area, eg the modern resorts of Avoriaz (cedar tile-hung buildings of grotesque appearance, 320 km of pistes (ski-routes) and 150 ski-lifts) and Flaine ('the impression of a huge war-time fortification with a honeycomb of boutiques, bars, night-clubs and a super-market', man-made snow if necessary). There are many other resorts.

**Grande Chartreuse Monastery, France (C Charrel)**

*a1   Chablais*
Les Hautforts 2464
Cornettes de Bise 2432
Dent d'Oche 2225

*a2   Giffre*
Mt Buet 3099 (fa 1770 JA de Luc)
Pic de Tenneverge 2987

*a3   Aiguilles Rouges*
Aig du Belvedere 2965
Le Brévent 2526, c-wy from Chamonix.
L'Index 2390, c-wy from La Praz by
   way of La Flégère 1877.

**b   Between the R Arve, the R Arly and
Lac Leman**

*b1   Mt Salève*, c-wy to point on ridge.
Le Grand Piton 1375

*b2   Chaîne du Bargy*
La Pte Blanche 2437

*b3   Chaîne des Aravis*
L'Étale 2484

**c   Bauges,** between Lac Bourget and
   Lac d'Annecy.
Pte d'Arcalod 2217

**Giffre** is contiguous with the Dents du Midi of Switzerland (qv). The **Aiguilles Rouges** face the Mont Blanc group across the valley of Chamonix. Le Brévent, La Flégère and L'Index are all celebrated viewpoints for Mont Blanc.

**Mont Salève** provides an excellent practice crag for climbers from Geneva.

The eastern edge of the **Grande Chartreuse** massif, above Grésivaudan, maintains an almost uniform height of 1900m over a length of 20km.

Inside the **Dent de Crolles** (2062) is the Trou du Glaz cave, the **longest in France** (31km). There is an entrance near the summit, two more on the sides, and another at the resurgence.

**Mont Granier** (1933) was the scene of a catastrophic landslide in 1248 involving 115 million m³ of material, which spread over 23 km. Some pieces were projected for 6km. There are crags now used by climbers.

Notable caves in the **Vercors** include the **Gouffre Berger, the third deepest in the world**. Having only one entrance, at the top, it provides the deepest cave trip. Bournillon Cave has an arch 80m high at the outlet.

**d  La Grande Chartreuse,** in a loop of
the R Isère.
Chamechaude 2082
Dent de Crolles 2062
Mt Granier 1933

**e  PN Vercors,** between R Isère &
R Drome.
Le Grand Veymont 2341
La Grande Moucherolle 2284
Mt Aiguille 2086 (fa 1492)

## 9  THE SOUTHERN PRE-ALPS

**a  Massif du Diois** – 1613

**b  Mont Ventoux Ra**
Mont Ventoux 1909, highest point in
Provence, rd to top, radio mast,
weather station.
Dentelles de Montmirail 627

**c  Massif des Baronnies**
Montagne de Laup 1757
Rocher St Julien 696, the 'Matterhorn
of Provence'.

**d  Montagne du Lubéron** (PNR Lubéron)
Le Grand Lubéron 1125

**e  Montagne Ste Victoire**
Pic des Mouches 1011

**f  Massif de la Ste Baume**
Signal des Béguines 1147

**g  Le Dévoluy**
Tête de l'Obiou 2790

**h  Montagne de Lure**
Signal de Lure 1826

**j  Pre-Alps de Grasse and de Digne**
Sommet du Cheval Blanc 2323
Les Monges 2115
Sommet de la Bernarde 1941

**k  Massifs des Maures & de l'Esterel**
La Sauvette 779

## 10  CORSICA

Mte Cinto 2706, **the highest point in
Corsica.**
Mte Rotondo 2622
Capo el Berdato 2586
Mte Spicié 2560
Capo Bianco 2554
Mte Mentone 2550
Mte Falo 2549
Pt Minuta 2547
La Moniccia 2536
Pta Rufi 2535
Paglia Orba 2525, the 'Matterhorn of
Corsica'.
Mte Padro 2393
Mte d'Oro 2389
Mte Renoso 2352
Capo Tafonato 2343

**Mont Aiguille** (2086) is a notably inaccessible looking obelisk, one of the earliest of such peaks to be assaulted. It was first climbed in 1492 at the command of Charles VIII by a party led by Anton de Ville. They stayed there for six days; a herd of chamois was found on the top (how did they get there(!) – these first human climbers needed ladders and 'subtils engins' to reach the top). The next ascent, this time without ladders, took place in 1834 (J. Liotard).

The boundary between the northern and **southern Pre-Alps** is climatic – oceanic in the north, Mediterranean in the south. The various massifs are again mostly limestone, but the folding here is more complex since the influences of the Alpine and Pyrenean mountain-building episodes are superimposed.

**Mont Ventoux**, a massive whale-back, was **one of the first mountains ever climbed**, by the poet Petrarch in 1336.

The **Dentelles de Montmirail** (627) and the **Rocher St Julien** (696) are cocked-hat shaped ridges accessible only to the climber. It was 1946 before the latter was completely traversed. There are hard rock climbs on the flanks of both.

The **Montagne Ste Victoire**, close to Aix-en-Provence, is for ever associated with Cézanne, who painted it many times. On top, the first Croix de Provence was erected in the 16th century by a Provençal seaman. The latest (20 m high) dates from 1875.

There is some climbing in the **Massif de la Ste Baume**. Only 20 km away are the **Mediterranean sea-cliffs** known as **the Calanques, a world famous climbing area**.

The outstanding feature of the **pre-Alps de Grasse and de Digne** is the 'Grand Canyon' of the River Verdon. Carved in limestone and 700 m deep, it was first explored in 1905. Now there are viewpoints accessible by road, way-marked footpaths and plenty of hard rock climbing.

The **Massifs des Maures and de l'Esterel** are lower mountains of ancient rocks on the Mediterranean sea-board between Toulon and Nice.

The island of **Corsica** is a block of granite with relatively high Alpine-type peaks. There are no glaciers, but snow and ice for part of the year add stature to the mountains. Part of the island is now PN Corsica; there are plenty of way-marked paths and some fine sea-cliffs.

**Capo Tafonato** has a hole 30 m wide and 10 m high piercing it below the south peak. The west face is **one of Europe's highest mountain walls**.

# The Pyrenees (F = France; S = Spain)

**WESTERN PYRENEES** (from the Atlantic coast to the C du Somport) (from W to E).

C de Roncesvalles (Puerto di Baneta) 1057 (rd, entirely in Spain)
Pic d'Orhy 2017
C de la Pierre St Martin (rd), the famous Gouffre de la Pierre St Martin, **deepest cave in the world,** is nearby.
Pic d'Anie 2504 (F)
Pic d'Ansabère 2376 (F/S)
Aigs d'Ansabère, well-known rock climbing area.
Visaurin 2668 (S)
Llana de la Garganta 2599 (S)
Pic d'Aspe 2645 (S)
C du Somport 1631 (int rd, ry tunnel below)

**WEST CENTRAL PYRENEES** (from C du Somport to Pont du Roi and R Garonne gap) (from W to E).

Pic du Midi d'Ossau 2884 (F) (fa 1582, or 1787 by a shepherd)
Peña Collarada 2883 (S)
C du Pourtalet 1792 (int rd)
*To the N the valleys d'Ossau and de l'Ouzom are linked by the:*
C d'Aubisque 1710
*Resuming now on the frontier ridge:*
Pic de Balaïtous 3144 (F/S) (fa 8/1825 Peytier & Hossand)
Pic de Vignemale 3298 (F/S) (fa 1837 H. Cazaux and B. Guillembet)
Pic de Tendeñera 2850 (S)

---

**Observatory on the Pic du Midi de Bigorre, Pyrenees (French Government Tourist Office)**

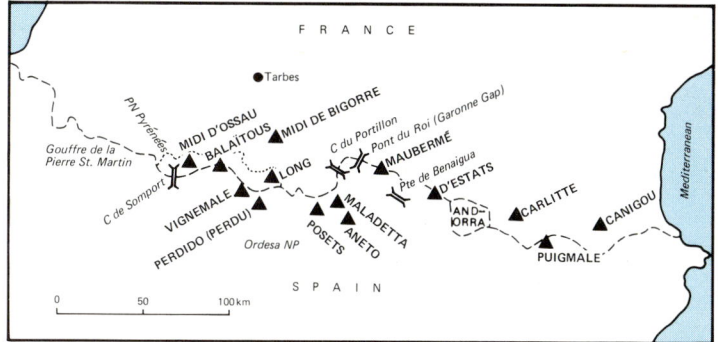

The Pyrenees

The **Pyrenees** form a mountain wall, 450km long, separating France from Spain. The range has two axes not quite aligned, so that there is a south to north 'kink' in the frontier and some overlap of the axes near the Col du Portillon. In the neighbourhood of the overlap the River Garonne crosses from Spain into France.

The highest point, Pic d'Aneto (3408), is in Spain; on the French side PN des Pyrénées Occidentales (2540 km²) and on the Spanish the Ordesa NP (22 km²) include some of the highest peaks and the finest scenery. There are a few small glaciers, totalling 33 km², also some considerable caves. The **Gouffre de la Pierre St Martin**, below the Pic d'Arlas, is **one of the deepest in the world** (1332 m); la Verna chamber is 200 m spherical. Further along the range the Grotte Casteret is **the highest ice cave in Europe.**

End-to-end scenic routes are available – for the motorist, the Route des Pyrénées; for the walker, the long-distance footpath GR 10; for the mountaineer, Haute Randonnée Pyrénéenne (Club Alpin Français).

Pic du Taillon 3146 (F/S)
*To the N, Luz and Ste Marie-de-Campan are linked by:*
C de Tourmalet 2114
*Resuming now on the frontier ridge:*
Marboré 3248 (F/S)
*and on a branch ridge S of the frontier ridge:*
Cilindro 3328 (S)
Mte Perdido (Mt Perdu) 3355 (S) (fa 1802 Rondo, Laurens and a shepherd; or 1802 R de Carbonnière)
*Resuming now on the frontier ridge:*
Pic de la Munia 3133 (F/S)
*Here there is a considerable area of high mountains to the N of the frontier:*
Pic Long 3192 (F)
Pic de Néouville 3091 (F)
Pic du Midi de Bigorre 2868 (F) (fa 1575)
*Resuming now on the frontier ridge:*
Port de Bielsa 2429 (rd tunnel under construction below)
Grand Bachimale 3174 (F/S)
*Here there is a considerable area of high mountains to the S of the frontier:*
Pico de Posets 3375 (S)
Espadas 3329 (S)
Gran Pico de Eriste 3053 (S)
Cotiella 2910 (S)
*Hereabouts to the N of the frontier ridge Arreau is linked to Luchon by:*
C de Peyresourde 1563
*Resuming now on the frontier ridge:*
Pic des Gourgs Blancs 3129 (F/S)
Pic des Spijoles 3065 (F)
Pic Periguère 3222 (F/S)
Pic de Maupas 3109 (F/S)
*A few km further on the frontier turns N to:*
C de Portillon 1308 (int minor rd)
*The line of the highest mountains, the S arm of the overlap, continues somewhat S of E to:*
Pico de la Maladetta 3312 (S) (fa 1817 Parrot)
Pico Maldito 3350 (S)
Pic d'Aneto 3408 (S) (fa 20/7/1842 Cte de Franqueville, P de Tchihatcheff & party), **the highest point in the Pyrenees.**
Pico Russell 3205 (S)
Tunnel de Viella (rd)
Besiberri 3030 (S)
Montarto 2830 (S)
Els Encantats 2747 (S)
Montseny 2881 (S)
*Resuming now on the frontier:*
Mt Bacanère 2194 (F/S)
Pont du Roi (int rd), in the valley of the Garonne where it crosses from Spain into France.

**EAST CENTRAL PYRENEES** (from Pont du Roi to Andorra) (from W to E)

Pic de Maubermé 2880 (F/S)
Pic de l'Homme 2722 (F/S)
Pic de Montvalier 2839 (F)
Port d'Aula 2260 (rd on French side)

Two important scientific sites are the observatory on the summit of the Pic du Midi de Bigorre and the solar furnace at Font Romeu.

The range is named for Pyrène, buried there by Herakles after being killed by wild beasts. Hannibal crossed in 218 BC on his way to the Alps and Italy. These mountains have since been traversed by several armies and by hosts of refugees in various wars.

Below the **Col du Somport** (1631) is a railway tunnel, 7·9 km long, opened on 18 July 1928.

**The PN des Pyrénées Occidentales** stretches from near the Col du Somport to beyond the Cirque de Gavarnie. The main park area of 457 km$^2$ includes the higher peaks, around it is a peripheral area of a further 2060 km$^2$.

The **Vignemale** (3298) will always be linked with the name of Count Henry Russell (1834–1909), an Irish catholic born in France, who climbed it 33 times, the last at the age of 70. Between 1880 and 1882 he organised the building of a grotto at 3200 m in the wall of rock at the head of the Ossoue glacier, in which he frequently spent many days at a time. A second was added in 1885 and a third in 1886. There were many visitors. In 1887 the glacier destroyed the lower grottoes, so another three were constructed at lower levels; finally in 1893 one was erected for him on the summit – a curious continuous link between a man and a mountain.

On the north side of the frontier ridge below the peaks of **Taillon** (3146) and **Marboré** (3248) is the famous Cirque de Gavarnie, a hollow, one kilometre in diameter, with a floor at 1650 m and a skyline of 3000 m. There is a waterfall of 460 m. Between the two peaks is the vertical-sided cleft of the Brèche de Roland, hewn by the sword of the legendary hero.

Parallel to the frontier ridge and 5 km south of the Brèche de Roland is the deep valley of **Ordesa Canyon** (Ordesa NP, 22 km$^2$), where there is high standard rock climbing on Tozal del Mallo (2283), Punta Gallinero, Mondaruego, etc.

Close to the **Pic de la Munia** (3133) on the north side of the frontier ridge are the Cirques d'Estaube and de Troumouse.

The **Pic du Midi de Bigorre** (2868) was climbed repeatedly by M. Plantade in the first half of the 18th century for scientific observations. Now there is a long-established observatory on the peak.

The glacier of the same name on the north side of the **Pico de la Maladetta** (3312) is **the largest ice field in the Pyrenees**.

North of the **Pic d'Aneto** (3408) the water from the Aneto glacier

Pic de Certescans 2840 (F/S)
Pic du Port de Sullo 3072 (F/S)
Pic de Montcalm 3077 (F)
Pic d'Estats 3141 (F/S)
Pic de Tristagne (Pic de Tristaina) 2878
Pic du Port 2903

## ANDORRA

Coma Redrosa 2946 (A/F)
Pic de Serrère 2911 (A/F)
Pic de Siguer 2905 (A/F)
Alt del Crio 2859 (A)
Pic Negre 2760 (A/S)
Pic de Casamanya 2736 (A)

**EASTERN PYRENEES** (Andorra to
  the Mediterranean) (from W to E)

 C de Puymorens 1915 (rd, Ariege
  valley to Segre valley and the Spanish
  frontier)
Pic Carlitte 2921 (F)
Pic Péric 2810 (F)
 C de la Perche (int rd, from the Têt
  valley to the Segre valley)
Puigmale 2912 (F/S)
Pic de la Donya 2714 (F/S)
Mont Canigou 2785 (F) (fa *c* 1280
  Peter III of Aragon)

# Spain and Portugal

## THE MOUNTAINS OF CATALO-NIA AND ARAGON (N of the R Ebro, the foothills of the Pyrenees)

**Sa del Cadi**
Cadi 2567
Tossa 2537

**Montserrat** – Turo de San Jeronimo
  1235

**Sa de Guara**
Tozal (or Punton) de Guara 2077

## THE CANTABRIAN MOUNTAINS

**Main chain**
Peña Prieta 2536
Pico de Curabocas 2517
Peña Espigüette 2450
Peña Ubiña 2417
Coriscao 2234
Mampodre 2197
Braña Caballo
Valdecebollas 2136
P de Arcenorio 2122
Catoute 2117
P Gildar 2083

sinks underground at the Trou de Toro, emerging on the north side of the range as the River Garonne.

**Els Encantats** (2747) is the best-known summit of the remote little district of Los Encantados. Hereabouts are the PNs Aigües Tortes and Lago de San Mauricio. Between this area and the parallel north arm of the overlap at the head of the River Garonne is the Pass of Bonaigua (2072).

**Andorra** is an independent principality of only 453 km² on the south side of the main ridge. It is linked to France by the Port d'Envalira (2407). A circuit of the country on the surrounding mountain walls makes a fine easy expedition. To the north near Foix in France is the show cave of Labouiche with an underground river.

Close to the **Col de la Perche** is Font Romeu, the site of the giant solar furnace of the CNRS (Centre National de la Recherche Scientifique). Sixty-three mirrors, servo-controlled to follow the sun, direct its radiation on to a parabolic reflector, 40 m high and 54 m wide, made up of 9500 mirrors. At the focal point one MW of thermal power is concentrated and temperatures of 3800°C can be realised.

**Mont Canigou** (2785) was one of the first mountains to be climbed (*c*1280). Peter III of Aragon, who made the ascent alone, is said by his chronicler to have thrown a stone into a lake there and set free an enormous dragon. It seems to have been a harmless specimen!

**Solar Furnace (R D Treble)**

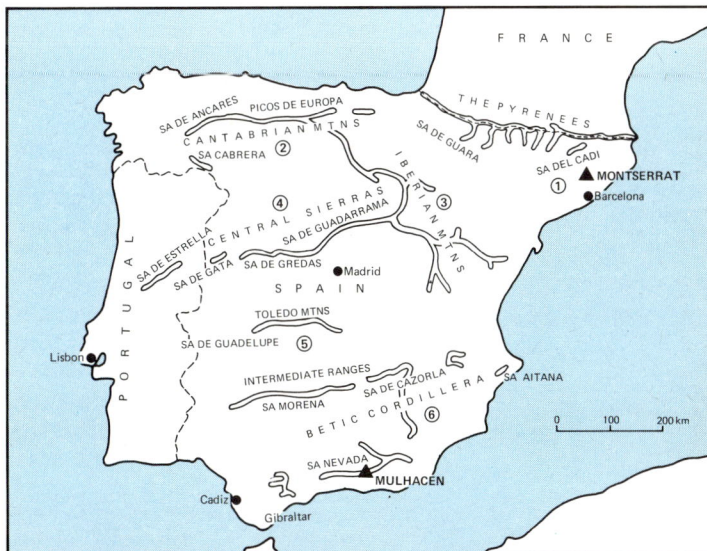

**Spain and Portugal**

Peña Sagra 2042
P de Tres Aguas 2035
Peña Labra 2006

**Picos de Europa**
Torre Ceredo 2648
Llambrion 2617
Peña Vieja 2613
Oriellos 2600
Peña Santa de Castilla 2596
Tesorero 2566
Naranjo de Bulnes (P de Urriello) 2516
    (fa 1904 Marquis v Villaviciosa;
    G Perez)
Tabla de Lechugales 2445
Cortes 2373

**Montañas de León**
El Teleno 2188

**Sa de Picos de Ancares** (Ancares NP)
P Miravalles 1969
P Cuiñá (Hórreo) 1958
Mustallar 1924
P de Tres Obispos 1848
Peñarrubia 1821
Peñalonga 1800

**Sa Cabrera**
Peña Treviñca 2124
Picon 2067
Moncalvo 2047

**THE IBERIAN MOUNTAINS** (NW
    to SE on the S side of the R Ebro)

**Sa de la Demanda**
Co de la Demanda (Sa de San Lorenzo)
    2305
San Millan (La Trigaza) 2132

**Sa de Cameros**
Cabezo del Santo 1857

**Sa de Urbión**
P de Urbión 2246

**Sa Cebollera** – 2141

**Sa de Moncayo** – 2265

**Sa de Albarracin**
Sa Alta 1856

**Sa de Gúdar**
Peñarroya 2020

**Sa de Javalambre** – 2020

**THE CENTRAL SIERRAS** (E & W,
    N of the R Tajo (Tagus))

**Sa de Guadarrama**
P de Peñelara 2444
Loma de Bailanderes 2440
Cabezas del Hierro 2397
Valdemartin 2291
Alto de las Guarramas 2275
La Picutilla 2236
Sa de la Maliciosa 2230
Peña del Oso 2209
Sa de la Mujer Muerta 2206

Mountain ranges are widespread throughout **Spain**. There is little snow and ice mountaineering, but there has been extensive development of the ski potential in many parts of the country. There is also a variety of rock climbing.

**Montserrat** is a rock climbers' and scenic paradise, only 60 km from Barcelona and close to the Costa Brava and Costa Dorada. A massif of conglomerate ($10 \times 5$ km) has been eroded into deep ravines with huge buttresses and pinnacles with impressive overhangs. There is a much visited monastery, well equipped with the paraphenalia of tourism – road access, cableways, cable-railway, shops, bars, etc.

The **Cantabrian Mountains** continue the line of the Pyrenees along the north coast of the peninsula. The Picos de Europa, which stand towards the north of the main line of the range, are the finest mountains in Spain – pointed rocky peaks offering plenty of climbing. They are limestone mountains reminiscent of some of the well-known Austrian mountain groups, such as the Karwendel. The western part of the Picos is a NP.

The railway from Leon to Oviedo through the Pass of Pajares has 58 tunnels in 42 km between points only 11 km apart.

Passing through the **Sierra Cabrera**, the Sierra de la Culebra Railway requires 182 tunnels, having a total length of 78 km, in the 173 km between Puebla de Sanabria and Carballino.

The **Sierra de Guadarrama**, rising 50 km north of Madrid, are 100 km long, and consist mainly of limestone and granite. These are the ancient Montes Carpetani; there is a megalithic monument near the top of the highest peak. In contrast, a lower summit, known as 'La Bola de Mundo' carries the antennae of Spanish television.

Madrid climbers practise on the crags of La Pedriza del Manzanares on the slopes of Cabezas del Hierro.

The **Sierra de Gredos** continues the line of the Sierra de Guadarrama to the west. This range is 150 km long, 20 to 30 km wide, and consists largely of granite. There is climbing, notably on the rock needles of Los Galayos.

The **Sierra Morena** is a lengthy ridge line rising to the north of the Guadalquivir River, having an average height of 750 m. It extends into Portugal as the Sierra de Monchique.

The **Betic Cordillera** embraces the mountains of south-east Spain, south of the River Guadalquivir.

The Sierra de Aitana offers mountains and rock climbing immediately behind the famed Costa Blanca.

The Sierra de Segura has been famous since Roman times, when it was called Mons Argentarius because of the deposits of silver-bearing lead.

Pinareja 2206
Montón de Trigo 2167
Siete Picos 2138
P de Najarra 2118
Peña del Aguila

**Sa de Gredos**
P del Almanzor 2592
La Mira 2416
Calvitero 2401
La Serrota 2294

**Sa de Gata**
La Alberca 1723

**Sa da Estrela**
Malhao da Estrela 1991, **the highest
point of the Portuguese mainland.**

**INTERMEDIATE RANGES** (between
the Rs Tajo (Tagus) and Guadalquivir)

**Montes de Toledo**
Corral de Cantos 1419

**Sa de Guadeloupe**
Carboneras 1443

**Sa Morena** – 1320

**THE BETIC CORDILLERA** (Cordillera Penibética) (S of the R Guadalquivir) (from E to W)

The Sierra Nevada are the highest mountains on the Spanish
mainland and in western Europe outside the Alps. There are no
glaciers, but permanent snow-fields ensure popular ski-ing
potential, while there is some rock climbing, notably on the
pinnacle of Pulpito de Carnales. **Mulhacen** (3481), **the highest point**,
carries the ruins of a summit observatory built in 1879. The range
is 110 km long and between base and summit there is a wide range
of flora from tropical to alpine. **The highest motor road in Europe**
crosses from north to south passing close to the summit of Veleta
(3401), the third highest peak.

**Sa de Aitana** – 1585

**Sa de Alcaraz** – 1798

**Sa de Segura**
La Sagra 2381
Tornajuelos 2128
Empanadas 2107
Las Cabras 2081
Calañas
Revalcadores 2001
Pinar Negro 1922
Blanquilla 1830

**Sa de Maria** – 2043

**Sa de los Filebres**
Ste Barbara 2269
Popa 2223
Calar Alto 2168

Dos Picos 2085
Padilla 2058

**Sa Nevada**
Mulhacen 3481 (fa 1840 Clementi), **the
highest point of mainland Spain.**
l'Alcazaba 3414
Veleta 3401, rd close to summit.
Cerro Delado 3182
Cerro del Caballo 3015
Chullo 2616

**Sa de Magina** – 2165

**Sa de Telox** – 1919

**GIBRALTAR**

The Rock 426

**Granada and the Sierra Nevada (Spanish National Tourist Office)**

# Spanish and Portuguese Islands

**BALEARIC ISLANDS**

**Majorca**
Puig Mayor 1445

The **Azores** lie on the mid-Atlantic Ridge at a point where it is
joined by an undersea ridge protruding from the Mediterranean.

Massanella 1350
Puig Tomir 1103

**Minorca**
El Toro 358

**Ibiza** – 409

**CANARY ISLANDS** (Islas Canarias)

**Tenerife**
P de Teide (V-recent eruption) 3718
(fa 1582 RE Scory), 'the Peak of
Hell', **the highest point in Spanish
territory.**

**La Palma**
Rocque de los Muchachos (V-recent
eruption) 2423

**Gran Canaria**
Pozo de las Nieves 1980

Most of the islands exhibit volcanic activity to a greater or lesser
extent. Pico (2351) is **the highest point of Portugal**.

| | |
|---|---|
| **Hierro** Pico Malpaso 1501 | **São Miguel** P da Vara 1103 |
| **Gomera** Garajonay 1487 | **São Jorge** P da Esperanza (V) 1053 |
| **Fuerteventura** Muda 689 | **Fayal** Caldeira (V-le 1957) 1043 |
| **Lanzarote** (V-shield type, le 1824) 565 | **Terceira** Caldeira (V) 1021 |
| **MADEIRA** | **Flores** Morro Gde 913 |
| P Ruivo 1862 | |
| **AZORES** | **Corvo** Caldeira (V) 718 |
| **Pico** | **Santa** Maria 587 |
| Pico (V-recent eruption) 2351, **the highest point on Portuguese territory).** | **Graciosa** Caldeira (V) 402 |

# Scandinavia

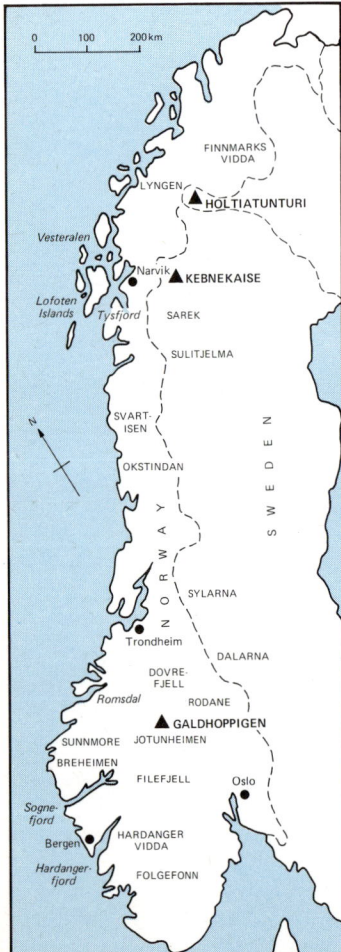

**Scandinavia**

Of the Scandinavian countries, **Norway** is particularly mountainous. Only one-fifth of the country is below 140m, half is above
450m, and there are considerable areas over 900m; 72 per cent
of the land surface consists of rock, mostly granite and slate. All
Sweden's mountains fall along the frontier and are therefore included in a general treatment.

The peninsula offers a possible ski expedition of some 2500km
from Banak in the north to Adneram (east of Stavanger) in the
south, crossing the highest points of Finland, Sweden and Norway
on the way. As a single trip it might take three months.

Oslo climbers practise on the rocks of Kolsas (west of the city);
Stockholm climbers at Häggsta (20km south-west); Gothenburg
climbers at Utby (6km north); Trondheim climbers at Korsvikke
and Innerfelnes.

The **Folgefonn glacier** (1661) is **the third largest ice field in Europe**
(220km$^2$).

The **Hardangervidda Plateau** with 7500km at 1000 to 1250m is
possibly **the largest plateau in Europe**. Now a well-established
walking area with 1200km of cairned routes, it will be a NP in
due course.

There is rock climbing in the **Finse-Filefjell** ranges. The Norsk
Alpincentre has been established in the Hemsedal.

Norway's highest mountains are in the **Jotunheimen**, 'the Home
of the Giants'. There are 250 peaks over 1900m and over 60
glaciers; the largest is the Holabre glacier (16km$^2$). These used to
be hunting lands, but the wild reindeer have now disappeared
and it has become a popular mountaineering area.

There is often dispute between Galdhoppigen (2469) and

## ROGALAND-SETESDAL

Vassdalsegga 1660
Mt Snonut 1606
Kjelatind 1476

## BLEFJELL AND DISTRICT

Gausta 1883
Borgsjåbrotet 1484
Skjerveggin 1381
Bletoppen 1341

## FOLGEFONN GLACIER – 1661

## HARDANGERVIDDA PLATEAU

Hardanger jøkule glacier 1876
Sandfloeggji 1706
Hårteigen 1681, a butte
Store Nup 1646

## STOLSHEIMEN (Vass and Mjølfjell ras)

Steganosi 1761
Fresvikbre 1660
Vassfjöro 1632
Tverrbotnanuten 1610
Øykjafonn 1604
Vosseskavl glacier 1579

## FINSE-FILEFJELL RAS

Hallingskarvet ridge 1933
Juklegga 1920
Storebotteggja 1830
Bleita 1820
Blaskavlen 1809
Reineskarvet ridge 1789
Suletind 1781
Storeskavlen 1729
Skogshorn 1728
Hornsnipa 1692, resembles a buck's
horn, long used as a landmark.

## SYNFJELLVIDDA MOUNTAIN PLATEAU

Hiemdalshø 1843
Sikilsdalshø 1783
Mt Skaget 1686
Langesuen 1595

## JOTUNHEIMEN

Galdhoppigen 2469 (fa 1850 L. Arne-
sen, S. Flotten, S. Sulheim), **the
highest point in Norway.**
Subsid pk 2369
Glittertind 2452 (fa 1841 by a survey
party), height has been measured up
to 2470 due to snow cover.
Store Skagastølstind 2405 (fa 21/7/1876
W. C. Slingsby alone)
Subsid pk 2345
Store Styggdalstind 2387
Subsid pk 2347
Skarstind 2373
Vetlepiggen 2368
Surtningssui 2368

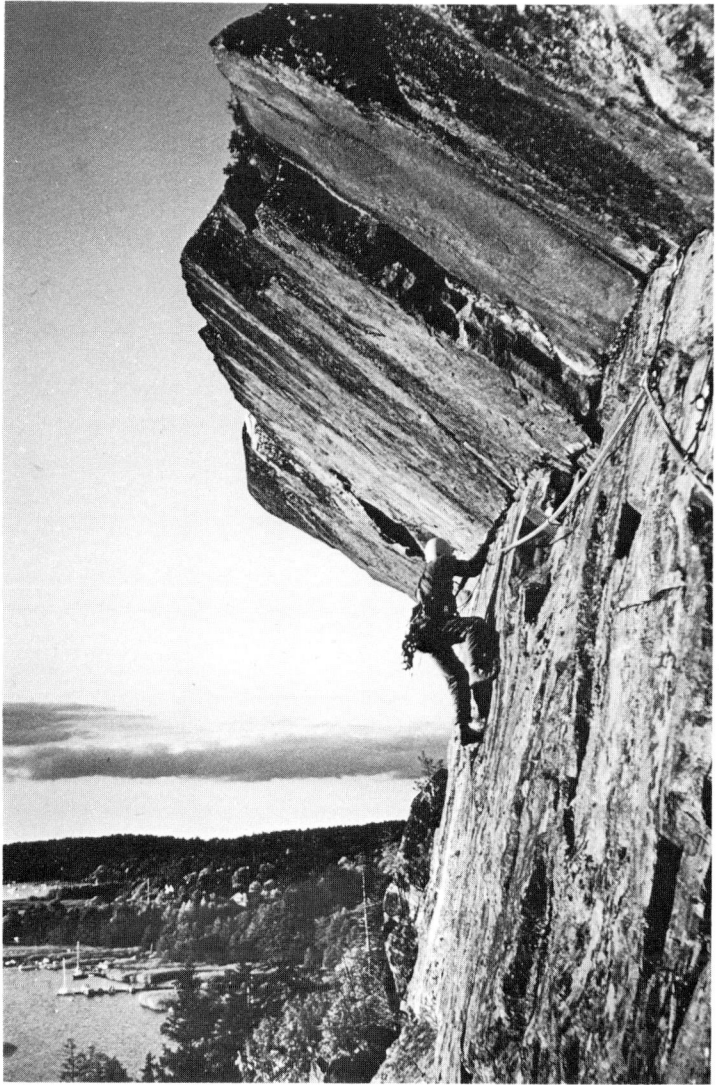

Rock climbing near Stockholm (L Cronlund)

Glittertind (2452) as to which is the higher. The latter varies a good deal due to fluctuating snow cover; the Norwegian authorities firmly favour the former.

**The Jostedalsbreen** (1500 km) in Breheimen is **the largest ice field in Norway** and **Europe's largest glacier**. The ice is 300 m thick with 24 glacier tongues descending on various sides. **Crossing** gives a fine mountaineering expedition, **first carried out** by W. C. Slingsby.

Rodane NP (572 km²) includes all the major peaks of the **Rodane mountains**. Wild reindeer, wolverine, etc live here permanently.

Subsid pk 2306
Store Memurutind 2364
Gjertvasstind 2352
Store Hellstugutind 2346
Subsid pk 2340
Storgjavtind 2343
Store Knutsholstind 2340
Tjørnholstind 2329
Leirhø 2328
Bukkehø 2314
Store Tverråtind 2309
Ymisfjell 2259
Mt Besshø 2258
Nautgardstind 2257
Store Smorstabbtind 2208

**BREHEIMEN**

Lodalskappa 2083
Suphellenipa 1825

**RODANE MOUNTAINS**

Rondslottet 2178
Storronden 2142
Høgronden 2114
Mt Storsmeden 2017
Sølnketten 1826

**ØSTERDAL-FEMUND AREA** (including Femundsmarke NP (385 km²))

Elgåhogna 1459
Mt Suuku 1415

**DALARNA** (Sweden)

Storevatteshogna 1204

**SUNNMØRE**

Jagta 1593
Skruen 1584
Slogjen 1559
Kolåstind 1463

**DOVRE-ROMSDAL RAS**

Snohetta 2286
Store Svånåtind 2215
Larstind 2196
Storekalken 1870
Vengetindane 1843
Trolltinder 1794
Kvanndalstind 1776 (fa 1885 W.C. Slingsby & party)
Breitind 1707
Dronninga 1568
Juratind 1562
Romsdalhorn 1554 (fa 1832), the Romsdal 'Matterhorn'.

**TROLLHEIMEN AND NORDMØRE**

Tarnfjell 1850
Trolla 1840
Vinnufjell 1823
Skardfjell 1741
Mt Snota 1689

*Above:* Galdhopiggen, the highest point of Norway (Aerofilms)

*Below:* Kebnekaise, the highest point of Sweden (G Malmström)

Blåhø 1680
Mt Trollhetta 1642
Dalatårnet 1394, the local 'Matter-
horn', the 'Tower of the Valley'

## SYLENE MOUNTAINS (Norway)/ SYLARNA (Sweden)

Helagsfjell 1796 (S)
Storsylen 1762 (N)
Sylarna 1710 (S)
Lillesylen 1704 (S)
Skarsfjell 1593 (S)
Storvigelen 1582 (N)
Skardorsfjell 1513 (N)

## GRESSÄMOEN NP (180 km²)

Mt Bugvassfjell 1009

## BORGEFJELL NP (1000 km²)

Kvigtind 1703

## OKSTINDAN MOUNTAIN RA

Oksskolten 1915, **the highest summit in N Norway.**
Okshornet 1907
Svartfjellet 1866
Tvillingtinden 1830
Okstind 1803
Steikvasstind 1751
Vesttind 1724
Okstjelltuva 1532
N Storfjell 1764 (Sweden)

## SVARTISEN

Skjelåtind 1640
Snøtind 1599
Istind 1577
Vestisen 1554
Østisen 1467

## SULITJELMA AND SALTFJELLET
(Norway/Sweden frontier)

Suliskongen 1913
Stortoppen 1830
Sulitjelma 1787
Ølfjell 1754
Nordsaulo 1745
Vardetoppen 1722
Mt Blåmannen 1571
Solvågtind 1561

## SAREK (Sweden)

Sarektjåkko 2098
Stortoppen 2089
Partjifjallen 2021
Akkavarre 2013
Spika 1976
Apartjakke 1914
Kavapakte 1906
Perikpakte 1789

## KEBNEKAISE (Sweden)

Kebnekaise 2117 (fa 1883 C. Rabot), **the highest point in Sweden.**
Kaskasatjåkko 2076
Kaskasapakte 2040
Kebnetjåkko 1976
Koupertjåkko 1970
Mormantjåkko 1970
Tuolpagorni 1960
Liljetoppen 1950
Tarfalatjåkko 1941
Mårmapakte 1939
Rullevarre (Riepovare) 1600
Nallo 1540

## BODO-TYSFJORD-NARVIK

Storstein fjell 1901
Hjornetoppen 1875
Kuinarcocka 1796
Kirken 1734
Sälkacocka 1694
Mt Caihnacocka 1596
Kongsbakktind 1576
Stortind 1537
Oallacocka 1490
Haugbakktind 1454
Stedtind 1381 (fa 1910 C.W. Ruben-
son, F. Schjelderup, A. Bryn), **'the most remarkable natural obelisk in the world'.**
Nordre Ippococka 1367
Fagernestoppen 1270
Rombakstotta 1243, the Narvik 'Mat-
terhorn'.
Strandaatind 862 (fa 1912 H. Jentoft,
C.W. Rubenson, F. Schjelderup),
'fantastic walls'.
Husbyviktind 806 (fa 1913), 'terrible
monolith'.

## LOFOTEN AND VESTERÅLEN ISLANDS

**Hinnøy**
Moysalen 1266

**Austvagøy**
Higravstind 1161 (fa 1901 J. Collie & party)
Gjeitgaljar 1084 (fa 1901 J. Collie & party)
Rulten 1062 (fa 1903 J. Collie & party)
Trolltinder 1045 (fa 1897)
Trakta 1025 (fa 1910)
Olsanestinder 1000
Blaafjeld 998 (fa 1901)

**Moskenesøy**
Hermansdaltind 1034

**Flakstad**
Stjernhodet 937

**Langoy**
Reka 607 (fa 1902 T. Ouston, H. Mundahl), 'a spectacular conical peak shaped like a huge obelisk'.

**TROMS AREA** (including Øvre Dividal NP (750 km²))

Kirkestind 1681
Kistefjell 1634
Rostafjell 1590

**SENJA ISLAND** (including Anderdalen NP (53 km²))

Kaperfjell 853

## LYNGEN PENINSULA

Jiekkevarre 1833, the 'Mont Blanc of the N'.
Store Jaegervarstind 1668 (fa 1898 W. C. Slingsby)
Najalvarre 1590
Anntind 1505

## NW FINLAND

Haltiatunturi 1324, **the highest point in Finland.**

**FINMARK MOUNTAIN PLATEAU** (including Stabbursdalen NP (96 km²))

Cuokka Rasša 1139
Muvravarre 618

The **Sunnmøre** is another rock climbing area, where Hjørundfjord and Geirangerfjord are notably scenic.

The wall of the Trolltinder above the **Romsdal** is one of the great rock walls of the world, with some climbs of very high standard. The west face of Hoarstadnebba is also a considerable precipice.

The **Svartisen** (400 km²) is **the largest ice field in north Norway**. Hereabouts too is the country's most important caving area; there are more than 100 caverns, of which Larshullet is 326 m deep.

**Sulitjelma** is an important mining area, yet remains fairly unspoilt. There will be a NP (1035 km²) on the Saltfjellet plateau in due course. There are caves here also.

**Mont Blåmannen** (1571) is the highest point of the Blåmannsisen ice field (124 km²).

The **Bodo-Tysfjord-Narvik** area is said to comprise 'fjords and some of the most unbelievable mountains in the world'. Notable peaks are Stedtind, Strandaatind and Husbyviktind.

**Lofoten** consists of six large islands and 76 more that are inhabited; **Vesterålen** consists of four large islands, parts of two others and numerous small ones. All present steep jagged rock peaks; 'if the ocean rose to the foot of the Chamonix Aiguilles, one would get a characteristic view of Lofoten peaks'.

For rock climbing practice, mountaineers from Copenhagen, **Denmark**, cross the straits to Sweden where there are some crags on the peninsula of Kulleberg, 20 km north of Hälsingborg.

## DENMARK

Yding Skovhøj 173, **highest point.**

# Germany and W Czechoslovakia

## THE BAVARIAN PRE-ALPS (Allgauer, Bayerische and Berchtesgaden Alps)

Zugspitze 2963 (c-wy to summit), **the highest point in W Germany.**
Watzmann 2714
Madelegabel 2646

## THE SCHWARZWALD (Black Forest) (on the right bank of the R Rhine between Basle and Karlsruhe)

The **Bavarian Pre-Alps** comprise the Allgauer, Bayerische and Berchtesgaden Alps, where the Austria/Germany frontier leaves a few major peaks on the German side. The **Zugspitze** (2963) **is the highest point in West Germany**; it can be reached by cableway. Near the summit is a hotel, also one of Europe's highest atmospheric sampling stations, measuring fall-out and pollution, electric charging and gradients.

The Watzmann (2714) throws down **the highest rock face in the Eastern Alps** (around 1800 m). The final stronghold of Adolf Hitler was sited in these mountains, but fortunately never used.

Feldberg 1493
Hornisgrinde 1164

## SCHWABISCHE AND FRANK-ISCHE ALB (Swabian and Franconian Alb (or Jura))

The Schwabische Alb (Lemberg 1015) sweeps NE on the left bank of the R Donau from the Rhine falls at Schaffhausen to Nordlingen. The Frankische Alb (Hessel Berg 770) continues the line round Nürnberg to the upper reaches of the R Main.

## THE W GERMANY/CZECHOSLO-VAKIA FRONTIER

**Bohmer Wald** (lines the frontier)
Grosser Arber 1457, **highest point on German side.**
Grosser Rachel 1452
Plöckenstein (Plechy) 1378, **highest point on Czech side.**
Boubin 1362

**Bayerische Wald** (parallel to the W)
Einodsriegel 1126

**Fichtelgebirge**
Scheeberg 1051

## CENTRAL GERMANY

**Hardt Mtns** 683, a continuation N of the Vosges, sandstone with thick forest cover.

**Hunsrück** 816, shales, slates and quartzites, well forested.

**Eifel** 760, an extension NE of the Ardennes.

**Odenwald** 626, a lower version of the Schwarzwald.

**Taunus** 880, well forested, ancient rocks.

**Westerwald** 657, signs of former volcanic activity.

**Rothargebirge** 840

**Vogelsberg** (Taufstein 773), **the largest continuous area of basalt in Europe.**

**Hohe Rhön** (Wasserkuppe 950)

**Harz Mtns**
Brocken 1142 (climbed 1697 by Peter the Great & 1777 by Goethe), rd to summit, hotel and observation tower.

**Thuringer Wald** (Grosser Beerberg 982), an 80km ridge with grassy summits and forested flanks.

## NW CZECHOSLOVAKIA

**Erzgebirge** (line the frontier with Germany)

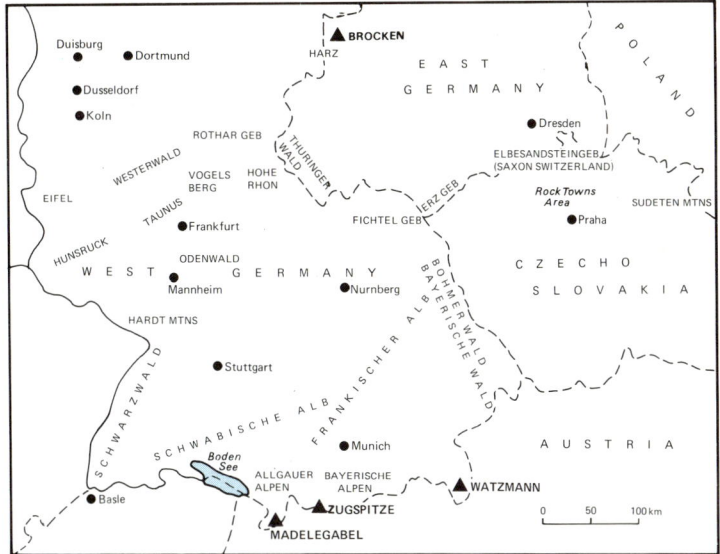

Germany and W Czechoslovakia

The **Schwarzwald** (Black Forest) faces the Vosges (qv) across the Rhine graben. The height and scenery are comparable and there is some rock climbing.

The **Schwabische and Frankische Alb** (Swabian and Franconian Jura) have a total length of 700km. Both are limestone and cut into blocks by dry valleys; deep gorges and steep slopes provide

Neuschwanstein, Swabia, Germany (Barnaby's)

Klinovec 1244 (c-wy and rd to summit,
  hotel and observation tower), **highest
  point on Czech side.**
Fichtelberg 1213, **highest point of E
  Germany.**

**'Saxon Switzerland'** (where the R Elbe
  flows through the Dresden Gate)

**Sudeten Mtns** (line the Czechoslo-
  vakia/Poland frontier) (from W to
  E)

**Krkonoše** (Giant Mtns) – Sněžkatl
  1603, Vysoké Kolo 1506

**Jeseniky** – Praděd 1492, Kral Sněžnik
  1423

*Right:* Watzmann and Berchtesgaden, Germany (Fremdenverkehr-
sverband München-Oberbayern
e.V. Photo: Ernst Baumann)

*Below:* Czechoslovak rocks (J
Jurak)

striking scenery for such low hills and dominating sites for castles
(eg Hohenzollern Hörnle (856), south of Tübingen). There are a
great number of climbers' crags in these ranges, which play an
important role in the training of German mountaineers.

There are a number of hill areas distributed across the **centre of
Germany**; all have networks of footpaths, while many offer crags
for the climber.

The **Eifel** (760) exhibits many volcanic phenomena – there are
some 50 cones, with lakes in former craters, also mineral and hot
springs. **Westerwald** (657) and **Vogelsberg** (773) also have volcanic origins.

The **Hohe Rhön** (950) was the cradle of gliding in Germany.

The **Harz Mountains**, a horst of ancient rocks ($80 \times 40$km) are
historically famous as a source of minerals, and home of the
celebrated 'Spectre' (see 'Optical Phenomena among Mountains'). It was the scene of Walpurgisnacht in Goethe's *Faust*.
**The first rack railway using the Abt System**, now used by three-
quarters of the world's mountain railways, was constructed here.
The summit is also reached by two roads and there is a hotel and
an observation tower.

An area of sandstones where the River Elbe flows through the
'Dresden Gate' is known as **'Saxon Switzerland'**. The rocks have
been eroded into isolated mountains and pinnacles giving many
steep faces for rock climbers. The hills have steep wooded sides
and heath-covered summits.

There is similar rock scenery across the frontier in Czechoslovakia to the north-west of Praha, where the collections of rock
towers are known as 'Rock Towns'. This too is a very important
rock climbing area.

# Italy (non-Alpine)

**THE NORTHERN APENNINES**
(Genoa to Ancona) (the Ligurian,
Etruscan and Emilian Apennines),
limestone.

Mte Cimone 2165
Mte Cusna 2121
Mte Prado 2054
Mte (Alpe di) Succiso 2017
Mte Giovo 1991
Corno alle Scale 1945
Alpe Tre Potenze 1940
Mte Maggiorasca 1803
Mte Bue 1780
Mte Ebro 1701

## MOUNTAINS OF SAN MARINO

Mte Titano 739

**THE APUAN ALPS** (on the W coast
behind Carrara)

Mte Pisanino 1946
Mte Cavallo 1889
Pania della Croce 1858
Pizzo d'Uccello 1782, the 'Matter-
horn' of the Apuans.

## THE TUSCAN MOUNTAINS

Colline Metallifere – Poggio di Mon-
tieri 1851
Mt Amiata 1749

**THE CENTRAL APENNINES** (Ancona to R Sangro) (the Umbro-
Marchigiano & Abruzzi Apennines),
limestone.

Mte Corno 2921 (fa 1794 Orazio Del-
fico), **the highest point of the Gran
Sasso massif, of the Apennines and of
non-Alpine Italy.**
Mte Amaro (Mtgna della Maiella) 2795
Pizzo Intermésoli 2646

**The Gran Sasso peaks, Italy (By
courtesy of the Italian State Tourist
Office)**

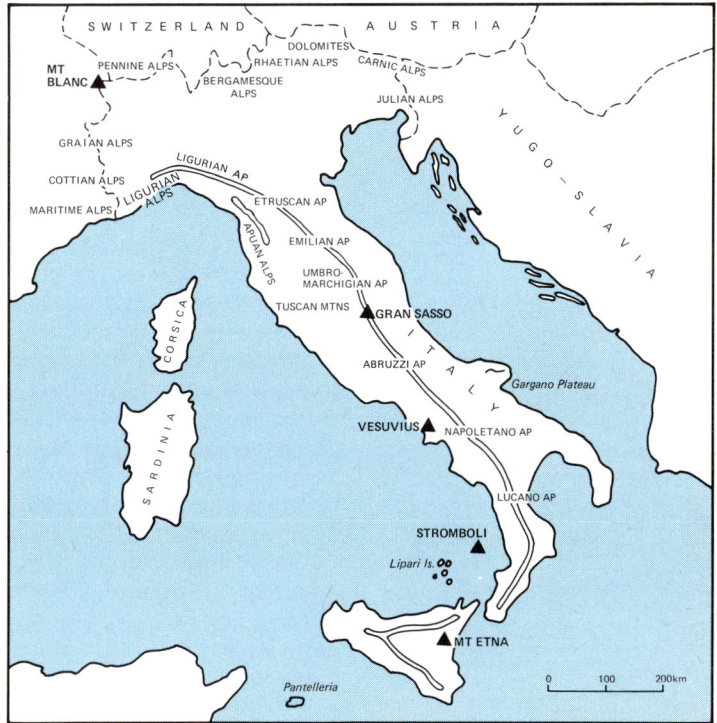

Italy

The **Apennines** run the whole length of Italy from Colle Altare in
the north-west down to the 'toe' in the south – more than 1000
km. The range nowhere provides a substantial barrier to communications and there are many road and rail crossings. A notable feature in the north is the Apennine Tunnel on the fast direct
line from Florence to Bologna. At 18·5 km, **the longest double-
track railway tunnel in the world**, it was started in 1920 and opened
in 1934. The excavators experienced tremendous difficulties –
flooding was a constant menace and the workings were often
saturated by methane gas – so that the final cost was 97 lives and
£5·1 million. The whole line was designed for high speeds and has
altogether 31 tunnels with a total length of 36 km.

The highest peaks are in the central Apennines, east-north-east
of Rome (Monte Corno (2921) in the Gran Sasso group). Close
to them the motorway from Rome to the east coast is only one of
a number of road crossings of the range.

On southwards the mountains are lower but wilder.

Carrara in the **Apuan Alps** is world famed as a source of marble –
the mountains are of marble with a capping of limestone.

At **Larderello** in the **Colline Metallifere** is **the world's first geothermal electricity generating plant**. Built in 1897, it has been con-

Mte Corvo 2643
Mte Camicia 2570
Mte Cefalone 2532
Mte Aquila 2498
Mte Velino 2487
Mte Vettore 2478
Mte Gorzano 2455
Mte Priore 2434
Pizo de Sevo 2422
Mte Sirente 2349
Mte Regina 2334
Mte Greco 2283
Mte Petroso 2247
La Meta 2241
Mte Terminillo 2213
La Terratta 2208
Mte Viglio 2156
Mte Simbruini 2156
Mte Argatone 2151

**THE SOUTHERN APENNINES** (R
  Sangro to the 'toe') (the Napolitan,
  Lucanian and Calabrian Apennines),
  limestone.

Sa Dolcedorme – Mte Pellino 2248
Mte Miletto 2050
Mte Serino 2007
Mte del Papa 2005
Mte Pelegrino 1986
Mte Alto 1956
Botte Donato 1929
Mte Cervati 1899
Mte Autore 1853
Mte Volturino 1836
Mte Cervialto 1809
Mte Terminio 1786
Mte Alburno 1742
Mte Lattari (Mons Lactarius) 1457
Vesuvius (V-le 1944) 1289 (rd & c-wy
  to summit).

tinuously refined and expanded and today makes a significant contribution to the nation's electricity. After World War II a further plant was set up in the Monte Amiata region, another area of ancient, but not present-day vulcanism.

Rock climbers from **Rome** practise on Monte Morra (north of Tivoli), Monte Leano (60km south-east), Monte Argentario (120 km north-west) and the sea-cliffs at Gaeta (120km south-east).

Sea-cliffs are also climbed upon at Gargano and on the Isle of Capri.

**Vesuvius** (1289), perhaps **the world's best known volcano**, has erupted at intervals throughout recorded history, notably in AD 79 when Pompeii and Herculaneum were destroyed by ashes and mud flow and thus preserved for voyeurs of the present age. The summit has always been a place of pilgrimage for the curious, and in 1880 Thomas Cook built a funicular railway to 1213m – the origin of the song 'Funiculi, funicula'. This was destroyed in 1944 and the ascent is now accomplished by bus and chairlift. There is an observatory at 608m.

The cone is partly enclosed by the semi-circular ridge of Monte Somma (1132), which is the remains of a former crater.

Two of the classes of volcanic eruption get their names here – Vesuvian, for the typical behaviour, and Plinian (after the Roman philosopher killed in the AD 75 disaster), for a similar but far more violent form.

On the coast, west of Naples, is a region of ancient vulcanicity called the Phlegraean Fields. In an area of 65km there are 19 craters, some with lakes, and the cones of Solfatara (which gave its name to a minor volcanic phenomenon), Monte Nuovi and Astroni.

# The Italian Islands

**Sicily**
Mt Etna (V-le 1979) 3308 (rd to 1916m,
  c-wy to observatory at 2943m).
La Madonie – Pizo Carbonara 1979,
  Mte S Salvatore 1910.
Mte Nebrodi – Mte Soro 1846

**Sardinia**
Mti de Gennargentu – P la Marmora
  1834

**Elba**
Mte Capanne 1019

**Ischia** (V-le 1300) 792

**Lipari Islands**
Salina 962
Stromboli (V-le 1971) 926
Filicudi 773
Alicudi 675
Lipari – Mte Velato (V-not recently)
  603
Vulcano (V-le 1888–90) 499

**Mount Etna** (3308) on Sicily is a volcano of massive volume. Greek mythology taught that the eruptions were produced by the fire-breathing Typhon with a hundred snake heads imprisoned below, or alternatively by the giant Enceladus chained in a cave and breathing fire and flame.

Empedocles (490–430 BC) threw himself into the crater. Emperor Hadrian (76–138 AD) claimed an ascent 'to see the sun rise', as did Emperor Trajan; there are Roman ruins just below the summit. Now there is a road to 1916m and a cableway to an observatory at 2943m.

A guide in an asbestos suit descended for several tens of metres into the crater by wooden ladders in 1974.

There is volcanic activity also in the **Lipari Islands**. **Stromboli** (926), wherein dwelt Aeolus, God of Winds, has given its name to a characteristic class of eruption phenomena. In September 1930 the whole island was lifted a metre and then fell back causing a tidal wave. Rocks weighing several tons were hurled consider-

Panarea 420
Vulcanello (V-not recently) 123

**Pantalleria**
Mte Grande 836, the whole island is the summit of a large underwater volcano.

**Malta** – 249

able distances. **Vulcano** (499), the mythical site of the forge of the God Vulcan, has given its name to all 'fire mountains'.

Since Italy forms part of the African tectonic plate, it is not surprising to find this typical plate boundary activity hereabouts. Somewhat further south the whole island of **Pantelleria** is the summit of a large underwater volcano.

# The Carpathians (W Czechoslovakia, S Poland, S W Ukraine, Romania)

## THE CARPATHIANS OF SLO-VAKIA

**Malé Karpaty** (Little Carpathians) (Zá-ruby 768), terminates in a castle-crowned cliff at Bratislava.

**Bilé Karpaty** (White Carpathians) (Vel'ká Javorina 970), steep sided valleys, heavily forested.

**Javorniky** (Javornik 1071)

**Beskydy** (Czech) (Lysá Hora 1324, Smrk 1280), between here and the Polish Beskydy is the Jablunkov Pass 553, the lowest crossing of the Carpathians.

**Beskydy** (Polish) (Babia Gora 1725)

**Malá Fatra** (Little Fatra) (Vel'ký Kri-vaň 1711), a 50 km range, there is a spectacular castle above Strečna.

**Vel'ká Fatra** (Great Fatra) (Ostredok 1592), granite mountains.

**Západny Tatry** (Western Tatra), a 40 km crest-line with 30 peaks over 2000 m, between Hucianska and Lili-owe Passes.
Bystrá 2248
Banikow Wierchy 2178
Starobociansky Wiercj 2175
Czerwone Wierchy 2123 (fa 1790s R. Townson)
Kasprowny Wierch 1988 (c-wy from Zakopane)

**Nizké Tatry** (Low Tatra), an 80 km ridge of smooth summits, rugged in places, forested slopes.
Dumbier (Gyömber) 2043
Chopok 2024 (c-wy to summit)
Dereše 2004
Chabenec 1955
Král'ova Hola 1948

**Vysoké Tatry** (High Tatra)
Gerlachovsky Štít 2663 (fa 1834 claimed for J. Still & hunters; certainly 1855 Z. Bošniaki & W. Grzegorzek), **the highest point in Czechoslovakia,** once called 'Stalin'.
Lomnický Štít 2634 (fa 16/8/1793 R. Townson), c-wy to summit from Slovakia.

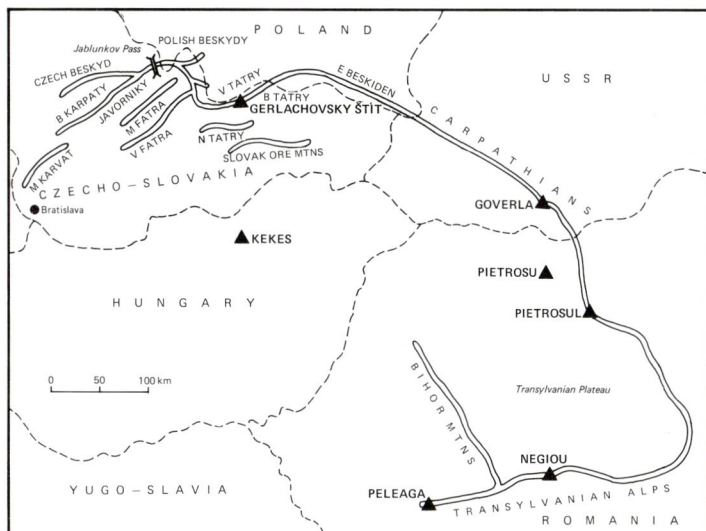

The Carpathians

The High Tatras, view from Gerlach (J Jurak)

L'adovy Štít 2628 (fa 8/1843 J. Ball)
Vysoká 2560
Kezmarsky Štít 2558
Rysy 2499 (fa 1840), **highest point in Poland.**
Kriváň 2496 (fa 26/8/1793 R. Townson)
Bradavica 2476
Ganek 2469
Slavkovsky Štít 2453
Batisovsky Štít 2448
Mieguszowiecki Pk 2438
Satzan 2432
Javorovy Štít 2424
Solisko 2404, c-wy to summit.

**Belanské Tatry**
Havran 2152
Szalony Weirch 2062

**Slovak Ore Mtns**
Stolica 1477
Polana 1458

**Zachodnie** (E Beskiden), extends the range to the USSR frontier.

## THE HILLS OF S POLAND

**Gory Swietokrzyskie** (Holy Cross Mtns) (Lysica Góra 611), named for the Monastery of the Holy Cross.

**Jura Krakowska,** a limestone scarp NW of Krakow.

## THE MOUNTAINS OF HUNGARY

Matra Ra – Kékes 1015

## THE UKRAINIAN CARPATHIANS

Goverla (Hoverla) 2058
Sevola 1818

## THE ROMANIAN CARPATHIANS
(Carpatii Orientali) (from the USSR frontier to the Predeal Pass) There are three longitudinal zones with newer rocks in the E, older rocks in the centre and a line of volcanic peaks, with craters still visible at the summits in the W.

Pietrosu 2305
Pietrosul 2102
Toroiaga 1939
Ceahlău 1904
Giumalău 1859
Hagimas 1793
Penteleu 1776

## TRANSYLVANIAN ALPS (Carpatii Meridionali)

This part of the range was once glaciated and shows arêtes, cirques and pointed peaks. 270 km long and 40 to 65 km wide.

**Predeal Pass to R Olt,** which breaks through a spectacular gorge, the

The **Carpathian range** begins at Bratislava where the River Danube (Donau, Duna) separates it from the Alpine ranges of Austria. It runs first north-east through Slovakia; then east close to the Czechoslovakia/Poland frontier into the USSR (Ukraine); then south-east and finally south into Romania; then west as the Transylvanian Alps and finally north again as the Bihor Mountains – a total length of close on 1500 km. Inside the last loop is the Transylvanian Plateau – 'the Dracula country'. A branch to the south faces the Stara Planina of Bulgaria across the Iron Gate (Portile de Fier) on the River Danube.

The **Vysoké Tatry** (High Tatra) are the high alpine mountains of both Czechoslovakia and Poland. On the Czech side is the Tatra National Park of 670 km$^2$. There are no glaciers, but many permanent snow-fields; these mountains offer mountaineering problems of a high standard amid fine peaks. An early explorer was an Englishman, Robert Townson, who made first ascents here as long ago as the 1790s.

The 80 km long ridge with 39 peaks and many lesser summits and pinnacles was first traversed in one expedition in 1957. The first winter traverse in the 1960s took 13 bivouacs.

**Gerlachovsky Štít** (2663) and **Rysy** (2499) are **the highest points of Czechoslovakia and Poland, respectively.**

The **Slovak Ore Mountains** were once famous as a source of minerals, only iron is now significant. They are rounded hills with igneous intrusions which sometimes present a bizarre landscape.

**Caraiman** is surmounted by a 33 m-high cross made from the metal of bridges destroyed during World War I.

Turnu-Rosu Pass, to reach the R Danube (from E to W)

Costila 2495
Caraiman 2325
Bocşoiu Pk 2492
Omu 2506
Papusa 2379
Ezeru 2468
Moldoveanu 2543
Negoiul 2548, **the highest point in Romania.**
Suru 2283

**R Olt to Jiu/Strei gap,** where there is another fine gorge through the range (from E to W)
Cindrelu 2245
Steflesti 2242
Mindra (Paringu) 2518
Surianu 2061

**Jiu/Strei gap to Timis/Cerna gap**
Peleaga (Verfu Pelaga) 2511
Gugu 2292

**BIHOR MOUNTAINS**

Cocurbeta 1849

**Mountain flowers (R D Treble)**

# Yugoslavia/Albania

## THE JULIAN ALPS

A fine group of rock peaks extending over the frontier from Yugoslavia into Italy (from W to E).

Mte Canin 2587 (fa 1750)
Jôf de Montasio 2754 (fa 1877)
Jôf Fuart 2666 (fa 1850)
Mangrt 2678 (fa 1794)
Jalovec 2643 (fa 1875)
Prisojnik 2541 (fa early-19th century)
Razor 2601 (fa 1842)
Špik 2472
Kanjavec 2568 (fa 1872)
Stenar 2501
Škrlatica 2738 (fa 1880)
Triglav 2863 (fa 26/8/1778 L. Willonitzer & party), **the highest point in Yugoslavia.**

**KARAWANKEN** (lining the Yugoslavia/Austria frontier)
Begunjščica 2603
Stol 2236
Peca 2126
Plešivec 1696

**KAMNIK ALPS** (parallel to the last, further S entirely in Yugoslavia) (from W to E)
Kočna 2539
Grintavec 2558
Skuta 2532
Planjava 2379
Ojstrica 2349

## POHORJE

Grni Vrh 1543
Vrh Kopa 1542
Veliki Vrh 1347

**ADRIATIC COAST RANGES** (Italy/Yugoslavia frontier to Yugoslavia/Albania frontier)
A confused tangle of mountains, invariably limestone, mostly trending parallel to the coast line (from NW to SE)

Snežnik 1796
Vaganski Vrh 1758
Klekovaca 1961
Dinaric Alps – Troglav 1913
Cinčer 2006
Vranica 2107
Vran 2074
Čursnica 2228
Treskavica 2088
Lelija 2032
Vlasulja 2339
Maǵlić 2387
Ljubišnja 2239
Durmitor 2522
Prutas 2398
Stožac 2140
Bjelastica 2135
Bjelasnica 2067
Komovi 2483

The **Julian Alps** have been called 'Kugy's Kingdom' after the great local mountaineer Julius Kugy. The highest point is Triglav (2863). This was the scene of fierce partisan fighting during World War II, commemorated by a memorial at the foot of the north wall of Triglav, which takes the form of a piton and karabiner five metres high.

In the **Karawanken**, Peca (2126) was the home of the mythical King Matthias, while Plešivec is topped by an ancient shrine. There are three passes over the frontier – Wurzen (1073), Loibl (1067) (summit tunnel) and Seeberg (1216).

**Jalovec, Julian Alps (AC Collection)**

Mokra Planina – Zljeb 2382
Kopaonik Planina – Suvo Rudiste 2017

## THE MOUNTAINS OF MACEDO-NIA AND S SERBIA (from N to S)

Ljuboten 2496
Koritnik 2394
Titov Vrh 2748
Rudoka Planina – Turcin Vrh 2702
Korab Ra – Zernonica 2725
Jakupica Ra – Solunska 2540
Jablanica (Jablanitza) 2257
Baba Ra – Perister 2600
Kaimakchalán (Kajmakčalan) 2523

## THE MOUNTAINS OF THE YUGO-SLAVIA/BULGARIA FRONTIER
(Osogovska Planina)

Ruen (Rujen) 2252
Carev Vrh 2085

## ALBANIA

Limestone mtns, inaccessible to West-ern mountaineers (from N to S).

Jezerce (Jezera) 2693, **the highest point in Albania.**
Daravica 2656
Mali-i-Shenjt 2110
Dêja (Mali-i-Dejes) 2246
Gur-i-Topit 2379
Mt Tomor (Mali-i-Tomorrit, Tomor-ica) 2418
Griba 2120
Mali-i-Kikes (Cika) 2045

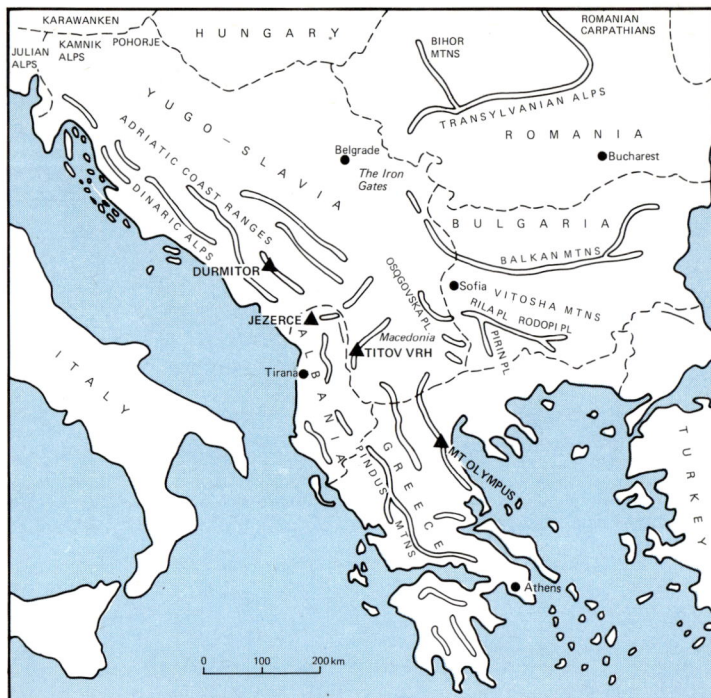

**The Balkans**

# Greece

## W AND CENTRAL PENINSULAR GREECE (extension of the coast ranges of Yugoslavia & Albania)

The main range is called the Pindos Mtns (from N to S).

Grammos 2520
Smólikas 2633
Peristéri 2295
Tringia 2204
Tzoumérka 2389
Karava 2184
Voutsikáki 2154
Timfristós 2316
Vardhousia Ori 2406
Gióna Oros 2510
Parnassos – Lia Koura 2458
Elikón Ór (Helicon) 1748
Kithairón Oros (Cithaeron) 1409

## E PENINSULAR GREECE (exten-sion of the Macedonian ranges of Yugoslavia) (from N to S)

Vernon (Vitsi) 2128
Vermion 2061
Siniatsikon 2112
Pieria Ori 2194

At **Meteora** in the plains of Thessaly, fourteen of a series of granite laccoliths, now exposed as rock pinnacles, are surmounted by monasteries, or monastic buildings. Visitors to some of these had at one time to be hauled up in a net or basket, though obviously someone had had to make the climb unaided in the first place. Now the rock faces are shared with climbers.

Mountains play a significant role in Greek mythology:
    In the flood story only Parnassos (not the highest mountain?) was left uncovered by the waters. Parnassos and Helicon were the home of Apollo and the Muses.
    Cithaeron was sacred to Dionysius.
    Mount Olympus was the home of Zeus, from which he flung thunderbolts. Pre-Christian remains have been found on the subsidiary peak Agios Antonios (2815), and Christian remains on Skilio (2911) and Profitis Ilias (2786).
    The Titans attempted to scale heaven by placing Ossa upon Olympus and Pelion upon Ossa. They were repelled by the arrows of Apollo.
    One of the twelve labours of Herakles was to capture a wild boar on Erimanthos.

*Top right:* **Monastery on Mount Athos (National Tourist Organisation of Greece)**

*Bottom right:* **Meteora, Greece (National Tourist Organisation of Greece)**

Mt Olympus – Mytikas Pk 2917 (fa recorded 1913 D. Baud-Bovy, F. Boissonas, K. Kakalos), **the highest point in Greece.**
Subsid pks 2911, 2786
Ossa (Kissavos) 1978
Pelion – Pliassidi 1547

**THE MOUNTAINS OF MACEDONIA AND THRACE** (NE Greece) (extensions of the Pirin & Rodopi Planina of Bulgaria)

Falakron 2229
Mt Koulá 2177
Áyion Óros (Hagion Oros, Mt Athos) 2033

**THE MOUNTAINS OF PELOPONNISOS** (from N to S)

Erimanthos 2224
Helmos (Aroánia) 2355
Killini (Sikionia) 2374
Taiyetos (Taygetos) 2407

**INSULAR GREECE**

**Crete** (from W to E)
Lévka Ori – Pakhnes 2452
          Troharis 2401
          Sokos 2397
          Suourikhti 2350
Idhi Óros – Mt Ida 2456
Dhíkti Ori – Mt Dikte 2148
Afentis Christos 2141

**Euboea** – Dhirfis 1743

**Cephalonia** 1620

**Samothrace** 1600

**Santorini** (Thira) (V-le 1956) 1315

**Chios** 1297

**Scarpanto** 1215

**Rhodes** 1215

**Samos** 1160

**Levkas** 1141

**Thásos** 1045

**Noxos** 1002

**Milos** (V) 751

**Nisiros** (V) 702

Killini was the birthplace of Hermes (Mercury), the messenger of the Gods.

Livy tells us that **Mount Haemus** in Thessaly was climbed by Philip Vth of Macedon in 181 BC, **one of the earliest records of a mountain climb.**

**Mount Athos** (2033) is the peninsular site of a series of monasteries; there are 20 monastic establishments, 14 dependencies and many hermitages, some of them caves. Many nations are represented. Women are not admitted, nor eunuchs, nor animals in case they be female!

# *Bulgaria*

**STARA PLANINA** (Balkan Mtns) (extending 550km from the Black Sea coast to the Yugoslavia frontier, then N through Serbia province to face the end of the Carpathians across the Iron Gate on the R Danube) (Mt Haemus)

**Sipka Pass to Botevgrad Pass**
Sipka Pass 1330 (rd)
Mazalet (Kadimlja) 2273
Botev (Yumrukchal, Ferdinand Yrak) 2376
Levski 2166
Vežen 2198

**Botevgrad Pass to Timok gap**
Botevgrad Pass 965 (rd)
Iskar Gorge (rd, Sofia to Danube valley; ry)
Kom 2016
Midžor 2169 (in Yugoslavia)

**RODOPI AND RILA PLANINA** (Rhodope Mtns) (a block of country between the coast of Thrace (NE Greece) and the Maritsa (Meriç) & Struma (Strimon) Rs)

**Ropodi Planina** (from S to N)
Snežanka 1926, c-wy to summit, TV mast
Prespa 2000
Goljam Perelik 2191
Goljam Persenk 2074
Karabalkan 1956
Goljam Sjutkja (Sütke) 2186
Belmeken 2646

**Rila Planina** (from E to W)
Slavovvrŭkh 2309
Jančovčal 2481
Mustacal 2633
Manco 2771
Musala (Moussalla, Rila Dagh) 2925, **the highest point in Bulgaria,** once called 'Stalin'.
Dimitrov 2901
Kadir Tepe 2785
Butschkaya Chal 2718
G Skakavec 2706
Maljovica (Malyovitza) 2729
Aigidik 2617
Eleni Vrŭkh 2791
Popova Shapka 2699

**Vitosha Mtns**
Cherni Vrŭkh 2286 (rd to summit)
Goljam Rezen 2277
Skoparnik 2226

**Pirin Planina** (from S to N)
Gocev Vrâh 2213
Kamenica (Kamenitza) 2822
Polezhan Pk 2850
Vikhren (El Tepe) 2915

The 550km **Stara Planina** terminates in the north at the Iron Gate on the River Danube. Here the river drops 26m in 150km of gorge with towering cliffs in places.

Near Belogradchik there is climbing on fantastic sandstone rocks.

There seems to be no firm basis for the claim made in a French encyclopaedia (but denied by the Bulgarian authorities) that there is a mountain **Dodow Wrch** (2957) above the famous Rila Monastery, which would, if indeed there, be the highest in the country.

**Cherni Vrükh** (Vitosha Mountain) (2286) is only 15km from Sofia, which is spreading over the foothills. The upper slopes are a National Park and Nature Reserve; there is a road to the summit and it is an important ski-ing site.

**Rila Monastery, Bulgaria (Bulgarian Tourist Organisation)**

# The Alps

Southward from the summit of Mont Blanc (FS Smythe – by courtesy of the Countess of Essex)

**LIGURIAN ALPS** (C d'Altare to C de Tende) (from E to W)

C d'Altare 495 (ry; Turin-Savona motorway)
Mt Antoroto 2144
C de Nava 947 (rd, Pieve di Teca-Ormea)
Pizzo d'Ormea 2476
Mte Mongioie 2630
Ca delle Saline 2612
Pte Marguereis 2650 (where the Italy/France frontier turns S)
Mte de Carsène 2383
C de Tende 1873 (int rd La Giandola to Borgo S Dalmazzo; ry tunnel below)

**MARITIME ALPS** (C de Tende to C de Larche)

**Frontier ridge from SE to NW**
C de Tende 1873 (int rd La Giandola to Borgo S Dalmazzo; ry tunnel below)
Ce de l'Agnel E 2852
Mt Clapier 3045 (fa 1818 Capt Cassato)
Ca della Maledie 3061 (fa 23/7/1895 E. & L. Maubert; J. B. & J. Plent)
Ce du Gélas 3143 (fa 1864 P de St Robert)
Ce de l'Agnel O 2927
C de Cerise 2543 (int path)
Mt Malinverne 2938
C de la Lombarde 2350 (int minor rd)
Mt Ténibre 3031
Tête de l'Enchastraye 2955
C de Larche (C della Maddalena, C de l'Argentière) 1995 (int rd Condamine to Vinadio)

**N of frontier ridge** (ie in Italy)
Ca Sud Argentera (Punta dell'Argen-

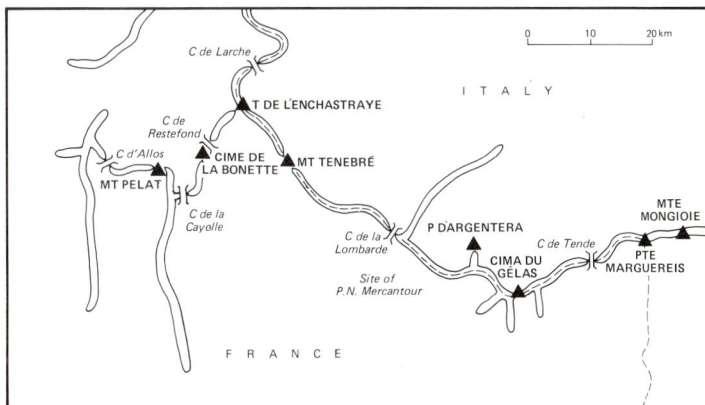

Ligurian and Maritime Alps

The Alps, the principal mountain range of Europe, are probably the best known high mountains of the world. They stretch for more than 1000 km from the Colle d'Altare, close to the Mediterranean coast in the Gulf of Genoa, to the Hochschwab in eastern Austria. The range is the source of many of Europe's greatest rivers – the Rhône and its left bank tributaries, the Rhine, the right bank tributaries of the upper Danube, the Drava, the Po and all its left bank tributaries. There are five basic watersheds – Mediterranean/Adriatic, Adriatic/North Sea, Adriatic/Black Sea, North Sea/Mediterranean and Black Sea/North Sea. The hydrographic centres of the Alps are the minor peaks of Wyttenwasserstock and Pizzo Lunghino; the slopes of each feed water into three distinct basins. The frontiers between the various Alpine countries – France, Italy, Switzerland, Austria, Liechten-

tera) 3297 (fa 18/8/1879 W. A. B.
Coolidge: C. Almer père et fils), **the
highest point in the Maritime Alps.**
Ca di Nasta 3108 (fa 27/9/1878 D.W.
Freshfield; F. Devouassoud – an
attempt on Argentera (*above*), the
whereabouts of which was not exactly known)
Alte Matto 3088

**S of frontier ridge** (ie in France)
Mt du Grand Capelet 2934
  C de Restefond 2802 (rd, Barcelon-
  nette to St Etienne de Tinée, now
  **the highest point in the Alps)**
Ce de la Bonette 2860
  C de la Cayolle 2327 (rd, Barcelon-
  nette to Guillaumes)
Mt Pelat 3051
  C d'Allos 2240 (rd, Barcelonnette to
  Colmars)

**COTTIAN ALPS** (C de Larche to C du
Mont Cenis: E of C du Galibier and
E of C d'Izoard)

**Frontier ridge from S to N**
C de Larche (C della Maddelena, C de
  l'Argentière 1995) (int rd, Conda-
  mine to Vinadio)
Bec de Chambeyron 3390
  C de Mary (int path, a possible way
  for Hannibal)
Bric de Rubren 3340 (fa 9/10/1875
  Arnaud, Reynaud, Faudon)
  C de la Traversette 2914 (int path,
  Queyras to upper Po valley)
Bric Froide (Pte Ramière) 3302
  C de Montgenèvre 1854 (int rd,
  Briançon to Cesana Torinese)
  C de Thures 2186 (int minor rd,
  Plampinet to Bardonecchia)
  For a short distance the int frontier
  forsakes the line of the ridge.
Mt Thabor 3181 (fa 1648 or 1694) –
  crowned by a chapel.
Rocne Bernaude 3225
  C de Fréjus 2542 (ry tunnel below)
Aig de Scolette 3508
Rognosa d'Etiache 3375
Pt de Ferrant 3376
Mt d'Ambion 3381
Dent d'Ambion 3382
  C du Mont Cenis 2083 (int rd,
  Lanslebourg to Susa)

**E of frontier ridge** (ie in Italy)
Mte Viso 3851 (fa 30/8/1861 W.
  Mathews & F. Jacomb; M. & J.B.
  Croz), **the highest point in the Cottian
  Alps.**
Visolotto 3348
Pta Caprera 3387

**W of frontier ridge** (ie in France)
Aig de Chambeyron 3411
  C de Vars 2111 (rd, St Paul sur Ubaye
  to Guillestre)
Grand P de Rochebrune 3325
  C d'Izoard 2361 (rd, Guillestre to
  Briançon)

stein and Yugoslavia invariably follow major or minor ridge
lines, crossed by famous international passes.

The range was formed originally in a collision between the
European and African tectonic plates. It is interesting to note that
Italy forms part of the African Plate. The level is considered to
have risen 2000m in ten million years.

Mont Blanc (4807), the highest in the range, was one of the
first snow mountains in the world to be climbed – in 1786. The
sport of mountaineering developed here during the 19th century,
as did the sport of downhill competitive ski-ing, somewhat later.
The tourist facilities are highly developed to an extent which has
now begun to deteriorate the amenities. There are rack- and
cable-railways, and cableways, also a large number of hotels and
mountain huts, so that climbing and ski-ing can take place all the
year round, while competent rescue services enable all sorts of
risks to be run.

Depending on the exact definition adopted, there are rather
more than 80 peaks over 4000m in the chain. **The first man to
reach the summits of all of them** was Karl Blodig (1859–1956), who
completed the task in 1911. Eustace Thomas (1869–1960) also
succeeded in 1929, having accomplished the feat within six years.

The **Ligurian Alps**, falling below the snow-line, are grassy and
forested with some crags and caves. The 60km ridge, mostly in
Italy, links the main chain of the Alps with the Italian Apennines.

**Fig. 1   The watersheds of the Alps**

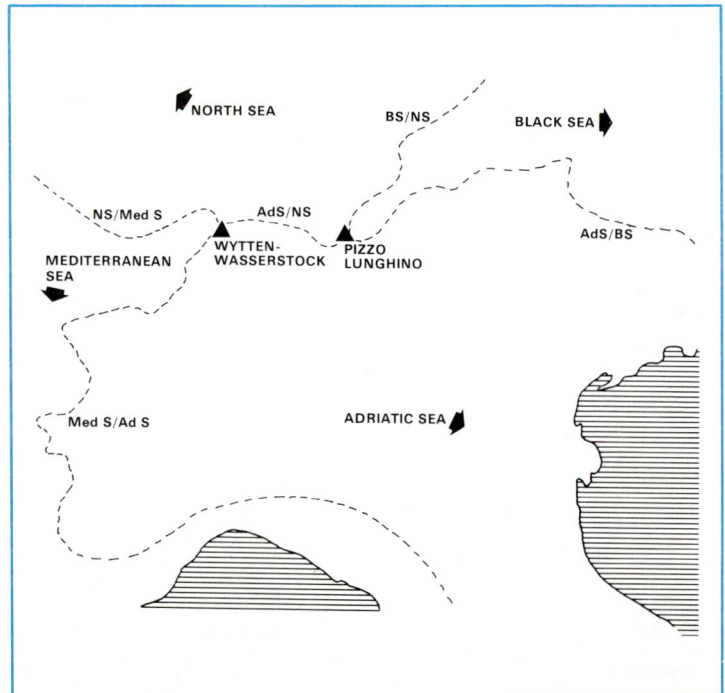

C du Galibier 2645 (rd, C du Lauteret
to St Michel de Maurienne)
C du Lauteret 2058 (rd, Le Bourg
d'Oisans to Briançon)

## CENTRAL DAUPHINÉ [Oisans;
Massif du Pelvoux; Ecrins Massif;
Parc National des Ecrins] (S & W of
the C du Galibier and the C du
Lauteret: S of the R Romanche: N
& W of the R Durance)

**Main N to S ridge** (from the N)
La Meije 3983 (fa 16/8/1877 E.B. de
  Castlenau; P. Gaspard père et fils;
  fa by a lady 24/8/1888 Miss K.
  Richardson)
Subsid pks 3890, 3974
Pic Gaspard 3883 (fa 6/7/1878 H.
  Duhamel; P. Gaspard père et fils,
  C. Roderon)
Le Pavé 3824 (fa 19/7/1879 W. A. B.
  Coolidge; C. Almer père et fils)
Roche Méane 3711
La Grande Ruine 3765 (fa 19/7/1873
  Miss M. Brevoort, W. A. B. Cool-
  idge; C. Almer père, P. Bleuer,
  P. Michel fils, C. Roth)
Subsid pk 3726
P. Bourret 3712
Roche Faurio 3730 (fa 20/6/1873 C.
  Taylor, R. & W. M. Pendlebury,
  Cox, Gardner; guides)
Barre des Ecrins 4101 (fa 25/6/1864
  A. W. Moore, E. Whymper; C.
  Almer père, M. Croz. fa by a lady
  6/8/1883 Mlle de Sireix), **the highest
  point in the Dauphiné.**
Subsid pk 4015
Pic Coolidge 3774 (fa 14/7/1877
  W. A. B. Coolidge; C. Almer père
  et fils)

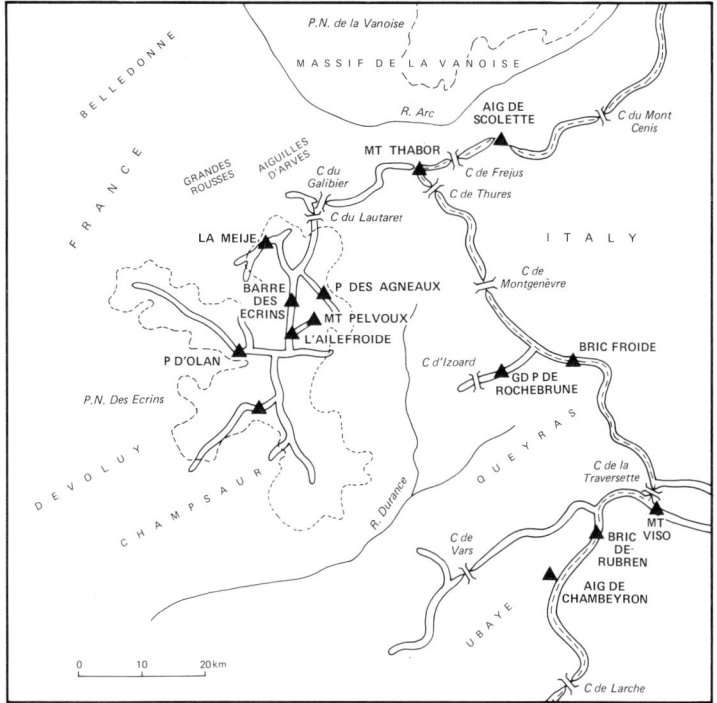

**Dauphiné and Cottian Alps**

**Drilling in Mont Cenis Tunnel, drawing by E Whymper (AC Collection)**

Ailefroide 3954 (fa 7/7/1870 W.A.B.
  Coolidge; C. Almer père, U. Almer,
  C. Gertsch)
Subsid pks 3928, 3848
Les Bans 3670 (fa 14/7/1878 W.A.B.
  Coolidge; C. Almer père et fils)

**W of main ridge**
Le Rateau 3809 (fa 11/7/1873 Miss M.
  Brevoort, W.A.B. Coolidge; C. Al-
  mer père, P. Bleuer, P. Michel fils,
  C. Roth)
Subsid pk 3766
l'Olan 3564 (fa 29/6/1877 W.A.B.
  Coolidge; C. Almer père et fils)
Subsid pk 3560
Tête de la Toura 2914, c-ry to a high
  point from les Deux Alpes.
P du Grand Galibier 3229, 2 km E of
  the Col, viewpoint towards Mont
  Blanc.

**E of main ridge**
P des Agneaux 3663 (fa 17/7/1873
  W.A.B. Coolidge; C. Almer)
Subsid pks 3646, 3631
P sans Nom 3914 (fa 10/7/1877 J. Col-
  grove, R. Pendlebury; G. & J.
  Spechtenhauser)
Mt Pelvoux 3946 (fa 30/7/1828 A.
  Durand; J.E. Matheoud, A. Lio-
  thard)
Subsid pks 3932, 3754
Serre Chevalier 2661, c-ry from Chante-
  merle by Briançon.

**MINOR MASSIFS OF THE DAU-
PHINÉ**

**Les Aigs d'Arves**
Main pk 3510 (fa 22/7/1878 W.A.B.
  Coolidge; C. Almer père et fils)
Subsid pks 3509, 3363

**Grandes Rousses Massif**
C de la Croix de Fer 2067 (rd, view-
  point to Mont Blanc)
P de l'Etendard 3468
P Bayle 3466
P du Lac Blanc 3327, c-ry from Alpe
  d'Huez

**Chaine de Belledonne**
Grand P de Belledonne 2978
Le Rocher Blanc 2928

**CENTRAL GRAIAN ALPS** (C du
  Mont Cenis to C de la Seigne)

**Frontier ridge** (from S to N)
C du Mont Cenis 2083 (int rd, Lansle-
  bourg to Susa)
Rocciamelone (Roche Melon) 3538 (fa
  1358 Rotario d'Asti)
Bessanese 3597 (fa 26/7/1873 M. Bar-
  atti; C. Cibrario)
Uia di Ciamarella 3676 (fa 31/7/1857
  A. Tonini; Ambrosini)
Levanna 3615 (fa 17/8/1875 L. Vac-
  carone, A. Gramaglia; A. & D.
  Castagneri)
Subsid pks 3591, 3558

The **Col de Tende** (1873) was crossed by invading Saracens in 906. The road was constructed in 1779–82; a tunnel bored a century later (1882), at 1320m, eliminated 69 bends on the old route to the watershed, where there are fortifications. The rail tunnel below is 8 km long.

The **Maritime Alps** present 80 km of rocky summits with a few small glaciers of bare ice and are notable for simultaneous views of high mountains on the one hand and the waters of the Mediterranean on the other.

The **Col de Larche** was crossed by Francis I in 1515 and later by armies in the 17th and 18th centuries. The road was started by Napoleon but not completed until much later in the 19th century; fierce fighting took place here in World War II.

The **Cottian Alps** are named for King Cottius, a comparatively minor figure; Coolidge suggested that they should have been named for Hannibal, since it is probable that he crossed somewhere along this 130 km stretch of the range. Its comparatively snow-less character has made it the scene of much fighting between France and the House of Savoy (eg the Catinat campaigns of 1692) and every pass has been crossed at some time by soldiers.

The **Col de la Traversette** (2914) was probably the route taken by Hannibal. The exact point where he crossed the Alps in 218 BC has long been a matter of controversy. A very thorough investigation of the problem has been carried out by Sir Gavin de Beer involving 'critical assessment of the meaning of texts as read in the oldest extant manuscripts' and 'the application of the principles of natural science to decide possible alternatives . . . narrowing the choice by rejecting some alternatives on the grounds of physics, meteorology, botany and other branches of science'. He concluded that the route was as follows: along the coast past Aigues Mortes, up the Rhône by way of Arles and Tarascon to the confluence of the Drôme, up the Drôme past Die, over the Col de Grimond, by Gap to the Durance valley, finally – either (probably) by Queyras to the Col de la Traversette, or (just possibly) by Ubaye to the Col de Mary.

Hannibal descended into Italy with 12 000 infantry, 8000 Spaniards, 6000 cavalry and all 37 of his elephants, this in spite of heavy losses due to terrain and hostile Gaulish tribes on the way.

As early as the 15th century a summit tunnel was bored here to facilitate passage.

The **Col de Montgenèvre** (1854), an outside possibility for Hannibal, was certainly crossed by Caesar when invading Gaul in 58 BC. Thenceforward there is a continuing history of crossings, for example by Charlemagne in 773, by Charles VIII in 1494 with considerable artillery and by French troops in World War I. The

The R Arc rises on the westerly glaciers and the R Isère a few km further N.

Tsanteleina 3605 (fa 9/8/1865 Blanford, Nichols & Rowsell; J.V. Favret)

Aig de la Grande Sassière 3747 (fa 1808 locals from Tignes)

Pte du Nantcruet 3610

C du Petit St Bernard 2188 (int rd, Bourg St Maurice to Pré St Didier)

C de la Seigne 2513 (int path)

**W of frontier ridge** (ie in France)

Pte de Ronce 3610 (fa 15/7/1784 De Lamonon *et al*)

Pte de Charbonnel 3750 (fa 17/7/1862 M.A. Boniface, M.A. Fodéré; J. & M. Personnaz)

Albaron 3638 (fa 2/9/1866 R.C. Nichols; J.V. Favret)

**E of frontier ridge** (ie in Italy)

Testa de Ruitor 3486 (fa 16/8/1858 G. Studer, Weilenmann, Bucher; J.B. Frassy or 1863 W. Mathews, T.G. Bonney; M. & J.B. Croz)

Grand Rousse 3607 (fa 4/8/1874 Gornet, Martelli, Barale; J.J. Maquignaz, S. Meynet)

Grande Traversière 3496

**EASTERN GRAIAN ALPS** [PN del Gran Paradiso; PN du Grand Paradis] (E of C del Nivolet, between the valleys of Aosta and Locana). A series of N–S ridges (listed from W to E)

C del Nivolet 2640 (path on N, rd on S, valley of Aosta to valley of Locana)

Becca di Moncair 3544 (fa 14/7/1881 G. Frasca; G. Blanchetti)

Ciaforon 3640 (fa 25/8/1871 F. Vallino; A. Blanchetti)

La Tresenta 3609 (fa 1867 M. Baretti; A. Blanchetti)

Grand Paradis (Gran Paradiso) 4061 (fa 4/9/1860 J.J. Cowell, W. Dundas; J. Tairraz, M. Payot), **the highest point in the Graian Alps.**

Piccolo Paradiso 3923 (fa 3/8/1869 P.G. Frassy; E. Jeantet)

Subsid pks 3921, 3868

Becca di Montandayné 3838 (fa 22/8/1875 L. Vaccarone, A. Gramaglia, Ricchiardi; A. & G. Castagneri)

Herbetet 3778 (fa 22/8/1873 L. Barale; A. & G. Castagneri)

La Grivola 3969 (fa 23/8/1859 J. Ormsby, R. Bruce; Z. Cachat, J. Tairraz, F.A. Dayne)

Roccia Viva 3650 (fa 4/7/1874 A.E. Martelli; J.J. Maquignaz, S. Meynet)

Torre del Gran San Pietro 3692 (fa 14/7/1867 D.W. Freshfield, C.C. Tucker, T.H. Carson, J.M. Backhouse; D. Balley, M. Payot)

Subsid pks 3651, 3618

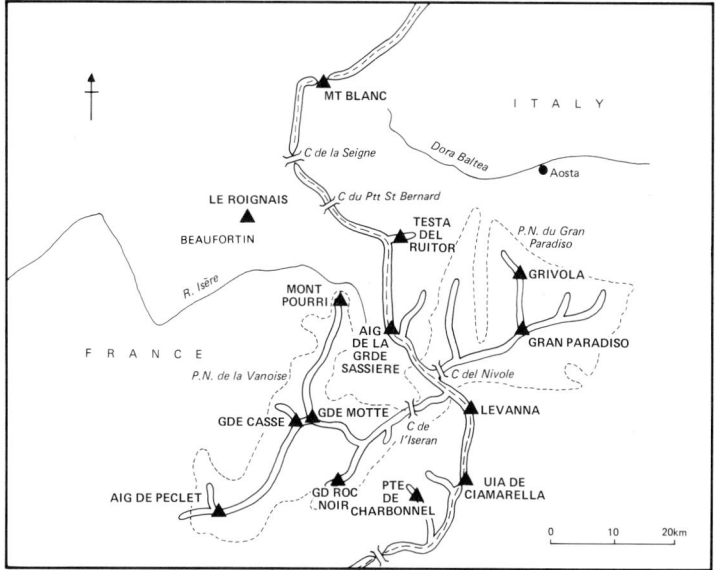

Graian Alps

road was built by Napoleon and finished in 1806; in the final stages of his campaigns he used it as a line of retreat after the Simplon and Mont Cenis routes had been cut off. There is a village close to the summit.

The tunnel below the **Col de Fréjus** (2542) was chronologically **the third railway crossing of the Alpine chain and the first by tunnel**.

The latter half of the 19th century saw tremendous advances in the science of tunnelling. Inspiration came from the mounting need to improve European rail communications by piercing the barrier of the high Alps. Previously lengthy tunnels had been constructed by digging in either direction from the bases of a series of vertical shafts, which facilitated both ventilation and the removal of excavated material. Such tactics could not be used to tunnel through mountain ranges, so that new solutions had to be found for the problems of aligning tunnels from either side of the range, of material removal, and of ventilation.

Until the 1850s travellers from France to Italy had to detrain at Modane and make a journey of 80km across the Mont Cenis Pass to Suse. A tunnel of 13·6km through the range to provide a direct link between the systems was started in 1857. At first, using existing manual methods, the tunnels only advanced by 23cm per day, which would have meant completion only by well into the next century. However, a mechanical drill driven by compressed air had recently been designed in the USA, for which Sommeiler, the engineer in charge of the Mont Cenis workings, developed a water-driven air compressor. After 1861 the daily rate of advance rose rapidly, reaching well over two metres per day by 1870. The two headings met on Christmas Day 1870, the

**WESTERN GRAIAN ALPS** [PN de la Vanoise; Tarentaise et Maurienne] (W of C de l'Iseran, between the valleys of the R Arc and R Isère)

**Main chain** (from E to W)
C de l'Iseran 2770 (rd, valley of the R Arc to valley of the R Isère)
Grand Motte 3656 (fa 5/8/1864 Blandford, Cuthbert, Rowsell; J.V. Favret)
Pte de la Grande Casse 3852 (fa 8/8/1860 W. Mathews; M. Croz, E. Favre)
Subsid pk 3795

final difference between them amounting to 30cm in level and 45cm in alignment. It had cost £3 million and 28 lives.

The **Col du Mont Cenis** (2083), another (unlikely) possibility for Hannibal's route, received its first recorded crossing in the 8th century, though it may have been used much earlier by the Romans. The hospice on the summit was founded between 814 and 825. It was often crossed in medieval times by royalty journeying about their domains. In the 15th century the practice of *ramassier* (or *glisser à la ramasse*) – tobogganing on wooden sledges guided by men called *marons* – was first introduced on the

Gran Paradiso, North Ridge (J G R Harding)

In the steps of Hannibal (R D Treble)

Dôme de Chasseforêt 3586
Dôme des Nants 3572
Dôme de l'Arpont 3611
Dent Parrachée 3684 (fa 1862 by surveyors)
Aig de Pectet 3562 (fa 1878 W.A.B. Coolidge and party)

**Spurs on the N side**
Dôme de la Sache 3608
Mt Pourri 3779 (fa 4/10/1861 M. Croz solo)
Croix de Verdon 2738, c-rys from Courcheval to NNE and from Méribel les Allues to WSW.

**Spurs on the S side**
Grand Roc Noir 3583
Pte de la Sana 3436

French side; this probably increased the 'tourist' attraction of the route.

Between 1803 and 1811 Napoleon replaced the mule track by a carriage road. A few years later J.M.W. Turner crossed during his Alpine travels, producing subsequently a picture which gives only a limited hint of the nature of the place. By mid-19th century there was an excellent stage-coach service, using mules for the steep zig-zag ascents with the passengers walking on the steepest turns; at the summit the mules were replaced by fast horses, the conductor during the descent screwing on mechanical brakes at the corners.

Between 1868 and 1871 a railway (110-cm gauge) employing the Fell system was operated across the Pass by English engine-drivers. This followed closely the line of the road from St Michel to Susa with 10 km of avalanche tunnels and curves as sharp as 40 m radius. Over a distance of 25 km the gradients averaged 1 in 15; the maximum speed permitted was 25 km per hour and the brake pads operating on the centre rail were renewed after every journey. It was dismantled after the completion of the tunnel below the Col de Fréjus.

The Col du Mont Cenis is the most direct route from the plains of France to the plains of Italy; the valley on the French side was heavily fortified in the 19th century. Later, with France and Italy in opposition, the Pass became an obvious scene of fighting in World War II. In 1947 the frontier was redrawn to give the French a tactical advantage for the future.

Close to the summit, in addition to the various buildings – hospice, barracks, hotel, etc – there is now a huge reservoir.

**Monte Viso** (3851) towers above the other mountains of the Cottian Alps and is thus outstanding in distant views from as far away as Monte Rosa (160 km). It is the source of the River Po and was noted as such long ago by Geoffrey Chaucer.

On the Ubaye side of the **Col de Vars** (2111) are the impressive earth pillars, the Colonnes Coiffées de Vars. The road was constructed in 1891.

Much of the high mountain area of the **Central Dauphiné** is included in the PN des Ecrins (central zone 530 km$^2$; peripheral zone 1500 km$^2$). This is a block of savage rocky mountains, which includes the most southerly of the 4000 m peaks of the Alps, Barre des Ecrins, and the last major peak to have been climbed there, La Meije (in 1877).

The **Central Graian Alps** fall on, or close to, the frontier ridge, 15 km of which form the boundary of PN Vanoise and some 6 or 7 km the boundary also of PN du Grand Paradis.

At the time of the first ascent in 1358, **Rocciamelone (Roche Melon)** (3538) was believed to be the highest point in the Alps. Rotario d'Asti, who made the climb, had been a prisoner of the Mo-

**MONT BLANC GROUP** (from C de
la Seigne to C des Montets, C de la
Forclaz, the Vallée d'Entremont
and Grand C Ferret)

**Frontier ridge**
C de la Seigne 2513 (int path)
Aig des Glaciers 3816
Aigs de Trélatête 3930 (fa 12/7/1864
A. A. Reilly, E. Whymper; M. Croz,
M. Payot, H. Charlet)
Subsid pks 3917, 3895, 3892
To the W is the Glacier de Trélatête,
its N wall formed by the Dômes de
Miage 3688.
C de Miage 3358 (snow pass)
Aigs de Bionnassay 4051 (fa 28/7/1865
E. Buxton, F. Crauford-Grove, R.
Macdonald; J-P. Cachat, M. Payot)
Dôme du Goûter 4304 (fa 1784 J.M.
Couttet, F. Cuidet)
Mt Blanc 4807 (fa 8/8/1786 M.G.
Paccard, J. Balmat), **the highest
point of France, of the Alps and of
Europe.**

hammedans and had vowed to set up an oratory there when freed.
There are chapel and statues on the summit, where an annual
religious fête and pilgrimage is held. The original bronze triptych
is carried there each year from Susa.

The **Col du Petit St Bernard** (2188) is one of the more likely routes
for Hannibal; it is lower and easier than some of the other favoured
lines and, moreover, there is a stone circle on the summit long
known locally as the **Cirque d'Hannibal**. It was certainly crossed
by Caesar returning from Gaul to Rome in 49 BC. There was a
hospice on the pass from the 11th century, but this was badly
damaged in a battle between French and Austro-Sardinians in
1794. On top there is also a statue of St Bernard (now the patron
saint of mountaineers) and the 6·5m Colonne de Joux. The
carriage road was not completed until 1871.

The highest mountains in the **Eastern Graian Alps** are all contained in PN Gran Paradiso (set up in 1922, 640 km$^2$) – formerly
hunting country for Italian royalty.

**Mont Blanc (C D Milner)**

Mer de Glace and Grandes Jorasses (C D Milner)

Aiguille du Dru, Mont Blanc (C D Milner)

Mt Maudit 4465 (fa 12/9/1878 W. Davidson, J. Seymour-Hoare; J. Jaun, J. Bergen)
Mt Blanc du Tacul 4248 (fa 8/8/1855 Hudson-Kennedy party)
C du Géant 3371 (int snow pass)
Aig du Géant 4013 (fa 20/8/1882 W. W. Graham; A. Payot, A. Cupelin)
Subsid pk 4009 (fa 29/7/1882 Four members of the Sella family; three members of the Maquignaz family)
Mt Mallet 3989 (fa 5/9/1871 L. Stephen, F. Wallroth, G. Loppé; M. Anderegg, J. Cachat, A. Tournier)
Aig de Rochefort 4003 (fa 14/8/1873 J. Eccles; M., C. & A. Payot)
Dôme de Rochefort 4012 (fa 12/8/1881 J. Eccles; M. & A. Payot)

**Mont Blanc group**

Much of the high mountain area of the **Western Graian Alps** is included in the PN Vanoise (central zone 530 km$^2$, peripheral zone 1440 km$^2$). It is a compact block of mountains having 107 summits over 3000 m. A number of high-level long-distance foot-paths facilitate access. The flora and fauna are particularly rich. Around the Dômes de Chasseforêt, des Nants and de l'Arpont a number of small glaciers combine to form the so-called Vanoise snow-field (11 × 5 km), **the second largest in the Alps.**

The **Mont Blanc Group** is an important massif of high mountains (25 peaks over 4000 m) divided between France, Switzerland and Italy, the frontiers meeting on Mont Dolent. The ridge runs sub-stantially west to east and is bounded in the north by the valley of

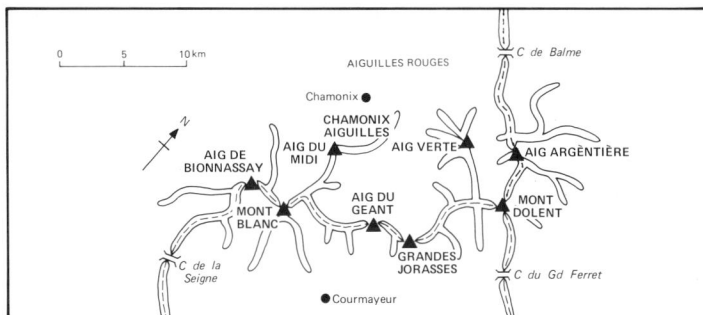

Grandes Jorasses (Pte Walker) 4208
(fa 30/6/1868 H. Walker; M. Ander-
egg, J. Jaun, J. Grange)
Subsid pks – Pte Whymper 4184 (fa
1865), Pte Croz 4110 (fa 1904), Pte
Marguerite 4065 (fa 1898), Pte Hél-
ène 4045 (fa 1898), Pte Young 3996
(fa 1904).
Aig de Triolet 3870 (fa 26/8/1874
J.A.G. Marshall; U. Almer, J.
Fischer)
Mt Dolent 3823 (fa 9/7/1864 A.A.
Reilly, E. Whymper; M. Croz, H.
Charlet, M. Payot), the meeting
place of France, Switzerland and
Italy.
Tour Noir 3836 (fa 3/8/1876 E. Javelle,
F.F. Turner; J. Moser, F. Tournier)
Aig d'Argentière 3900 (fa 15/7/1864
A.A. Reilly, E. Whymper; M. Croz,
H. Charlet, M. Payot)
Aig du Chardonnet 3824 (fa 20/9/1865
R. Fowler; M. Balmat, M.A. Du-
croz)
Aig du Tour 3544 (fa 17/8/1864 C.
Heathcote; M. Andermatten)
C de la Forclaz 1527 (rd, Trient to
Martigny), the Châtelard to Barber-
ine c-ry nearby has a gradient of 1
in 1·15.

**N of frontier ridge** (ie in France)
Aig du Goûter 3863
Between is the great basin of the
Taconnaz and Bossons glaciers,
which provided the route of the first
ascent on the one side and on the
other is even today the usual route to
Mont Blanc.
Aig du Midi 3842 (fa 5/8/1856 J.A.
Dévouassoud; J. & A. Simond)
Aig du Plan 3673 (fa 7/1871 J.E. Coles;
M. & A. Payot)
Aig de Blaitière 3522 (fa 8/8/1874 E.
Whitwell; C. & J. Lauener)
Aig de Grépon 3482 (fa 5/8/1881 A.F.
Mummery; A. Burgener, B. Venetz)
Aig des Grandes Charmoz 3444 (fa
9/4/1885 H. Dunod, P. Vignon; F. &
G. Simond, J. Desailloux, F. Folli-
guet)
Between this and the next S–N ridge
lies the celebrated ice stream of the
Mer de Glace
Les Courtes 3856 (fa 4/8/1876 H.
Cordier, T. Middlemore, J. Oakley
Maund; J. Anderegg, J. Jaun, A.
Maurer)
Les Droites 4030 (fa 7/8/1876 as for
Les Courtes above, less J. Anderegg)
Aig Verte 4122 (fa 29/6/1865 E. Whym-
per; C. Almer, F. Biner)
Subsid pks 4023, 3982, 3791
To the E is the Glacier d'Argentière
Aigs du Dru – Grand 3754 (fa 12/9/
1878 C.T. Dent, J. Walker; A.
Burgener, K. Maurer)
Aigs du Dru – Petit 3733 (fa 29/8/1879
J.E. Charlet-Straton, P. Payot, F.
Folliguet)
Aig des Grands Montets 3292, c-ry
from Argentière to 3250m.

the Arve and in the south by Val Veni and Val Ferret. An important walkers' route, the Sentier International du Tour du
Mont Blanc, encircles the massif with 160km of way-marked
paths. The group, its summit much higher than any other Alpine
peak, forms the backdrop to the view from many surrounding
elevations, including points of easy access, both near and far.

The **Col de Miage** (3358) was the scene of a record fall (and survival) of 550m by J. Birkbeck in 1861.

**Mont Blanc** (4807) has always attracted attention as the highest
mountain in the Alps and in Europe. H.B. de Saussure, Swiss
savant, offered a prize in 1760 for the first ascent. After a number
of attempts this was won by J. Balmat and M. G. Paccard on 8
August 1786; for many years controversy raged as to which of
them carried the dominant responsibility; modern scholarship
has come down on the side of Paccard, who was the local doctor
in Chamonix. De Saussure followed a year later (accompanied
by Balmat, a valet and 17 other guides), spending four-and-a-half
hours on the summit making scientific observations. Six days
after de Saussure came Colonel Beaufroy for **the first British
ascent**. All of them climbed from Chamonix up the spur known
as the Grands Mulets between the glaciers de Tacconaz and des
Bossons. **The following 'firsts' ensued:** by a lady (Marie Paradis)
1808; by a German (Rodatz) 1812; by Americans (Van Renssel-
aer and Howard) 1819; by a Frenchman (Comte de Tilly) 1834;
second by a lady (Henriette D'Angeville) 1838; by an English lady
(Mrs T. Hamilton) 1854; in winter (Miss I Straton) 1876. The
honour of taking **the first photographs on the summit** (around
1860) is in dispute between J. Tairraz, who founded the famous
Chamonix photographic firm, and A. Bisson.

As early as 1921 an aeroplane landed at 4250m and parachutists dropped to the summit in August 1961.

There are also bizarre records. The first ascent by a dog took
place in 1837. The lame, one-legged and no-legged, have climbed
it, as have the blind. In July 1960 a crowd from a certain sect
camped high on the mountain to await their scheduled 'end of the
world', but they had mistaken the date!

On the south (Italian) side there are steep faces between long
ridges – Brouillard (in 1901), Innominata (in 1919) and Peuterey (in 1927).

The **Col du Géant** (3371) was occupied for 16 days in 1788 when
de Saussure camped there while making scientific observations.

Below at an altitude of 1220m is the Mont Blanc road-tunnel
linking Chamonix with Courmayeur. It was begun in 1959 and the
headings joined in 1962; 11·6km long and 4·8m high, the
carriage-way is 7m wide.

Above the range the cableway crosses from Aig du Midi to
Pte Helbronner.

The north face of the **Grandes Jorasses** (4208) is one of the three

C des Montets 1461 (rd, Chamonix to
  Vallorcine)

**S of frontier ridge** (ie in Italy)
Mt Blanc de Courmayeur 4748, **the
  highest peak in Italy.**
Part of Mont Blanc (Monte Bíanco) is
  in Italian territory up to the 4760 m
  contour.
P Luigi Amédeo 4469 (fa 20/7/1901
  G. F. & G. B. Gugliermina; J.
  Brocherel)
Mt Brouillard 4069
  Between here and the next ridge to
  the E are the glaciers du Brouillard
  and du Fresnay.
Aig Blanche de Peuterey 4107 (fa
  31/7/1885 H. Seymour King; E. Rey,
  A. Supersaxo, A. Anthamatten)
Aig Noire de Peuterey 3772 (fa 5/8/
  1877 Lord Wentworth; E. Rey,
  J-B. Bich)
  Between here and the frontier ridge
  lies the Glacier de la Brenva.
Grand C Ferret 2537 (int path)

**DENTS DU MIDI GROUP** (between
  the R Rhône, Martigny to Lac
  Leman, and the France/Switzerland
  frontier) (from N to S)
Dents du Midi 3257 (fa 1784 J. M.
  Clément)
Subsid pks 3212, 3187, 3180, 3165, 3164
Tour Sallière 3219 (fa 18/7/1858 H.
  Marguerat, J. Oberhauser, J. Rey,
  E. Gonet)
Grand Mt Ruan 3047 (fa 1875 G.
  Béranek)
P de Tenneverge 2987 (fa 6/10/1863 A.
  Wills; C. Gurlie)

**PENNINE ALPS** (between Grand C
  Ferret and the Simplon Pass)

**Frontier ridge** (Grand C Ferret to C de
  Fenêtre) (from W to E)
  Grand C Ferret 2537 (int path)
  C du Grand St Bernard 2469 (int rd,
  Martigny to Aosta)
Mt Velan 3734 (fa 31/8/1779 L. J.
  Murith; Moret, Genoud)
C de Fenêtre 2805 (int snow pass, Val
  de Bagnes to Val Pellini)

**N of frontier ridge, W of Val des Bagnes**
  (from S to N)
Grand Combin 4314 (fa 30/7/1859
  C. StC. Deville; D., E. & G. Balleys,
  B. Dorsaz)
Subsid pks 4243, 4184, 4142, 3833
Maisons Blanches – Grande Aig 3682
  (fa 23/6/1874 M. Maglioni; D. Bal-
  leys, N. Knubel)
Combin de Boveyre 3663 (fa 1888 R.
  Broft; J. & O. Balleys)
Tournelon Blanc 3707 (fa 5/7/1867 F.
  Hoffmann-Merian; J. Fellay, S. Bes-
  sard)
Combin de Corbassière 3716 (fa 14/8/
  1851 G. Studer; J. von Weissenfluh,
  J. B. Fellay)
Petit Combin 3672 (fa 27/7/1890 C. de
  la Harpe, E. V. Viollier; J. Bessard)

great north faces of the Alps, and the scene of modern climbing
of high standard.

The **Aiguille du Midi** (3842) is **the highest point in the Alps reached
by téléphérique**. It rises from Chamonix 2900 m in 4500 m.

A hang glider descent to Chamonix from the summit has been
made in 25 min.

The mountains, known as the **Chamonix Aiguilles**, provide high
standard rock climbing on good granite. They provided the
scene for the renaissance of French mountaineering after World
War I. In addition to the main peaks listed there are several more
minor peaks and pinnacles on the ridge, which was first traversed
from end to end in 1939.

In 1964 the guide, Armand Charlet, recorded his 100th ascent of
the **Aiguille Verte** (4122).

The frontier ridge of the **Pennine Alps**, running roughly west to
east, throws out a number of long spurs to the north towards the
Rhône valley, so that most of the area falls in Switzerland. There
are many 4000 m peaks and glacier basins.

The **Col du Grand St Bernard** (2469) was known before the times
of the Romans and was used extensively by them. The hospice on
the summit existed by 859. It has been an important crossing place
throughout the ages and a favourite route for pilgrims to Rome.
Napoleon came this way in 1800, on the way to Marengo, with
large numbers of troops and considerable artillery. The big guns
were dragged over encased in hollowed-out tree trunks.

The through carriage-way was not completed until 1905. In
1958 a tunnel was started at 1830 m to cut off the upper part of
the route. It was completed in 1962 at a cost of £12 million and
17 lives; it is 5·9 km long (with another 13 km of covered-in
access roads), 4·5 m high and has a carriage-way 7·5 m wide. The
Col is closed for some months in the winter, but the tunnel is
always open. There is a statue of St Bernard here.

The Pass is associated with the famous breed of dog, said to
succour travellers lost on the crossing. Barry, who rescued 40
during his lifetime – 1800–14 – now stands guard in the Nature
Museum in Berne. He could predict avalanches and lead the
monks to buried victims. Now in his honour the lead dog is
always known as Barry.

The first ascent of the **Dent Blanche** (4357) was made by the
south ridge. The east-north-east ridge (ascended in 1882) is
known as the Arête des Quatres Ânes, from a remark by one of
the guides – 'Nous sommes pourtant quatre ânes d'être montés
par ici'. The west ridge was the scene of the accident in which the
famous British rock climber, Owen Glynne Jones, was killed in
1899.

**Frontier ridge** (C de Fenêtre to C d'
  Hérens) (from W to E)
C de Fenêtre 2805
La Singla 3714 (fa 22/7/1867 C. Schroe-
  der; S. Bessard)
Subsid pks 3704, 3691
  C Collon 3130
Mt Brule 3621 (fa 7/8/1878 A. Cust &
  porter)
Dents de Bouquetins 3838 (fa 6/9/1871
  A.B. Hamilton; J. Anzevui, J.
  Vuignier)
Subsid pks 3779, 3670
Tête Blanche 3724 (fa 15/8/1849 G.
  Studer, M. Ulrich, G. Lauterburg;
  J. Madutz, N. Inderbinnen, A. & J.
  Binner)
  C d'Hérens 3480 (snow pass, Val
  d'Hérens to Nikolaithal)

**N of frontier ridge, between Val des
Bagnes and Val d'Arolla** (from S to N)
La Ruinette 3875 (fa 6/7/1865 E.
  Whymper; C. Almer, F. Biner)
La Serpentine 3795
Pigne d'Arolla 3796 (fa 9/7/1865 A.W.
  Moore, H. Walker; J. Anderegg)
Mt Blanc de Cheilon 3869 (fa 11/9/1865
  J.J. Weilenmann; J. Fellay)

The **Matterhorn** (4477), the most famous mountain of the Alps and perhaps indeed of the world, rises on the frontier ridge between the Zermatt valley and the Italian Val Tournanche. From northerly and easterly aspects it presents the near-perfect mountain shape, a steep-sided pyramid. From the west it appears more as a blunt wedge, while on the south side the familiar shape gives way to a bulky mass of walls and towers.

The first ascent in 1865, by a party led by Michel Croz and including Edward Whymper, was the culmination of a series of attempts over a period of five years. The struggle against the mountain had become an obsession with Whymper who was often accompanied in the earlier years by a local guide J. A. Carrel of Val Tournanche. In 1865 a number of prominent Italian mountaineers decided that the first ascent should really fall to Italy and they engaged Carrel to lead their attempt. Learning of their plans, Whymper hurried back to Zermatt to find two parties there planning an assault by the Hornli ridge – Charles Hudson and the completely inexperienced Douglas Hadow with the famous Michel Croz and Lord Francis Douglas with the guide Peter Taugwalder and his son. They joined forces and within a few days the mountain had been climbed, proving straightforward

**Matterhorn (AC Collection)**

**Zinal glacier with Obergabelhorn, Matterhorn and Dent Blanche; Switzerland**

Mt Pleurer 3703 (fa 13/7/1866 E. Hoffmann-Burkhard; J. Fellay, S. Bessard, J. Gillioz)
Aigs Rouges d'Arolla 3646 (fa 23/6/ 1870 J. H. Isler; J. Gillioz, Bruchez)
Subsid pks 3595, 3584
Mt Gele 3023, c-wy from Verbier

**Matterhorn from Gornergrat (E Pyatt)**

in spite of its appearance of inaccessibility. From the summit the toiling Italians were observed still far below.

Tragedy struck during the descent. A slip by Hadow knocked Croz from his holds; in turn Hudson and Douglas were dragged from theirs, but the rope broke between them and Peter Taugwalder so that he, his son and Whymper were saved. It is the most controversial incident in the history of mountaineering; the actual sequence of events, responsibility for various parts of the action, statements by Whymper, speculation by his friends and enemies have led to prolonged arguments, to lengthy articles and even to books. The exact details never seem likely to be resolved.

The Italians reached the top from their side, the south-west ridge, two days later. The Zmutt (north-west) ridge fell in 1879. The Furggen (south-east) ridge has a very steep section just below the summit which defied all efforts until 1941, by which time the techniques of climbing had advanced sufficiently to overcome them. As early as 1899 the celebrated Italian mountaineer, Guido Rey, having been stopped by this steep part, descended the mountain, climbed it by the ordinary route and had himself lowered down the Furggen ridge to his previous highest point, thus completing the exploration without actually doing the climbing.

The first ascent by a lady was made on 22 July 1871 by Miss Lucy Walker, who was accompanied by her father, aged 63, and F. Gardiner and guided by Melchior Anderegg. She forestalled by only a few days Miss Meta Brevoort, who had to be content with the first traverse by a lady, up by the Hornli ridge and down by the Italian.

**N of frontier ridge, between branches of the Val d'Hérens** (from S to N)

L'Evêque 3716 (fa 5/8/1867 A. Baltzer, C. Schroder)

Mt Collon 3637 (fa 31/7/1867 G.E. Forster; H. Baumann, J. Kronig)

Douvres Blanches 3664

Aig de la Tsa 3668 (fa 21/7/1868 P. & J. Vuignier, P. Beytrison, G. Gaspoz, P. Quinodoz)

Dent de Perroc 3676 (fa 31/8/1871 A.B. Hamilton, W.R. Rickman; J. Anzevui, J. Vuignier)

Subsid pk 3651

**S of frontier ridge**

Becca Luseney 3506

**N of frontier ridge, between Val d'Hérens and Val de Zinal** (from S to N)

C d'Hérens 3480

Dent Blanche 4357 (fa 18/7/1862 T.S. Kennedy, W. Wigram; J-B. Croz, J. Kronig)

Grand Cornier 3962 (fa 16/6/1865 E. Whymper; C. Almer, M. Croz, F. Biner)

Pte de Bricolla 3658 (fa 30/7/1879 C. Socin; E. Peter)

**Frontier ridge** (C d'Hérens to Theodule Pass)

C d'Hérens 3480

Tête de Valpelline 3802 (fa 1866 E. Whymper; F. Biner)

Dent d'Hérens 4171 (fa 12/8/1863 W.E. Hall, F. Crauford Grove, R.S. Macdonald, M. Woodmass; M. Anderegg, P. Perren, J-P. Cachat)

Obergabelhorn (B R Goodfellow/ AC Collection)

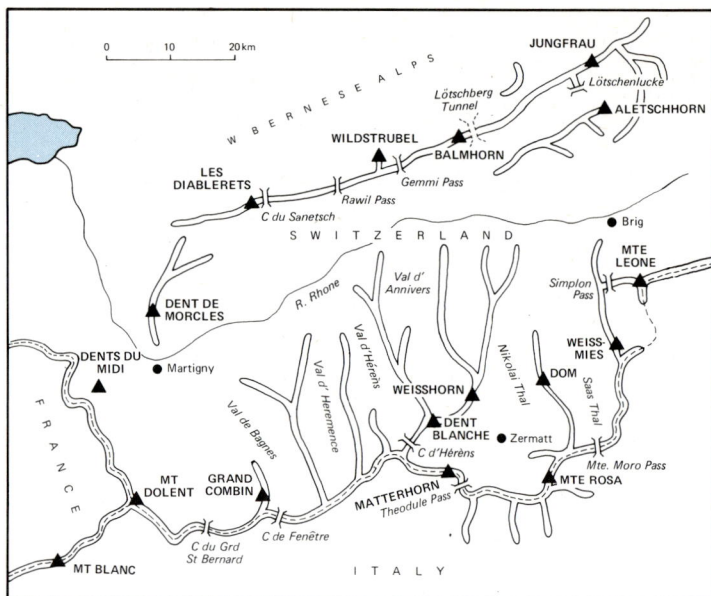

Pennine and W Bernese Alps

The first ascent in winter months, rendered considerably more difficult by the increased snow and ice cover and by the shortness of the days, took place in 1894.

There are records without number. It has been climbed by the blind and by the crippled (on one artificial leg and even on two). A cat has been up, making the descent in a rucksack. The Hornli Ridge has been climbed in 63min, and up and down by a guide and his client in three hours. The four ridges have been traversed, up Furggen and down Hornli, up Zmutt and down the Italian inside 24 hours.

The striking outline of the Mountain has led to the proliferation of the use of the name for similar outlines elsewhere. At least four North American mountains are actually called *Matterhorn* – in British Columbia, Oregon, Nevada and Colorado. Many other mountains are proudly dubbed the *Matterhorn* of their particular country, varying from tiny hills like Roseberry Topping (Cleveland, England) (322) and Cnicht (Wales) (690) through Mount Aspiring (New Zealand) (3035) and Mount Assiniboine (Canada) (3618) to mountains higher than the original, like Ushba (Caucasus) (4710), Shivling (Indian Himalaya) (6543) and Ama Dablam (Nepal Himalaya) (6856). The similarity is often detectable from only one viewpoint, the mountain may be cocked-hat shaped rather than pyramidal.

The **Theodule Pass**, now approached over the Theodule glacier on the north side, was formerly snow-free. Roman relics were found near the summit, while a Stone Age axe was found close to the Gandegg Hut in May 1959. It weighed close on a kg and was 30cm long and 8cm wide.

Matterhorn (Mt Cervin) 4477 (fa 14/7/
  1865 E. Whymper, C. Hudson, D. R.
  Hadow, Lord F. Douglas; M. Croz,
  P. Taugwalder père et fils)
Theodule Pass 3317 (int snow pass,
  Zermatt to Breuil-Cervinia)

**N of frontier ridge, between Val de
Zinal and Nikolaithal**
Pte de Zinal 3789 (fa 1871 E. Javelle,
  G. Beraneck Jnr; J. Martin)
Mt Durand 3713
Ober Gabelhorn 4063 (fa 6/7/1865
  A. W. Moore, H. Walker; J. Ander-
  egg)
Wellenkuppe 3903
Trifthorn 3728 (fa 5/7/1865 Lord F.
  Douglas; P. Taugwalder, P. Inäbit)
Pte de Mountet 3877
Zinalrothorn 4221 (fa 22/8/1864 L.
  Stephen, F. Crauford-Grove; M. &
  J. Anderegg)
Besso 3668 (fa 1862 J-B. Epiney; J.
  Vianin)
Schalihorn 3974 (fa 10/8/1864 J.J.
  Hornby, T. H. Philpott; C. Almer,
  C. Lauener)
Subsid pk 3955
Weisshorn 4505 (fa 19/8/1861 J. Tyn-
  dall; J.J. Bennen, U. Wenger)
Bieshorn (Bishorn) 4153 (fa 18/8/1884
  G. S. Barnes, R. Chessyre-Walker;
  J. Imboden, J.M. Chanton)
Brunegghorn 3833 (fa 1853 J. & F.
  Tantignoni; H. Brantschen)
Tête de Millon 3693
Diablons 3609

**Frontier ridge** (Theodule Pass to Monte
Moro Pass) (from W to E)
Theodule Pass 3317 (int snow pass)
Klein Matterhorn 3883 (fa 13/8/1792
  H. B. de Saussure; J. M. Couttet & 6
  others)
Breithorn 4165 (fa 13/8/1813 H. May-
  nard; J. M. Couttet, J-G., J-B. &
  J-J. Erin: 2nd by Sir J. Herschel, the
  astronomer, in 1821)
Subsid pks 4141, 4160
Schwarztor 3781
Pollux 4091 (fa 1/8/1864 J. Jacot; P.
  Taugwalder, J-M. Perren)
Zwillings Pass 3861
Castor 4226 (fa 23/8/1861 W. Mathews,
  F. W. Jacomb; M. Croz)
Felikjoch 4068
To the N is the Gorner Glacier.
Lyskamm 4527 (fa 19/8/1861 J.F.
  Hardy and party; 5 guides)
Lysjoch 4260 (reached by Italian
  peasants in 1778)
Mte Rosa (Dufourspitze) 4634 (fa
  1/8/1855 J.G. & C. Smyth, E.J.
  Stephenson; V. Lauener, J. & M.
  Zumtaugwald), **the highest point in
  Switzerland.**
Subsid pks – Nordend 4609 (fa 1861),
  Grenzgipfel 4596 (fa 1848), Zum-
  steinspitze 4563 (fa 1820), Signal-
  kuppe 4556 (fa 1842), Parrotspitze
  4436 (fa 1817), Ludwigshohe 4341
  (fa 1822). The cols between are all
  over 4000 m.

**The high passes across the frontier ridge** above this part of the Gorner glacier were used by escaping prisoners-of-war during the 1939–45 conflict. Large numbers of ill-fed, poorly clothed and shod escapees made the attempt, often without any mountaineering experience. Some won through, many perished. So-called guides abandoned their charges at the frontier, where the real mountaineering difficulties begin, but the Swiss maintained rescue patrols on their side and saved whomsoever they could.

The **Klein Matterhorn** (3883), a tiny summit which mimics the shape of its namesake nearby, will eventually be reached by cableways from Zermatt by way of Trokenersteg on the one side and from Breuil-Cervinia via Testa Grigia on the other.

**Monte Rosa** (4634), a massive mountain with many summits, is the highest point of Switzerland. It was ascended in 1894 by Winston Churchill and by many celebrated others, before and since. One hundred women took part in a mass ascent of the Signalkuppe in the summer of 1960 as a tribute to Mme Claude Kogan and Mlle van der Stratten, lost that year on Cho Uyo in the Himalaya.

**Monboso** on the south slopes of **Monte Rosa**, climbed over by Leonardo da Vinci, cannot be positively identified.

Worthy of note is the ridge of low peaks forming the north wall of the **Gorner glacier** above Zermatt: Riffelhorn (2927), on the summit of which Bronze Age artefacts have been found; Gorner-grat (3136), summit of a rack railway from Zermatt (opened 1898, **second highest in Europe, highest point reached in the open**); Stockhorn (3534) (a cableway from Gornergrat reaches 3413m, **the highest in Switzerland**).

The **Dom** (4545), **the highest mountain entirely in Switzerland**, and the adjacent heights are called collectively the Mischabelhorner.

The **Antrona Pass** (2844) was once the main route from the Valais to Milan and relics of the paved track remain in places. It was replaced eventually by the carriage road over the Simplon Pass.

The first recorded crossing of the **Simplon Pass** (2005) was in the 13th century, when the original hospice was constructed. For some centuries the Antrona Pass was preferred, though this one was used during the 17th century by considerable commercial traffic organised by the entrepreneur G. de Stockalper from Brigue. The position changed abruptly when Napoleon built the carriage-way in 1801–5, with a large barracks on the summit. The whole cost 12 million francs. A scheme for a road tunnel from Berisal to Varzo, put forward in 1961, has not yet been implemented. The col is sometimes closed in winter months, but a car ferry is available through the railway tunnel.

A tunnel beneath the Simplon Pass to link the valley of the Rhône with the cities of north Italy was first planned in 1895. It

Jägerhorn 3970
 Alt Weissthor 3576
Ca di Jazzi 3804 (fa 8/1851 G.M.
 Sykes; M. Zumtaugwald)
Neu Weissthor 3645
 Mte Moro Pass 2776 (int path, Saas-
 tal to Val d'Anzasca)

**N of frontier ridge, between Nikolaithal
 and Saastal** (from S to N)
Strahlhorn 4190 (fa 15/8/1854 E.J.
 Grenville, C. Smyth; F.J. Anden-
 matten, U. Lauener)
Fluchthorn 3802
Rimpfischhorn 4199 (fa 9/9/1859 L.
 Stephen, R. Liveing; M. Anderegg,
 J. Zumtaugwald)
Allalinhorn 4027 (fa 28/8/1856 E.L.
 Ames; F.J. Andenmatten, Imseng)
Alphubel 4206 (fa 9/8/1860 L. Ste-
 phen, T.W. Hinchliff; M. Anderegg,
 P. Perren)
Täschhorn 4491 (fa 30/7/1862 J.L.
 Davies, J.W. Hayward; J. & S.
 Zumtaugwald, P.J. Summermatter)
Dom 4545 (fa 11/9/1858 J.L. Davies;
 J. Zumtaugwald, J. Kronig, H.
 Brantschen)
Lenzspitze 4294 (fa 8/1870 C.T. Dent,
 A. & F. Burgener)
Nadelhorn 4327 (fa 16/9/1858 J. Zim-
 merman; A. Supersaxo, B. Epiney,
 F. Andenmatten)
Subsid pk 4241
Hohberghorn 4219
Ulrichshorn 3925 (fa 10/8/1848 Ulrich,
 J.J. Imseng; F. Andermatten, J.
 Madutz, S. Biner, M. Zumtaugwald)
Dürrenhorn 4035 (fa 7/9/1879 A.F.
 Mummery, W. Penhall; A. Bur-
 gener, F. Imseng)
Balfrin 3796 (fa 6/7/1863 Mr & Mrs
 R.S. Watson; J. Imseng, F. Anden-
 matten, J-M. Claret)
Gr Bigerhorn 3625

**Frontier ridge** (Mte Moro to the Simplon road) (from S to N)
 Antrona Pass 2844 (int path)
Portgengrat (P d'Andola) 3654 (fa
 7/9/1871 C.T. Dent; A. & F. Burgener)
 Andola Pass 2417 (int path)

**W of frontier ridge, between Saastal
 and Simplon Pass** (from S to N)
Weissmies 4023 (fa 8/1855 J.K. Heus-
 ser; P.J. Zurbriggen)
Lagginhorn (Laquinhorn) 4010 (fa
 26/8/1856 E.L. Ames, 3 other Eng-
 lishmen; J.J. Imseng, F. Andenmatten, 3 other guides)
Subsid pk 3906
Fletschhorn 3996 (fa 28/8/1854 M.
 Amherdt; J. Zumkemmi, F. Klausen)
 Simplon Pass 2005 (rd, Brigue to
 Domodossola; the frontier is well
 down on the S side; rail tunnel
 below)

The summit of the Simplon Pass with one-time barracks and commemorative
obelisk, Bernese Oberland beyond (E Pyatt)

was decided to use two linked tunnels rather than a double track
tunnel; work commenced in August 1898 and an average advance
of 5·5 m/day was achieved. As excavation proceeded the temperature at the working face rose steadily, eventually reaching over
55°C. Work on the Swiss side had to be abandoned and sealed off,
while that on the Italian side continued until the headings were
joined in February 1905. The differences amounted to a mere 9
cm in level and 20 cm in direction. Now at last the trapped hot
water was able to run away down the southern half of the tunnel,
enabling work to be resumed from both ends. The first tunnel was
opened on 1 June 1906, the second not until 1922. At 19·8 km this
is **the longest railway tunnel in the world**.

The **Western Bernese Alps** consist of a ridge running from west-
south-west to east-north-east parallel to the Rhône valley on its
north side.

The **Gemmi Pass** (2316) was known as early as 1250; the path was
improved in 1740. A hospice erected on the summit later became
an inn.

The first ascent of the **Stockhorn** (2190) in 1536 was made in
search of dragons, none were found. Now there is a cableway to
the summit from Erlenbach.

The first ascent of the **Balmhorn** (3709) is noted as the only
occasion in Alpine history where a father, son and daughter
participated jointly in such an event.

In the **Bernese Alps – Main Group** the simple ridge structure of
the west section becomes more complex, first with parallel ridges,
then later breaking out to a knot of mountains around the big ice
basin called Concordia Platz. This is the wall of the Alps above
the Swiss plain and can be seen from many a viewpoint, being
quite outstanding even from as far away as the Jura.

There are a great many glaciers, notable among them the
Aletsch, **the longest in the Alps**, fed by the glaciers round Concordia Platz.

**WESTERN BERNESE ALPS** (Lac Leman to Gemmi Pass)

**Main ridge** (from W to E)
Dent de Morcles 2969
Grand Muveran 3051
Les Diablerets 3209 (fa 19/8/1850 G. Studer, M. Ulrich; J. Madutz, J.D. Ansermoz)
Oldenhorn 3123
 Sanetsch Pass 2243 (rd on S side only)
Wildhorn 3248 (fa 10/9/1843 G. Studer; M. Schäppi & a shepherd)
Rawil Pass 2429 (path)
Wildstrubel 3243 (fa 16/8/1856 E. von Fellenberg; J. Tritten)
Gross Lohner 3049 (fa 9/1875 F. Ogi; C. Hari)
 Gemmi Pass 2316 (path, c-wy to top on S side)

**NORTHERN FOOTHILLS** (from SW to NE)
Tornettaz 2541
Gifferhorn 2542
Albristhorn 2762
Männlifluh 2652
Stockhorn 2190 (fa 1536 Johann Müller), c-wy to summit
Niesen 2362, c-ry to summit

**BERNESE ALPS – MAIN GROUP** [Bernese Oberland] (Gemmi Pass to Grimsel Pass)

**Gemmi Pass to Lötschenlücke (N of Lötschental)** (from W to E)
Altels 3629 (fa 1834 by peasants)
Balmhorn 3709 (fa 21/7/1864, F., H. & Miss L. Walker; J. & M. Anderegg)
Lötschen Pass 2690 (path Kandertal to Lötschental; ry below in Lotschberg Tunnel)
Doldenhorn 3643 (fa 30/6/1862 E. von Fellenberg, A. Roth; C. Lauener, J. Bischoff, K. Blatter, G. Reichen)
Blumlisalphorn 3664 (fa 27/8/1860 L. Stephen, R. Liveing, J.K. Stone; M. Anderegg, P. Simond, F. Ogi)
Morgenhorn 3613 (fa 14/8/1869 H. Baedeker; U. Lauener, J. Bischoff)
Gspaltenhorn 3437 (fa 10/7/1869 E. Foster; H. Baumann, J. Anderegg)
Lauterbrunnen Breithorn 3782 (fa 31/7/1865 E. von Fellenberg; P. Michel, P. Egger, J. Bischoff, P. Inäbit) (J.J. Hornby, T.H. Philpott: C. Almer, C. Lauener arrived at the summit only a few minutes later)
Grosshorn 3762 (fa 8/9/1868 H. Dübi, E. Ober; J. Bischoff, J. Siegen)
Mittaghorn 3893 (fa 19/7/1878 C. Montandon, A. Ringier, A. Rubin)
Lötschenlücke 3204 (snow pass, Lötschental to Concordia Platz)

**S of Lötschental, S of Lötschenlücke and Concordia Platz, W of Aletsch glacier** (from E to W)
Drieckhorn 3811 (fa 26/8/1868 T.L. Murray-Brown; P. Bohren, P. Schlegel)

The **Lötschen Pass** (2690) was well known by the 14th century, and for a long time was more important than the Gemmi Pass. Battles were fought on, or near, the summit in 1384, 1419 and 1656.

Work began in 1906 on the Lötschberg Tunnel below the Pass, designed to shorten the route between central Switzerland and the Simplon Tunnel. Serious difficulties were encountered on the north side of the range, where the working ran from solid rock into a deep silt-filled fault below the Kander valley. A huge mass of sand, mud and boulders filled the tunnel for some 1600m back from the working face and 25 tunnellers lost their lives. A wall was built shutting off the head and the tunnel diverted round the fault; the north-bound tunnel was deflected correspondingly, the two meeting in March 1911. The 14·6km tunnel was opened to traffic on 15 July 1913.

The **Jungfrau** (4158) is a mountain of three ridges – north-east to Jungfraujoch, south over the Rothalhorn to Lauitor, west to the Silberhorn. Prominent in the view from Interlaken and points north, it was one of the first major mountains of the Alps to be climbed (in 1811). In 1940 the guide Fritz Steuri Snr completed his 100th ascent.

The **Jungfraujoch** (3474) is the saddle between the Jungfrau and the Mönch and looks out to the south over the wide ice field of Concordia Platz. It is served by a rack railway from Kleine Scheidegg (see Fig. 17 – p. 49), completed in 1912 after twelve years work, which cost 12 million francs and 28 lives. The line

**Central Switzerland**

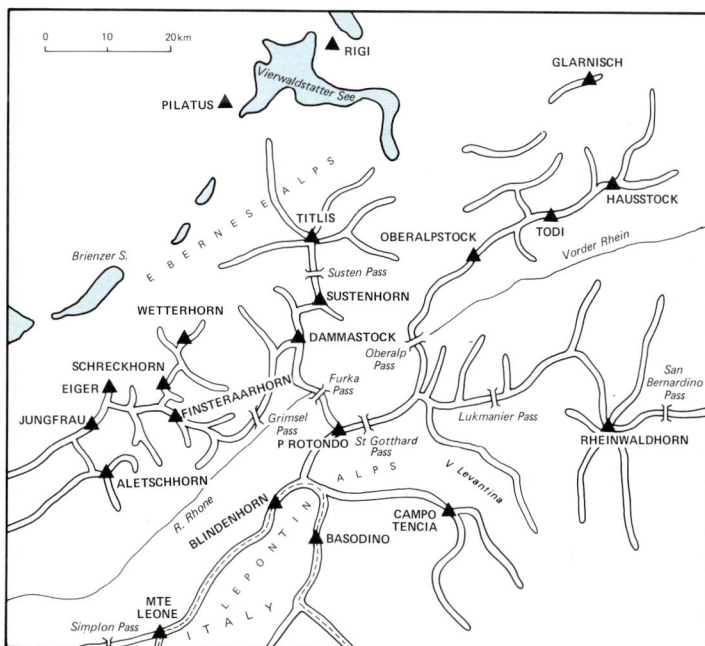

Klein Drieckhorn 3641
Aletschhorn 4195 (fa 18/6/1859 F.F.
  Tuckett; J.J. Bennen, P. Bohren, V.
  Tairraz)
Geisshorn 3740
Sattelhorn 3741 (fa 26/8/1883 K.
  Schulz; A. Burgener, J. Rittler)
Distelhorn (Distlighorn) 3748
Schienhorn 3797 (fa 30/8/1869 G.J.
  Häberlin; A. & J. von Weissenfluh)
Nesthorn 3824 (fa 18/9/1865 H.B.
  George, A. Mortimer; C. & U.
  Almer)
Lötschentaler Breithorn 3785 (fa 28/6/
  1869 G.J. Häberlin; A. & J. von
  Weissenfluh)
Bietschhorn 3934 (fa 13/8/1859 L.
  Stephen; J. & A. Siegen, J. Aebiner)

**Around Concordia Platz from Lötschen-
lücke to Grünhornlücke** (from W to
E)
Ebnefluh 3964 (fa 27/8/1868 T.L.
  Murray-Brown; P. Bohren, P.
  Schlegel)
Gletscherhorn 3983 (fa 15/8/1867 J.J.
  Hornby; C. Lauener)
Jungfrau 4158 (fa 3/8/1811 J.R. & H.
  Meyer and party)
Jungfraujoch 3474
Mönch 4099 (fa 15/8/1857 S. Porges;
  C. Almer, U. & C. Kaufmann)
Eiger 3970 (fa 11/8/1858 C. Barrington;
  C. Almer, P. Bohren)
Trugberg 3933 (fa 13/7/1871 E. Burk-
  hardt; P. Egger, P. Schlegel)
Gr Fiescherhorn 4049 (fa 23/7/1862
  H.B. George, A.W. Moore; C.
  Almer, U. Kaufmann)
Subsid pk 4025

---

**The North Wall of the Eiger,
Switzerland, with the Finsteraarhorn
beyond**

enters the mountain at Eiger-gletscher Station and climbs inside the Eiger to Eigerwand Station, where there is a view across this notorious rock wall and down to Grindelwald. It climbs, doubling back, to Eismeer Station, with views across the far side of the range to the Schreckhorn and the Lauteraarhorn, and so to 3454m at Jungfraujoch terminus. The extension of the line to the summit of the Jungfrau, mooted in 1962, is not being proceeded with at present. This, **the highest railway in Europe**, employs the Strub (a modified Abt) rack.

On the Joch there are hotel, restaurant, observatory, laboratory, terraces, ice-rink – linked by tunnels and galleries. There is access to the snows of the col and to the upper edge of the northerly snow and ice fields (ski-ing and sledge rides). **The first aeroplane landing on a mountain** took place here – R. Ackermann in a Haefels DH3 in July 1919.

The **Eiger** (3970) is a mountain of three ridges – north-east (Mittellegi) not climbed until 1921, south to the Eigerjoch and the Mönch, west which in combination with the west face was the route of the first ascent. Barrington, who accomplished this in 1858, had never been to the Alps before, but he took the lead when the guides wanted to turn back.

The face between the north-east and west ridges is the Nordwand, the third of the great North Faces of the Alps. First climbed in 1938 by A. Heckmair, W. Vorg, H. Harrer and F. Kasparek and in winter in 1961 by T. Hiebeler, W. Almberger, A. Kinshofer and A. Mannhardt, it has been the scene of considerable international rivalry and tragedy. While engaged in these struggles the mountaineers provide 'entertainment' for onlookers with telescopes at the Kleine Scheidegg.

The north face of the **Wetterhorn** (3701), above Grosse Scheidegg, is one of the most photographed in the Alps. The first téléphérique in Switzerland was opened here in 1908 from just below the foot of the Upper Grindelwald glacier to 1677m at the west corner of the north face. Closed in 1914, it never re-opened. Winston Churchill climbed the mountain in 1894.

The **Grimsel Pass** (2165) provides the easiest route from the Bernese plains to the Valais. It was crossed by armies in 1211 and again in 1419; the hospice near the summit was probably there in the 14th century. Until 1895, when the carriage-way was built, there was only a mule track (part of a route which continued over the Gries Pass into Italy). There are splendid views on the east side over the snout of the Rhône glacier.

The **Schilthorn** (2974) is reached by a cableway from Stechelberg via Murren. This rises 2092m over 6931m in four stages – the lower two accommodating 100 persons/cabin and the upper two 80 persons/cabin.

The railway up the **Brienzer Rothorn** (2350), opened in 1892, and still by popular demand operated by steam, climbs 1681m (max

Gr Grünhorn 4043 (fa 7/8/1865 E. von Fellenberg; P. Egger, P. Michel, P. Inäbit)
Subsid pk 3913
Grünegghorn 3860
Grünhornlücke 3305

**S and E of Grünhornlücke** (from W to E)
Fiescher Gabelhorn 3876 (fa 8/8/1884 P. & C. Montandon)
Schönbühlhorn 3854 (fa 13/7/1884 L. Kurz and party)
Gr Wannehorn 3905 (fa 6/8/1864 G. Studer, R. Lindt; K. Blatter, P. Sulzer)
Eggishorn 2930, c-wy from Fiesch to 2878 m.
  Continuing now ESE of the Gr Fiescherhorn and N of the Walliser Fiescherfirn
Aggassizhorn 3953 (fa 7/9/1872 W. A. B. Coolidge; U. Almer, C. Inäbit)
Finsteraarhorn 4274 (fa claimed for 16/8/1812 A. Volker, J. Bortis, A. Abbühl, certainly 10/8/1829 J. Leuthold, J. Währen), **the highest point in the Bernese Oberland.**
Studerhorn 3638 (fa 5/8/1864 G. Studer, R. Lindt; K. & J. Blatter, P. Sulzer)
Oberaarhorn 3638 (fa 23/8/1860 L. Stephen; M. Anderegg)
 Finsteraarjoch 3330 links with the Finsteraarhorn (*above*).
Gr Schreckhorn 4078 (fa 14/8/1861 L. Stephen; C. & P. Michel, U. Kaufmann)
Gr Lauteraarhorn 4042 (fa 8/8/1842 E. Desor, M. Girard, A. Escher vd Linth; J. Leuthold, D. Brigger, J. Madutz, Fahner, M. Bannholzer)
Klein Lauteraarhorn 3737
 Lauteraar Sattel 3156 links with the Gr Schreckhorn (*above*).
Berglistock 3657 (fa 26/9/1864 C. Aeby; P. Egger, P. Inäbit)
Mittelhorn 3704 (fa 8/7/1845 S. T. Speer; J. Jaun, K. Abplanalp *et al*)
Wetterhorn 3701 (fa 31/8/1844 M. Bannholzer, J. Jaun)
 To the NE above Rosenlaui is a famous rock climbing area – the Engelhorner and the Wellhorner.
 Grimsel Pass 2165 (rd, Haslital to the Rhône valley)

**NORTHERN OUTLIERS**

Schilthorn 2974, c-wy to summit
Faulhorn 2681
Schwarzhorn 2928
Brienzer Rothorn 2350, r-ry to summit

**LEPONTINE AND ADULA ALPS**
 (Simplon Pass to Splügen Pass) (from W to E)

**Main Ridge**
 Simplon Pass 2005
Mte Leone 3553 (fa 1850 G. Studer *et al*)

gradient 1 in 4) to a summit of 2249 m, **the greatest level difference of any Swiss rack line**. The summit of the mountain is 100 m higher.

The **Lepontine and Adula Alps** are bounded in the north by the Rhône valley and the Vorder Rheintal, with ridges running down towards Italy in the south. After the **Gries Pass** (which dates back to medieval times) the south slopes are also Swiss (Canton of Ticino) until the Splügen Pass is reached. The road over the **Nufenen Pass** was only constructed after World War II.

The **Wyttenwasserstock** is one of the hydrographic centres of the Alpine chain, sending water to the Rhine (North Sea), the Rhône (Mediterranean) and the Po (Adriatic).

The **St Gotthard** (2090) is a simple pass across the watershed ridge, but the approach valleys on both sides are narrow and rugged. To the north are other passes linking with various parts of Switzerland – the Oberalp to the Rhine valley, the Furka to the Rhône valley, the Susten to the Bernese Oberland and the Klausen to Glarus. The descent on the south side was at one time by a tremendous series of hair-pin bends; however, the top 25 can now be avoided by using a new road.

The Pass was known in the 13th century; a mule track was opened in 1293; the summit chapel and hospice were first mentioned in 1331. It is named (and no one knows why) after a bishop who died in 1038. By the 14th century this was an important route from Switzerland to Italy.

The Schöllenen Gorge on the north approaches to the Pass was the big obstacle; it was passed for many years by a 60 m wooden terrace (frequently renewed) suspended by chains from the face of the flanking mountain above the torrent. A narrow tunnel was bored at this point in 1707. Lower down the famous Devil's Bridge spanned the river; the original fell in 1888, its successor has been by-passed by the modern road.

The Pass was the scene of fierce fighting early in the 19th century between the French and the Austro-Russian alliance under General Suvarov, a monument to whom now graces the Schöllenen Gorge.

The first through carriage-way was completed in the 1820s and improvements and refinements have since been repeatedly made. While undoubtedly an international route of considerable commercial and tourist importance, the effect on the environment is devastating – to call these monstrosities 'cathédrales de notre temps' is an arrogant mis-statement of their true place in our civilisation.

The successful completion of the Mont Cenis tunnel in 1870 encouraged the next venture – a line beneath the St Gotthard Pass to link Zurich with Milan. Work started on 13 September 1870 under Louis Favre, who had accepted a contract with onerous penalty clauses which soon led him into difficulties. In spite of new machinery the working conditions were terrible and the

The new St Gotthard road, Switzerland

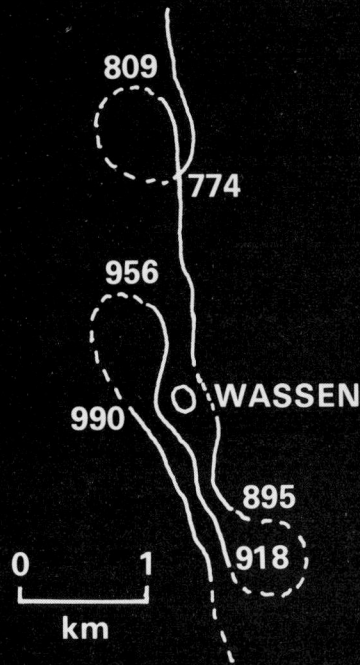

Fig. 2    Loop tunnels on the St Gotthard line

Devil's Bridge on the St Gotthard road, Switzerland

accident and death rates high. Favre, worn out and ruined, died in 1879 some eight months before the headings met (28 February 1880), with differences of only 5 cm in level and 40 cm in direction. The tunnel, 15 km long and double track, was finally opened to goods on 1 January 1882 and to passenger traffic six months later. It had cost £2·3 million and 310 lives.

The present-day traffic density amounts to 11 million tonnes of freight and 7·5 million passengers per annum and there are plans for a St Gotthard Base tunnel, at lower level and longer, from Amsteg to Giornico, or alternatively for the construction of relief tunnels further east.

A feature of the approach lines, here appearing **for the first time**, was **the use of spiral tunnels** to gain height on the way up from the valleys, while not exceeding the slope appropriate to adhesion working.

Further east there are three more important ways over the main chain. The **Lukmanier Pass** (1917) was first crossed as early as 700 and continued to be used in medieval times. Gradually it was more and more overshadowed by the St Gotthard route. The road was constructed in 1871–2. The **San Bernardino Pass** (2063) dates back at least to the Middle Ages. The summit of the road, first built in 1813–23, is now avoided by a 6·6 km tunnel at around 1600 m. The road over the **Splügen Pass** (2113) was constructed during 1818–23; until that time it had always been overshadowed

Helsenhorn 3272
  Albrun Pass 2409 (int path, Rhône
    valley to Valle Antigorio)
Ofenhorn 3235
  Blindenhorn 3374 (fa 5/9/1866 Sedley-
    Taylor; Tännler, Guntren)
  Gries Pass 2460 (int path, Rhône
    valley to Valle Antigorio)
On the int frontier are:
  Passo di San Giacomo 2313 (int path,
    Ticino to Valle Antigorio)
  Basodino 3273 (fa 3/9/1863 Zanini,
    Gaud, Padovani, Scuella; Josi)
Continuing the line of the watershed:
  Nufenen Pass 2478 (rd, Valais to
    Ticino)
Wyttenwasserstock
Pizzo Rotondo 3192
  St Gotthard Pass 2090 (rd, Göschenen-
    Reuss valley to Airolo-Ticino; ry in
    tunnel below)
  Lukmanier Pass (Passo de Luco-
    magno) 1917 (rd, Disentis-Vorder
    Rheintal to Olivone-Ticino)
Scopi 3199
Piz Medel 3210
Piz Terri 3149
Güferhorn 3393
Rheinwaldhorn 3402 (fa 1789 Placidus
  à Spescha)
Zapporthorn 3152
  San Bernardino Pass 2063 (rd, Hinter
    Rhein to Val Messoco)
Piz Tambe 3279
  Splügen Pass 2113 (int rd, Splügen-
    Hinter Rheintal to Chiavenna-Val
    San Giacomo)

**The Mountains of Ticino**
Campo Tencia 3072

**EASTERN BERNESE ALPS** (Grim-
  sel Pass to Reuss valley) (from S to
  N)
  Furka Pass 2431 (rd, Rhône valley to
    Reuss valley; ry below)
Galenstock 3583 (fa 18/8/1845 E.
  Desor, D. Dollfuss-Ausset, D. Doll-
  fuss; H. Wähun, H. Jaun, M. Bann-
  holzer, D. Brigger)
Gletschhorn 3305
Dammastock 3630 (fa 28/7/1864 A.
  Hoffman-Burkhardt; A. von Weis-
  senfluh, J. Fischer)
Fleckistock 3417 (fa 21/7/1864 A.
  Raillard, L. Fininger; A. Zgraggen,
  K. Blatter)
Hinter Tierberg 3447 (fa 1901 L.S.
  Powell *et al*)
Subsid pks 3444, 3418, 3300
Gwachtenhorn 3425 (fa 8/1861 R.W.
  Elliot-Forster, L. Hardy Dufour; J.
  von Weissenfluh)
Sustenhorn 3504 (fa 17/8/1841 G.
  Studer; J. & H. von Weissenfluh)
Salbitschijen 2981, on a spur N of
  Göschenertal is an important rock
  climbing mountain.
  Susten Pass 2224 (rd, Haslital to Reuss
    valley)
Titlis 3239 (fa 1739 by monks)

by the Septimer Pass. Of the three this is the sole international link.

The funicular railway in the **Ticino** between Piotta and Piora (Lake Ritom) is rated as being **the steepest in the world** with a gradient of 1 in 1·125.

The principal passes of the **Eastern Bernese Alps** are as follows: the **Furka Pass** road, which gives fine views of the Oberland mountains and the Rhône glacier (the rail tunnel below is 2 km long); the **Susten Pass** road, constructed after World War II, which has a 400 m tunnel at the summit.

The narrow gauge **Furka-Oberalp Railway, one of the most remarkable in Switzerland,** was one of the last to be built (1926). In the 110 km between Brigue (650) and Disentis (1130) it crosses the passes of Furka (2 km tunnel at 2160 m) and Oberalp (2044), climbing by mixed Abt System (*see* p. 47) rack and adhesion, with liberal use of spiral tunnels and ample provision of avalanche sheds. Parts of the line are closed in winter; one particular bridge in the Steffenbach Gorge is dismantled every year to prevent its destruction by avalanche. In between the passes, the line descends to Andermatt (1436), where it passes directly over the St Gotthard line 335 m below.

**Pilatus** (2122) rises above the western end of the Lake of Lucerne (Vierwaldstatter See). An insignificant lake close to the summit was long thought to have been the burial place of Pontius Pilate and ascents were forbidden for fear of supernatural retaliation; in 1387 six priests were punished for making an attempt. However, the top had certainly been reached by early in the 14th century.

There were two ascents in 1518 – one by the Duke of Wurtemburg, the other by a company of Swiss scholars, including Oswald Myconius whose pupil Conrad Gesner (1516–65) was to become one of the earliest of real mountaineers. Gesner, a botanist who became a professor at the University at Zurich, vowed 'to climb at least one mountain every year'; he reached the top of Pilatus in 1555, 'crawling up, clutching the turf'. It was not until 1585 that Pastor Johann Muller and some of his congregation finally dispelled the Pilate legend.

Soon after the completion of the Rigi Railway in 1871 it was decided to treat Pilatus likewise. The line had to climb 1629 m in a distance of 4 km, ie at an average gradient of 1 in 2½, with the steepest parts 1 in 2, making it **the steepest in the world**. The rack system used by Riggenbach on the Rigi (*see* p. 47) was not considered suitable for this high slope and the engineer, Locher, invented a different form of rack which is still widely used to this day. This consists of flat bars of steel with teeth on either side which are gripped by horizontal pinions on the engine; no flanges are needed on the running wheels (*see* p. 47). The coaches and the platforms are stepped to conform to the gradient. The 80 min journey of the original steam locomotives is now accomplished in 30 min by electric traction.

Gr Spannort 3199
Uri Rotstock 2928 (fa 1795 J.E. Muller)
Pilatus 2122 (fa before 1307), r-ry to summit
Rigi 1800, r-ry to summit
Stanserhorn 1898, c-ry to summit

**GLARNER ALPS** (Reuss valley to the bend of the R Rhine, N of the Vorder Rheintal) (from SW to NE)

**On or near main ridge**
Oberalp Pass 2044 (rd & ry, Reuss valley to Vorder Rheintal)
Bristenstock 3072
Piz Giuf 3096
Oberalpstock 3328
Dussistock 3256
Gr Windgälle 3188 (fa 31/8/1848 G. Hoffman; M. Tresch)
Gr Scheerhorn 3294
Claridenstock 3267 (fa 1863 E. Rambert; E. Streiff, G. Stussi)
Tödi 3620 (fa 1/9/1824 P. Curschelles, A. Bisqualm; Placidus à Spescha halted at Porta à Spescha and sent the others on)
Subsid pks 3601, 3434
Piz Urlaun 3371 (fa 1793 Placidus à Spescha)
Bifertenstock 3426 (fa 1863 A. Roth, G. Sand, A. Raillard; H. Elmer)
Salbsanft 3029
Kisten Pass 2638 (path, Glarus to Vorder Rheintal)
Hausstock 3158
Panixer Pass 2407 (path, Glarus to Vorder Rheintal)
Vorab 3028
Segnes Pass 2627 (path, Glarus to Vorder Rheintal)
Piz Sardona 3056
Ringelspitz 3247

**On parallel ridge to the N**
Klausen Pass 1948 (rd, Reuss valley to Linthal)
Glärnisch 2914

**ALBULA ALPS** (Splügen Pass to Flüela and Maloja Passes) (from W to E)
Splügen Pass 2113 (int rd, Hinter Rheintal to Val San Giacomo)
Piz Timun 3210
Pizzo Stella 3162
Piz Gallegione 3136
Piz Duan 3131
Maloja Pass 1815 (rd, Ober Engadin to Val Bregaglia)
Piz Platta 3392 (fa 7/11/1866 B.E. de Beurnonville, Baltzer, Gadient; S. Hartmann)
Pizzo Lunghino 2784
Piz Lagrev 3164
Julier Pass 2284 (rd, Oberalbstein to Ober Engadin)
Piz Picuogl 3336
Piz d'Err 3378
Piz d'Aela 3340
Tinzenhorn 3172

**Pilatus (Aerofilms)**

Pilatus is a highly accessible and important place of pilgrimage and famous ascenders include Queen Victoria in 1868.

The **Rigi** (1800) is a celebrated viewpoint close to Lucerne, popular for seeing the sun rise. There is a large summit hotel, as well as plenty of other tourist accommodation and facilities. The first rack railway in Switzerland was constructed here in 1871 by Niklaus Riggenbach; initially the line, which started at Vitznau in the Canton of Lucerne, had to terminate at Staffel since the summit was in the Canton of Schwyz, which refused permission. Soon afterwards Schwyz began their own line from Arthgoldau to Staffel, at the same time building the line from Staffel to the summit which they leased to Lucerne. Riggenbach's original steam locomotives, which had an inclined chassis (so that the boiler could be near vertical on gradients of 1 in 5), climbed the 1311m in 80min. One is preserved in the Swiss Transport Museum in Lucerne. Today, electric locomotives make the climb in 35min.
Many celebrated names among European travellers in the last century visited this summit, including Queen Victoria in 1868. The view embraces a circumference of 800km.

Above the **Segnes Pass** (2627) a natural hole 22m high, the Martinsloch, pierces the range. On the north side the village of Elm was the scene of a disastrous rock-fall in 1881, when 115 of the inhabitants were killed.

The face of **Glärnisch** (2914) above Klönthaler See is **one of the highest and steepest in the Alps**, 1980m at an angle of 45°.

The **Maloja Pass** (1815) crosses a branch ridge which joins the

Piz Nair 3030, reached from St Moritz
by two c-rys and a c-wy.
Piz Ot 3246
  Septimer Pass 2310 (path, W foot of
  Julier Pass to W foot of Maloja Pass)
  Albula Pass 2312 (rd, Albula valley
  to Ober Engadin; ry in tunnel below)
Piz Kesch 3418 (fa 1864 D.W. Fresh-
  field, H. Walker, R.M. Beachcroft
  et al)
  Scaletta Pass 2606 (path, Davos valley
  to Ober Engadin)
Piz Vadret 3229
Schwarzhorn 3147
  Flüela Pass 2383 (rd, Davos valley to
  Ober Engadin)

**BERNINA ALPS** (Lago di Como and
  Maloja Pass to Reschen Scheideck
  and Stelvio Passes)

**Frontier ridge to Bernina Pass** (from
  W to E)
Piz Badile 3307 (fa 27/7/1867 W.A.B.
  Coolidge; F. & H. Dévouassoud)
Piz Cengalo 3370 (fa 25/7/1865 D.W.
  Freshfield, C.C. Tucker; F. Dé-
  vouassoud)

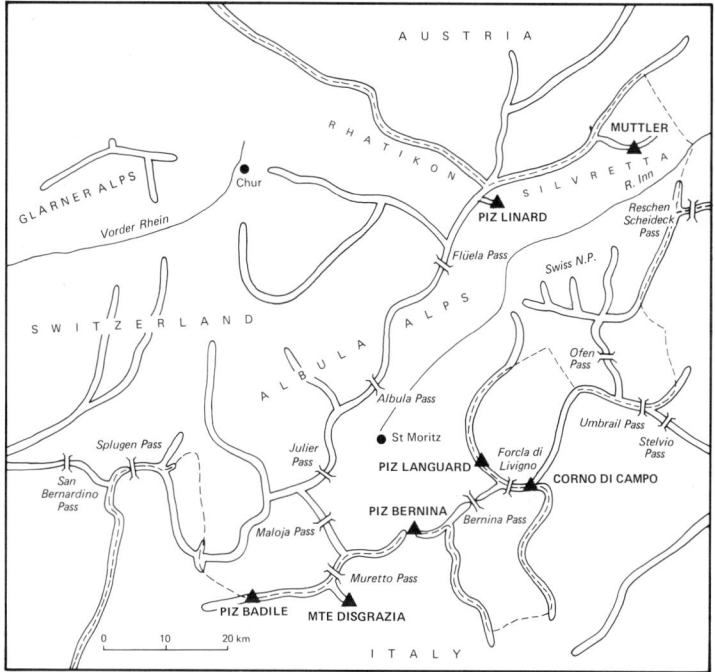

Eastern Switzerland

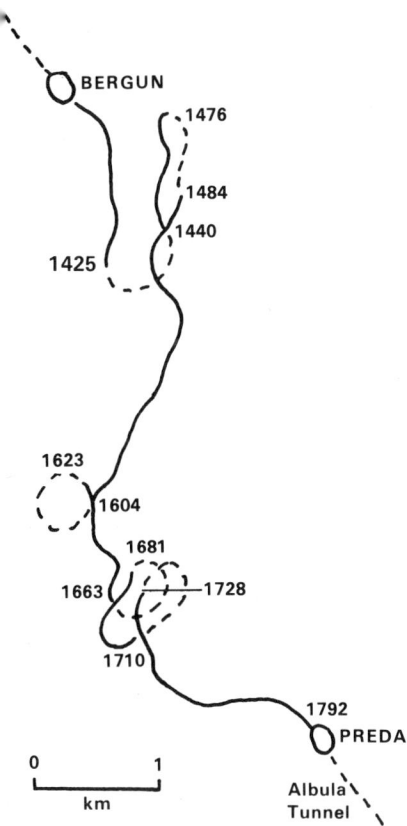

0                    1
    km

**Fig. 3   Loop tunnels in the northern
approach to the Albula Tunnel**

Albula and Bernina Alps. It has a long easy slope on the north-
east side and an abrupt drop with hairpin bends on the south-
west, leading eventually into Italy.

**Pizzo Lunghino** (2784) is another hydrographic centre peak,
sending water to the Po (Adriatic), the Rhine (North Sea) and
the Danube (Black Sea).

The first ascent of the **Piz d'Aela** (3340) made by a group of guides
from Pontresina, without clients, brought down on them the
wrath of the Establishment.

Of the passes of the **Albula Alps** the **Septimer Pass** (2310) is by
far the most important historically, having been in use since
Roman times. After the carriage roads were built over the
Splügen, Julier and Maloja Passes, it declined rapidly. The
**Albula Pass** road is an inferior alternative to the **Julier Pass**,
which now carries the principal road. The **Albula rail tunnel, the
highest of the principal Alpine tunnels** at 1823 m, was opened in
1903. It is 6 km long; the line is metre gauge. On the north side
the height is gained in a confined space by the ingenious use of
loop tunnels. The carriage road over the **Flüela Pass** (2383) was
built in 1866–7; on the far side of the Inn valley the **Ofen Pass**
(in use since the Middle Ages) continues in a direct line into Italy.

The **Parsenn Railway**, a funicular from Davos to Weissfluhjoch,

Pizzo del Ferro Cent 3289 (fa 1876 L. Held solo)
*From this point the Sciora chain of rock peaks runs N to:*
Ago di Sciora 3201 (fa 4/6/1893 A. von Rydzewsky; C. Klucker, M. Rey)
*Resuming the main ridge line:*
Ca di Castello 3392 (fa 31/7/1866 D.W. Freshfield, C.C. Tucker; F. Dévouassoud, A. Fluri)
Punta Rasica 3305 (fa 27/6/1892 A. von Rydzewsky; C. Klucker, M. Barbarin)
Terrone group 3349, 3333, 3233, 3290, all hard rock peaks
Mte Sissone 3335 (fa 10/8/1864 D.W. Freshfield, J.D. Walker, R.M. Beachcroft; F. Dévouassoud)
Ca di Rosso 3366 (fa 30/7/1867 W.A.B. Coolidge; F. & H. Dévouassoud)
Passo di Muretto 2562 (int path)
Piz Fora 3363 (fa 24/8/1875 E. Burkhardt; B. Cadonau)
Piz Tremoggia 3441 (fa 1859 J.J. Weilenmann solo)
Piz Glüschaint 3593 (fa 7/8/1875 E. Burkhardt; H. Grass)
Piz Roseg 3937 (fa 28/6/1865 A.W. Moore, H. Walker; J. Anderegg)
Piz Scerscen 3967 (fa 13/9/1877 P. Güssfeldt; H. Grass, C. Capat)
Piz Bernina 4049 (fa 13/9/1850 J. Coaz; J. & L. Ragut Tscharner), **the highest point of the Bernina Alps.**
Piz Zupo 4008 (fa 9/7/1863 L. Enderlin, P. Seradi & a chamois hunter)
Piz Palü 3905 (fa – probably K.E. Digby; P. Jenny and porter)
Piz Cambrena 3603 (fa 5/8/1863 V. Cruzemann; G.M. Colani)
Bernina Pass 2323 (rd, Postchiavo to Ober Engadin; ry in the open)

**Mountains S of the frontier ridge** (ie in Italy)
Mte Disgrazia 3678 (fa 24/8/1862 L. Stephen, E.S. Kennedy; M. Anderegg, T. Cox), an outstanding isolated summit.
Sasso d'Entova 3323
Sasso Moro 3102

**Mountains N of frontier ridge** (ie in Switzerland)
Il Chapütschin 3391 (fa 23/7/1850 J. Coaz & party)
Piz Corvatsch 3451 (fa 1850 J. Coaz & party)
Piz Morteratsch 3751 (fa 11/9/1858 C.G. Brügger, P. Gensler; K. Emmermann, A. Klaingutti)
Piz Tschierva 3560 (fa 18/8/1850 J. Coaz & party)

**Frontier ridge W of Val Postchiavo** (from N to S)
Piz di Verona 3453 (fa 6/7/1865 D.W. Freshfield, F.F. Tuckett, E.N. Buxton; F. Dévouassoud, P. Michel, J.B. Walther)
Piz Scalino 3323 (fa 22/6/1866 F.F. Tuckett *et al*)
Ca Painale 3248

was the first mountain line specifically constructed for skiers. It has now been extended by cableway to the Weissfluh.

The **Bernina Alps** are the most easterly of the high mountain groups having peaks over 4000 m.

The range from Piz Badile to the Passo di Muretto is known as the **Bregaglia**, a very important rock climbing area.

The **Bernina Pass** road, sometimes closed in winter, is said to give the finest mountain and glacier views of any road pass in the Alps. The railway (metre gauge) summit, in the open at 2257 m, is **the highest through-line in Europe.**

The **Umbrail Pass** road is **the highest in Switzerland**. In the Middle Ages, and for many years afterwards, it was used in preference to the Stelvio Pass, until the carriage-way was built on the latter (1820s). The Umbrail road only dates from 1900–1.

The road over the **Stelvio Pass** is **the third highest in Europe**. The carriage-way was built in 1820–5, magnificently engineered with 40 to 50 hairpin bends on each side. It now carries very heavy commercial traffic through magnificent mountain scenery, one of the finest view roads in Europe.

**The Swiss National Park** is a closely preserved wilderness, with access only on foot.

The summit hospice on the **Reschen Scheideck Pass** was founded in 1140. The village of Reschen (Resia) high up on the Pass is inhabited all the year round.

**Säntis** is served by a cableway from Schwägalp; to the north on the Kronberg is **the longest cableway span between pylons in Switzerland** of 2220 m.

The **Arlberg Pass** was crossed as early as 945. A mule path was available from early in the 14th century, with a hospice opened in 1385. The carriage-way was built sporadically between 1785 and 1824. A 14-km road tunnel, **the longest in the world**, was opened here in December 1978.

The railway tunnel, 10·25 km long and completed in 1884, is at 1315 m **the highest main-line rail summit in Europe**.

The school of rock climbing which grew up amongst the limestone mountains of the Austrian and south German Alps has played a significant part in the history of mountaineering. The high standards of achievement on rock led the world, possibly in the latter years of the last century, certainly in the 1920s and 1930s. Climbing on these great crags under winter conditions produced mountaineers who were able to make significant advances in the higher mountains of West Europe.

**Frontier ridge – from above Bernina Pass to Stelvio Pass** (from S to N)
Il Paradisino 3305
 Forcla di Livigno (int rd on Italian side, path on Swiss side, Postchiavo to Unter Engadin, but crossing a corner of Italy.
Piz Lagalb 2963 (3 km WSW) has a c-wy from Curtinatsch to 2902 m.
Mt Gotschen 3108 (fa Placidus a Spèchsa)
Piz Languard 3262, just N of the ridge.
Piz Quattervals 3165
 A 3·5 km road tunnel, open all the year, links Zernez with Livigno.
Piz Murtaröl 3180
 Umbrail Pass 2501 (int rd, Val Venosta to Valtellina, but crossing a corner of Switzerland)
 Stelvio Pass 2756 (rd, Val Venosta to Valtellina)

**S & E of the above frontier ridge section**
 (ie in Italy)
Cima Lago 3299
Cima di Piazzi 3439

**Ofen Pass to Reschen Scheideck Pass**
 [Swiss National Park]
Ofen Pass 2149 (rd, Unter Engadin to Val Mustair and Val Venosta)
Piz Sesvenna 3205
Piz Pisoc 3174
Piz Tavrü 3168
Piz Plavnadadaint 3166
Piz Nuna 3124
Piz Lischanna 3105
 Reschen Scheideck Pass (Passo di Resia) 1508 (int rd, Inn valley to Val Venosta)

**SILVRETTA AND RHÄTIKON ALPS** (the Liechtenstein/Swiss and Austria/Swiss frontiers from St Gallen to Unter Engadin)
Säntis 2504, c-wy from Schwägalp to 2483 m.
Churfursten 2385
 Here is a low gap traversed by R Rhine, rd and ry. The ridge line continues along the frontier of Liechtenstein, **the highest point** of which is Grauspitze 2599.
Schesaplana 2964 (fa 1730)
Rotbuelspitze 2955
Drusenfluh 2828
Madrisahorn 2826
Sulzfluh 2817
Valüla 2815
Gr Litzner 3109 (fa 27/7/1866 J. Jacot; C. Jahn, A. Schlegel)
Silvrettahorn 3244
Piz Buin 3312 (fa 14/7/1865 J. Specht, J. Weilenmann; J. Pfitscher, F. Pöll)
Plattenhorn 3221
Piz Linard 3411 (fa 1835 O. Heer; J. Madutz)
Fluchthorn 3399 (fa 1861 J. Weilenmann; F. Pöll)
Muttler 3294 (fa 29/7/1859 J. Weilenmann)

Notable among the great crags and faces are the Wettersteinwand (Wetterstein Gebirge), the La[l]idererwand (Karwendel Gebirge), Fleischbank and Predigstuhl (Kaisergebirge), the Watzmann (Steinernes Meer), the Dachstein and the Eisenerzer Alpen. There are many others in almost every mountain group.

Limestone implies caverns, and caverns at high level become filled to a large extent with ice. Among the notable ice caves of Austria are the **Dachstein** and **Eisriesenwelt**, each open to the public and approached by cableway.

The practice crag for Viennese climbers is the **Peilstein** between Raisenmarkt and Neuhaus at the end of the Weiner Wald.

The road over the **Timmelsjoch** (2509) was only completed in 1970. There are hotels on the Austrian side at 2195 m.

The **Brenner Pass** (1370) is **the lowest pass across the main range of the Alps**. The recorded history started in 15 BC. It was used by the Romans in their conquests and then by barbarians returning the compliment. It was frequently fought over in subsequent centuries as the Tyrol periodically changed hands. A track was constructed in 1314–7 and gradually improved for carts and carriages. The carriage-way dates from 1772, but the post-war period has seen this upgraded (?) to a toll motorway. Usually open all the year round, this now carries very heavy traffic indeed. Forty-four bridges and 10 km of viaducts on the Austrian side have been described as 'a permanent civil engineering exhibition'. In this technological bonanza the major bridges are of three different structural types and include the highest in Europe,

**Churfursten, peaks above the clouds (Barnaby's)**

**AUSTRIAN AND S GERMAN ALPS**
(N of Inn and Pinzgau valleys) (from
W to E)

**Allgauer Alps**
Hohes Licht 2687
Madelegabel 2646
Hochvogel 2594
Nebelhorn 2224, c-wy to summit

**Lechtaler Alps**
Valluga 2811
Freispitze 2887
Wetterspitze 2898
Parseierspitze 3038 (fa 23/8/1869 J.
  Specht; P. Seiss)
Schlenkerspitze 2821
  Arlberg Pass (rd, Inntal to Rheintal
  in long tunnel; also ry tunnel)
  Bielerhöhe Pass 2038 (rd, parallel to
  and S of the Arlberg)

**Verwall group**
Kuchenspitze 3170 (fa 1877 A. Mad-
  lener; J. Volland)
Riffler 3160 (fa 1864)

**Samnaun group** (an extension of the
  Silvretta Alps above)
Hexenkopf 3038

**Wetterstein Gebirge**
Zugspitze 2963 (fa 27/8/1820 K. Naus,
  Maier; G. Deuschl), **the highest point
  in Germany** and the only German
  peak with glaciers. C-ry to summit.
Hochwanner 2746
Hochblassen 2697
Dreitorspitze 2673 (fa 1871 Barth)
Schusselkarspitze 2538 (fa 1894 O.
  Schuster; H. Moser)

**Karwendel Gebirge** (a series of ranges
  running E to W – starting with the
  most southerly)
Gr Solstein 2542
Hafelekarspitze 2334, c-wy to summit
Kleine Solstein 2641
Praxmarerkarspitze 2642
Bettelwurfspitze 2725 (fa 1870 Barth)
Odkarspitze 2743
Birkkarspitze 2749 (fa 1870 Barth)
Kaltwasserkarspitze 2733 (fa 1870
  Barth)
Laliderspitze 2583
Grubenkarspitze 2661
Lamsenspitze 2508 (fa 1843)
Eastkarwendelspitze 2537

**Kitzbühler Alpen**
Kreuzjoch 2559
Gr Rottenstein 2362
Hahnenkamm 1655, c-wy from Kitz-
  bühel.

**Kaisergebirge** (from W to E)
Tuxeck 2226
Treffauer 2304
Ellmauer Haltspitze 2344 (fa 1826
  Hauzensteffel solo)
Kleine Haltspitze 2119

with piers of 146m. Another example of 'cathédrales de notre temps'.

The rail route was opened in 1867, **the second crossing of the chain.**

The **Hohe Tauern** is the home of the legendary ice dwarfs.

The **Grossglocknerstrasse** is a modern Alpine highway opened in 1935. It is one of the finest view roads in the Alps, notably at special viewpoints at the ends of two spur roads. However, a staggering provision of car-parks and other tourist facilities have turned it into an environmental eyesore. The summer tourist traffic is heavy, but it is usually closed in some winter months. There is a 300m summit tunnel and a toll.

The nearby Pasterze glacier is the third longest in the Alps.

The **Radstädter Tauern** (1738) dates from Roman times. Further south the route also crosses the Katschberghöhe (1641).

Laliderspitze, Karwendel, Tyrol (By courtesy of the Austrian National Tourist Office)

Hintere Karlspitze 2283
Vordere Karlspitze 2261
Totenkirchl 2193 (fa 16/6/1881 G. Merzbacher; M. Soyer)
Fleischbank 2187 (fa 11/7/1886 C. Schöllhorn; T. Widauer)
Predigtstuhl 2115 (fa 30/6/1895 P. Scheiner; J. Tovonaro)
Ackerlspitze 2331 (fa 1826 Thurwieser)
Maukspitze 2227

**Loferersteinberge** – 2634

**Steinernes Meer** (Berchtesgadener Alpen)
Watzmann 2713 (fa 1801 V. Stanig)
Hochkonig 2938 (fa 1826 P. Thurweiser)
Schönfeldspitze 2650
Höhe Göll 2519

**Dachstein Gebirge**
Höhe Dachstein 2996 (fa 18/7/1834 P. Thurweiser; A. & P. Gappmaier), c-wy to summit.
Torstein 2947 (fa 1819 J. Buchsteiner, G. Kalkschmied)
Gr Bischofsmütze 2455 (fa 1879)
Manndkogel 2277 (fa 1892 L. Purtscheller)

**Totes Gebirge**
Hoh Priel 2514 (fa 1817)
Schermberg 2451, has a N face of 1400m, one of the highest in Europe.

**Eisenerzer Alpen** (Gesauseberge, Ennstaler Alpen)
Hochtor 2372 (fa 1871)
Gr Ödstein 2355 (fa 1877)
Festkogel 2272
Sparafeld (Reichenstein) 2245
Gr Pyhrgas 2244
Gr Buchstein 2224

**Hochschwab**
Hochschwab 2277

**AUSTRIAN ALPS** (S of Inn and Pinzgau valleys) (from W to E)

**Otztaler Alpen and Alpi Venosti** (from N to S)
Höhe Geige 3395 (fa 1853 J. Ganahl)
Watzespitze 3533 (fa 1869 A. Ennemoser solo)
Wildspitze 3772 (fa 1848 L. Klotz)
Subsid pk 3770
Hinterer Brochkogel 3635 (fa 1858 A. Wachtler; L. Klotz)
Ramolkogl 3551 (fa 1862 J.J. Weilenmann)
Weisskogl (Weisskugel) 3736 (fa 1861 J.A. Specht; L. Klotz)
Fineilspitze 3516 (fa 1865 F. Senn; C. Granbichler, J. Gstrein)
Similaun 3607 (fa 1834 T. Kaserer; J. Raffeiner)
Hintere Schwarze 3628 (fa 1867 E. Pfeiffer; B. Klotz, J. Scheiber)
Hochwilde (Höhe Wilde) 3480 (fa 1858 J. Ganahl with locals)

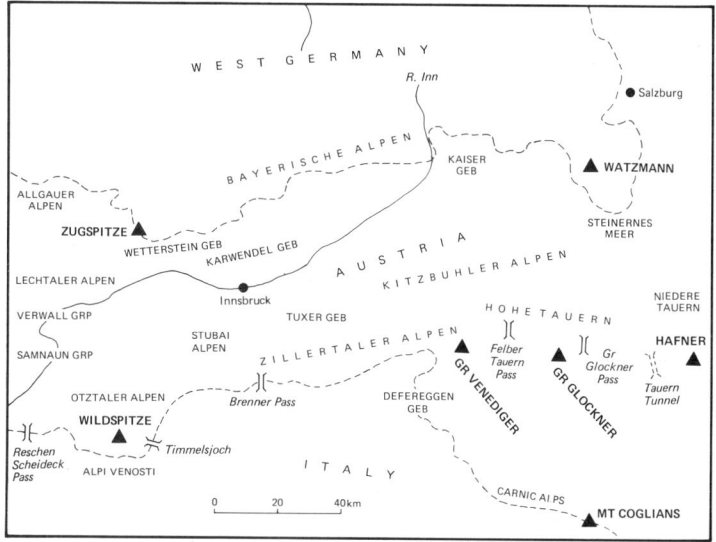

**Austria**

The railway across the **Semmerling Pass** (980) was **the first to cross the main chain of the Alps** in 1854.

The **Carnic Alps** are separated from the Defereggen Gebirge by the River Drau, the gap also providing passage for road and railway.

Roman inscriptions have been found near the summit of the **Plocken Pass** (1363).

Much of the **Ortler group** lies within the PN dello Stelvio. This area was the scene of fierce fighting in World War I when the summit of the Ortler was occupied by Austrians and the summit

**Heiligenblut and the Gross Glockner (C D Milner)**

Punta Saldura 3433
Gaislachkogel 3044, c-wy from Sölden.
  Timmelsjoch (Passo di Rombo) 2509
  (int rd, Otztal to Val Venosta)

**Stubai Alpen**
Zuckerhütl 3507 (fa 1863 J. A. Specht;
  A. Tanzer)
Schrankogl 3500 (fa 1840 J. B. Schöpf)
Sonklarspitze 3475
Ruderhofspitze 3471
Wilder Pfaff 3470
Wilde Freiger 3426
E Seespitzen 3419
Pflerscher Tribilaun 3096, fine rock
  climbing
Kalkkogel 2750, the Stubai dolomites.
  Brenner Pass 1370 (int rd, Inn valley
  to Adige valley; ry in the open)

**Tuxer Alpen**
Olperer 3480 (fa 1867 P. Grohmann;
  Samer, Gainer)
Patscherkofel 2247, c-wy to summit
  from near Innsbruck.

**Zillertaler Alpen**
Gr Pilastro (Hochfeiler) 3510
Hochfernerspitze 3463 (fa 1878 R.
  Seyerlein *et al*)
Thurnerkamp 3422 (fa 1872 W. Hud-
  son, C. Taylor, R. Pendlebury; G.
  Spechtenhauser, Jaseler)
Schrammacher 3416 (fa 19/8/1847 P.
  Thurwieser; S. Jörgl, G. Jaggl)
Gr Löfflerspitze 3382 (fa 1843)
Fusstein 3337 (fa 5/9/1880 R. Starr;
  Lecher, J. Eberl)
Zsigmondyspitze (Feldkop) 3080 (fa
  25/7/1879 E. & O. Zsigmondy)
Grundschartner 3066 (fa 14/7/1877 F.
  Löwl; Bliem)

**Defereggen Gebirge**
Hochgall (Collalto) 3435 (fa 1854 H.
  von Acken & survey party)

**Hohe Tauern** (main watershed ridge)
  (from W to E)
Dreiherrenspitze 3505 (fa 2/11/1866
  Ploner père & fils, M. Dorer)
Gr Venediger 3660 (fa 3/9/1841 J.
  Schwol, F. Scharler)
  Felbertauern Road Tunnel below, 5·2
  km long at 1600m.
Muntanitz 3232
Gr Glockner 3798 (fa 28/7/1800 Hor-
  asch, curé of Dollach), **the highest
  point in Austria.**
Glockerin 3425 (fa 18/9/1869 K. Hoff-
  mann, J. Studl; T. Groder, J. Schëll)
Hochtenn 3371 (fa 1869 Cardinal
  Count Schwarzenberg)
Gr Wiesbachhorn 3564 (fa 18th cen-
  tury by peasants), this lies N of the
  main ridge.
Gr Glockner Pass 2503 (rd, Pinzgau
  valley to Drau valley – the Gross-
  glocknerstrasse)
Hohe Sonnblick 3106
Stubner Kogel 2246, c-wy from Bad-
  gastein.

of Cevedale by Italians. Engagements between specially trained troops took place at all altitudes and casualties were high on both sides.

Climbers from Milan practise on the crags and pinnacles of the **Grigna** (2148) above Lecco.

The **Dolomites** present the most striking rock peaks of the Alps, their huge rock walls and pinnacles providing climbing of the highest standard. The rock, a limestone with magnesium carbonate, was named for the Marquis de Dolomieu (1750–1801), a French geologist. The Brenta Dolomites are south-west of Bolzano; the main area of some $65 \times 80$ km is east of Bolzano. Over the centuries the ownership of these peaks has passed back and forth between Austria and Italy, so that many have alternative names in either tongue.

The **Pordoi Pass** (2239) is one side of a square of passes around the Sella group, all with fine mountain views. The Campolungo Pass (1875) links the east side of the Pordoi with Val Badia, the Passo Gardena (2121) links Val Badia with Val Gardena, the Passo di Sella (2214) links Val Gardena with the west side of the Pordoi, the valley of Avisio.

**Monte Cristallo** (3216), **Monte Piano** (2324) and **Sorapis** (3205) were the scene of prolonged fighting in World War I. Many relics still remain on the ground, including a chapel and a summit cross on Monte Piano.

**The Dolomites of Lienz, Austria (C D Milner)**

Tauern Railway Tunnel below, 8·5
km long at 1183 m.
Ankogel 3251 (fa 18th century)
Hochalmspitze 3355 (fa 1855 K. von
Malta & party)
Hafnereck 3061 (fa 1825 Lt Gorizutti
& survey party), the most easterly
snow peak of the Alps.

**Hohe Tauern** (S of watershed ridge)
Rotor Kulm 3296
Petzeck 3283
Reisseck 2959
Polinck 2780
Hochkreuz 2704
Lienzer Dolomites – Sandspitze 2863 –
face the Carnic Alps across the
Lesachtal.

**Niedere Tauern**
Radstädter Tauern 1738 (rd, Ober-
ennstal to Drau valley)
Hochgolling 2863 (fa 1791)
Hochwildstelle 2747 (fa 1801)
Sölker Pass 1790 (rd)
Greimberg 2474
Gr Bosenstein (Pölsenstein) 2449
Hochreichart 2417
Zinken 2398
Schneeberg 2075, r-ry to 1795 m from
Puchberg Semmerling Pass 980

**CARNIC ALPS** (Austria/Italy frontier)
(from W to E)
Mte Cavallino (Pfannspitze) 2686
Mte Peralba (Hochweissenstein) 2693
(fa 1854 Schönhuber & survey party)
Mte Coglians 2781 (fa 1865 P. Groh-
mann)
Ca dei Preti 2708
Mte Duranno 2668
Plocken Pass 1363 (int rd, Ober Gail-
tal to Tagliamento valley)

## THE ALPS OF NORTH ITALY

**Ortler group** (Stelvio Pass – N, Val
Venosta – E, Tonale Pass – S, Val
Tellina – W)
Stelvio Pass 2756 (rd, Val Venosta to
Val Tellina)
Ortler (Ca Ortles) 3899 (fa 27/9/1804
M. Gebhard; J. Pichler, J. Leitner, J.
Klausner)
Mte Zebru 3740 (fa 1866 J. von Payer;
J. Pinggera)
Gran Zebru (Königsspitze) 3859 (fa
24/8/1854 S. Steinberger solo)
Ca Vertana (Vertainspitze) 3544 (fa
1865 J. von Payer; J. Pinggera)
Cevedale 3778 (fa 7/9/1865 J. von
Payer; J. Pinggera)
Subsid pks 3764, 3687
Gioveretto 3438 (fa 1868 J. von Payer;
J. Pinggera)
Mte Vioz 3644 (fa 1867 J. von Payer;
J. Pinggera)
Pta San Matteo 3684 (fa 1865 F.F.
Tuckett, D.W. Freshfield; 2 guides)
Pta Tresero 3602 (fa 1865 as Pta San
Matteo)
Tonale Pass 1883 (rd, Adige valley to
Val Tellina)

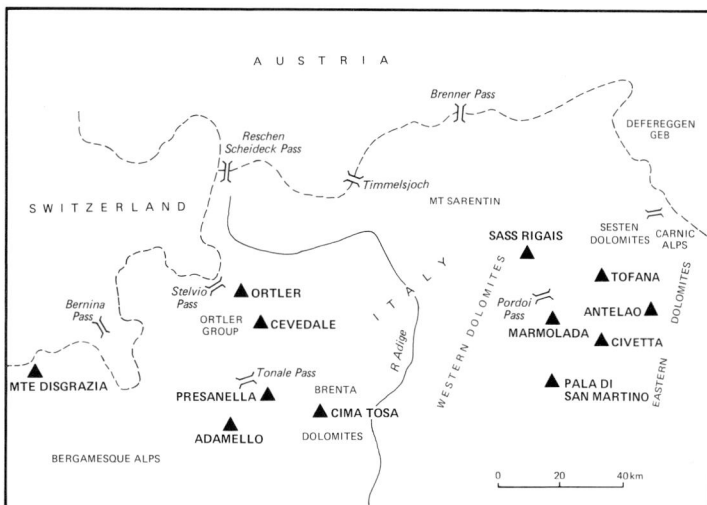

**Presanella-Adamello group**
Presanella 3556 (fa 17/9/1864 D.W.
Freshfield, M. Beachcroft, J. Walker;
F. Dévouassoud & porter)
Mte Madrone 3283
Adamello 3554 (fa 15/9/1864 J. von
Payer; Caturani)
Care Alto 3462
Mte Fumo 3418

**Bergamesque Alps** (Alpi Orobie)
Pizzo di Coca 3052
Pizzo Redorta 3037

## THE DOLOMITES

**Brenta Dolomites**
Ca di Brenta 3155 (fa 1871 F.F.
Tuckett, D.W. Freshfield; F. Dé-
vouassoud)
Torre di Brenta 3014
Campanile Basso (Guglia di Brenta)
2877 (fa 18/8/1899 O. Ampferer, K.
Berger)
Crozzon di Brenta 3135 (fa 8/8/1884 K.
Schulz; N. Nicolussi)
Cima Tosa 3176 (fa 1865 J. Ball, R.
Forster)

**Western Dolomite Alps** (W of Val
Badia and Val Cordevole) (from N
to S)
Saas Rigais 3027
Sassolungo (Langkofel) 3181 (fa 1869
1869 P. Grohmann; F. Innerkofler,
Sacher)
Fünffingerspitze (Pta delle Cinque
Dita) 2996 (fa 8/8/1890 J. Santner;
R. Schmitt)
Grohmannspitze (Sasso Levante) 3126
(fa 1880 M. Innerkofler solo)
Sella (Piz Boè) 3151 (fa 1864 P. Groh-
mann et al)

Vajolet Towers – Stabeler 2805 (fa
1892), Winkler 2800 (fa 1887) and
Delago 2790 (fa 1895), named for
the first ascenders.
Rosengartenspitze (Catinaccio) 2981
(fa 31/8/1874 C. Tucker, T. Carson;
F. Dévouassoud)
Passo di Pordoi 2239 (rd)
Marmolada (Marmolata) 3342 (fa 28/
9/1864 P. Grohmann; A. & F.
Dimai), the only glaciated peak in
the area and **the highest point of the
Dolomites.**
Ca di Vezzana 3191
Cimone della Pala 3185 (fa 3/6/1870 E.
Whitwell; C. Lauener, S. Siorpaës),
the 'Matterhorn' of the Dolomites.
Pala di San Martino 2987 (fa 1878
Meurer, Pallavicini; A. Dimai, S.
Siorpaës)
Ca di Canali 2846
Sass Maor 2816 (fa 4/9/1875 H. Beach-
croft, C. Tucker; B. Della Santa, F.
Dévouassoud)
Ca della Madonna 2741 (fa 2/8/1886
G. Winkler, A. Zott)

**Eastern Dolomite Alps** (listed clock-
wise round the Val d'Ampezzo)
Civetta 3218 (fa pre1867 di Silvestri
solo)
Mte Pelmo 3168 (fa 19/9/1857 J. Ball
with a local hunter)
Croda da Lago 2716 (fa 1878 F. Silber-
stein, P. Föschels; A. & P. Dimai)
Cinque Torri (Torre Grande) 2366
Passo di Falzarego 2107 (rd)
Tofana 3243
Subsid pks 3237, 3225
Croda Rosa (Hohe Gaisl) 3139 (fa 1870
E. R. Whitwell; C. Lauener, S. Sior-
paës)

**Italian Alps**

**Dolomite mountains – Tre Cime di Lavaredo (C D Milner)**

Mte Cristallo (Cristallospitze) 3216 (fa
14/9/1865 P. Grohmann; A. Dimai,
S. Siorpaës)
*Extending E from the last:*
Mte Piano 2324
Tre Cime di Lavaredo (Drei Zinnen) –
Ca Grande 2999 (fa 21/8/1869 P.
Grohmann; F. Innerkofler)
Subsid pks 2972, 2856
Zwölferkofel (Croda di Toni) 3094
*Resuming the circuit of the Val
d'Ampezzo:*
Passo Tre Croci 1809
Sorapis 3205 (fa 1864 P. Grohmann;
A. Dimai, F. Lacedelli)
Marmarole 2932
Antelao 3263 (fa 18/9/1863 P. Groh-
mann; F. & A. Lacedelli)

**Dolomite village (R D Treble)**

# Africa

Groot Spitzkopf, Namibia (F K Elliott)

## MOROCCO

**Er Rif**
J Tidighine (Tidiguin) 2496
J Taghzout 2459

**Moyen Atlas**
Caberral – Adrar Bou Naceur 3343
J Moussa ou Salah 3190
Bou Iblane 3103

**Haut Atlas**
Toubkal 4165 (fa 12/6/1923 Marquis de Segonzac, V. Berger, H. Dolbeau), **the highest point in Morocco.**

**Village below the Atlas Mountains, Morocco (Moroccan Tourist Organisation)**

**The Atlas Mountains** of north-west Africa comprise a number of ranges, the most impressive of which is the Haut Atlas of Morocco **(highest point – Toubkal** (4165)) running for 650km from the Atlantic sea board by Agadir to the Algerian frontier. Here there are snow-capped peaks and valuable mineral resources. Parallel to the north is the Moyen Atlas **(highest point – Adrar Bou Naceur** (3343)). To the south-west rises the Anti Atlas, 250km long with a volcanic landscape of plateaux, craters and weird peaks. Close to the Mediterranean sea board in the west are the lower mountains of Er Rif. Further east in Algeria there are more coastal ranges, while inland the Atlas Saharie extends the Haut Atlas on as far as Tunis.

It is a popular area for mountaineering and ski-ing, particularly amongst the French.

**North Africa**

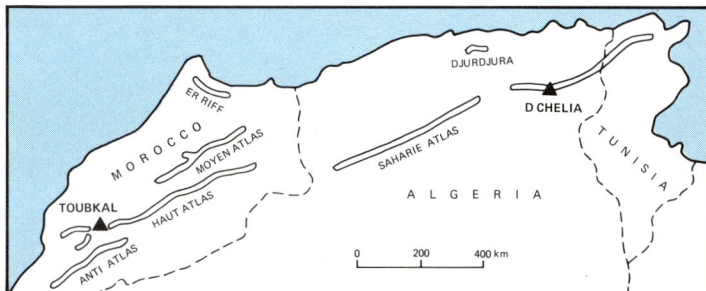

Timesguida n'Ouanoukrim 4089
Ras n'Ouanoukrim 4083
J Ighil (M'Goun) 4071
Afella n'Ouanoukrim 4040
Akiud Bu Imrhas 4030
Biiguinnoussene 4002
Iferouane 4001
Tazarharht 3980
Aksoual 3910
Anrhemer 3893
Iguenouane 3875
Ouanoums 3870
Aguelzim 3860
Azrou n'Tamadout 3860
Tadat 3837
J Tignousti 3825
J Rhat 3789
J Ouaougoulzat 3770
Tadaft 3764
Ayachi 3751 (fa 1900 M de Segonzac)
Bu-Uzzal 3740

**Anti Atlas**
Adrar-n-Aklim 2531
J Lekst 2350
J Siroua 3304, **highest point of a vol-
canic ra which links the Anti Atlas
with the Haut Atlas.**

**ALGERIA**

**N coast ras**
Dj Chélia 2326
Djurdjura – Ras Timedouin 2305 (fa
1886 D.W. Freshfield)

**Atlas Saharie**
The Ksour, Dj Amour and Ouled-Nail
Mtns rise 1800m above the desert to
the S. Highest point – Dj Aissa 2236.

**Ahaggar (Hoggar)**
Mt Tahat 2918 (fa, recorded, 1931 Dr
Wyss Dunant), **the highest point in
Algeria.**
Tehoulag Nord 2800
Ilamane 2758 (fa 1935 Housser, Bos-
sard)
Assekrèm 2728
Tamahagueni 2680
Saouinan 2650
Amdjer 2567
Tafedge 2525
Taridalt 2500
Aûkenet 2500 (fa 1953 Martin, Pierre,
Syda)
Tahararete 2462
Dj Tellerteba (Insakare) 2455
In Acoulmou 2369
Garet el Djenoun 2327 (fa 1953 Coche,
Frison Roche)
Mt Serkout 2306
Haggarhene 2170
Aokasset 2159 (fa 1938 R. Jaquet, J.
Rouget)
Mt Afao 2158
Escarnaied 2137
Akar Aka 2132
Adaouda 2100 (fa 1938 A. Jacquet)
Elfadoul 2065
Tasabat 1933

**The volcanic mountains of north-west Africa** fall into two distinct and widely-spaced groups. Both are in the heart of the Sahara Desert – the Ahaggar of southern Algeria, approximately midway between the Gulf of Guinea and the Mediterranean, and further east Tibesti on the Libya/Chad frontier, 1200km south of the Mediterranean coastline. Both exhibit fantastic landscapes of wind-eroded volcanic features – lava fields, plugs, rock spires, craters and so on, reminiscent of similar scenery elsewhere, such as in the deserts of south-western USA.

The Ahaggar mountain massif is some 250km across and rises to over 2900m (**highest point – Tahat** (2918)) above a plateau with a general level of 2000m. It was almost unknown before the beginning of the present century. Much of the mountaineering exploration has been carried out by the French, for whom this was until recent times colonial territory. Iharen (1732) and Ilamane (2758) are noteworthy isolated rock peaks, specially difficult of access. The peak of Garet el Djenoun (2327) (Mountain of Goblins) is a holy mountain, which features in many legends and is feared and avoided by the local Touaregs. The first recorded ascent took place in 1935 but it had probably been climbed long before that since neolithic pottery has been discovered high on the slopes.

The Tibesti massif, which occupies 100000km$^2$, was first sighted by Europeans in 1869 (G. Nachtigal's expedition). It was thoroughly explored around the turn of the century by J. Tilho. **The highest point (Emi Koussi** (3415)) lies on the rim of an ancient crater 20km wide and 1200m deep. Close to Pic Tousside is a crater 5km across and 750m deep filled with white sodium carbonate (*see also* Oldoinyo Lengai in the Great Rift Valley – p. 125). Since these mountains are still so very remote, little mountaineering has been done so far.

Visits by foreign mountaineers to the troubled Peninsula of **Sinai** are likely to be severely limited. There are fine rocky peaks with obvious climbing prospects and indeed climbing holidays here used to be advertised a few years back.

The history goes back to Exodus and, in the last century-and-a-half, large numbers of travellers have attempted to fit the local geography into its woolly descriptions. However, it is tradition rather than history that has placed **Moses on Gebel Musa** (Mount Sinai) and **credits him with a first ascent**.

**The highest point of the area is Gebel Katherina** (2637). Close by at a height of 1500m is St Katherine's Monastery which was built in the 6th century. Thenceforward very large numbers of pilgrims came to visit the various holy relics hereabouts – the Holy Mount, the site of the 'burning bush', the summit of Gebel Katherina (on which the saint's bones were miraculously found) and the summit of Gebel Musa (which is reached by a pathway of 3000 steps).

Among the foothills of **Gebel Faraid** (Pentadactylus) alongside the Red Sea coast of Egypt is a natural flying buttress, looking

Assekrar 1902 (fa 1938 Dr Wyss
   Dunant)
Issekrar 1896 (fa 1951 Martin, Pierre,
   Syda)
Aharen 1852
Immerous 1820
Iharen 1732 (fa 1953 Coche, Frison
   Roche)
Adriane 1709

## NIGER

**Aïr (or Azbine)**
Mt Gréboun 2000

## TUNISIA

Dj Chambi 1544

## LIBYA

The highest point (2286m) is on an extension of the Tibesti (*see:* Chad below)
in the SW of Cyrenaica.

## EGYPT

**Sinai**
G Katherina 2637, **the highest point of
   Egypt**
G Um Shamar 2586 (fa 1862 T. Yorke,
   T. Prout)
G Loda 2555
G el Thabt 2438
G Rimhân 2412 (fa 1933)
G Musa (Mt Sinai) 2286 (fa Moses)
G Sabbâgh 2266
G Serbal 2070
G Abu Qadis 1981 (fa 1894)

**Red Sea coast**
G Shâyib el Banât 2187 (fa said to have
   been made by Salima, a Bedouin
   girl, around 1870), **the highest point
   of a group of granite peaks.**
G Hamata (El Samak el Mulak) 1977
   (fa 1905)
G Qattar 1963
G Shendib 1912
G Umm Anab 1766
G Dâra 1751
G Gharib 1751 (fa 1823)
G Abu Hammamid 1747
G Umm Harba 1714
G Faraid (Pentadactylus) 1366
Berenice's Bodkin 1232, a volcanic plug

## CHAD

**Tibesti**
Emi Koussi 3415 (fa claimed for Eng-
   lish party in 1957), **the highest point
   of the Chad Republic.**
Tarso Taro 3325
Pic Tousside 3265
Tarso Tieroko 3179 (fa 1965 claimed
   for an English party)
Tarso Emissi 3150
Aigs Sisse, give some rock climbing.

**Guera massif** – 1800

from afar like a gigantic cup handle through which daylight can
clearly be seen. It stands close on 10m above a sloping hillside, is
several tens of metres wide and appreciably more slender at the
lower end than at the upper.

**Lake Bosumtwi** (the Ashanti Crater) in Ghana is **the largest known
meteor crater in the world** (95km diameter).

**Mount Cameroun** (4070), beside the Gulf of Guinea, first climbed
by Sir Richard Burton in 1861, was climbed also in 1893 by Mary
Kingsley, the Victorian lady explorer whose fame rests on one
comparatively short, though very enterprising, journey made in
this part of Africa.

The **Great Rift Valley** is a tremendous graben-type fault line which
runs for 10000 km from Lebanon to Mozambique. It was described
by one of its earliest discoverers as **the one earth feature which
would be visible from the moon**. Now recent satellite pictures have
borne out his prediction. The line follows the Jordan Valley and
the Red Sea, then through Ethiopia and Kenya into Tanzania
and Mozambique. The 2500km of trench from the Red Sea to
Lake Manyana in Tanzania are a striking feature of African
scenery; the width is some 50km with escarpments on either hand
rising upwards of 500m. **The geology** and the main details of the
**geomorphology were first worked out by J.W. Gregory in the
1890s**. The relation between the Great Rift Valley and the world
system of tectonic plates and ridges does not seem to be completely
clear at present.

    The down-faulting which produces a graben is invariably
associated with volcanic activity and there is certainly plenty of
it hereabouts. The principal volcanoes are Oldoinyo Lengai,
which blew up in 1966 (the lava, rich in alkalis, is black when it
emerges, but turns white when converted by moisture in the air
to sodium carbonate), Longonot, and Teleki. There are many
others lesser. Between Longonot and Lengai are the highly
alkaline lakes of Magadi and Natron, the latter the home of **the
world's largest colony of flamingoes**.

    After crossing the Kenya/Ethiopia frontier at the north end of
Lake Rudolf (itself with two volcanic islands) the Great Rift
Valley runs north-eastwards through the Ethiopian massifs,
spreading out beyond Addis Ababa to a width of 500km, the
Danakil Depression. This is an area of great volcanic activity,
since it lies at the junction of the Great Rift Valley and a mid-
ocean ridge which runs up the Gulf of Aden. There are salt lakes,
brightly coloured rocks and depressions of $-116$ and $-150$m,
where rock temperatures of 160°C are said to have been recorded.
The river Awash does not flow to the sea but ends inland at
Lake Abbe, a geomorphological feature only confirmed by the
explorations of Thesiger in the 1930s. Previous travellers had been
deterred by the reputation for ferocity towards strangers of the
local Afar tribesmen.

## SUDAN

**North-west**
J Uweinat 1911

**Red Sea coast & Ethiopia/Sudan frontier**
J Oda 2260
J Erba 2218
J Asoteriba 2217
J Elba 1594
J Hamoyet
J Kessala, an inselberg 850m above Kessala offering a hard rock climb to surmount.

**Western desert**
Marra Mtns – J Gimbala 3071

**Southern frontier** (with Uganda)
Kinyeti 3187, **the highest point in Sudan.**
J Lotuke 2963
J Dongotona 2623
J Kedong 2445

## ETHIOPIA AND SOMALIA

**a    Amaro Mtns** – Mt Delo 3600

**b    Gugé (Gughe) highlands** – 4200

**c    Batu ra**
Dimtu 4433
Saneti 4315
Tigrita 4302
Guramba 3367

**d    Mtns around & W of Jima**
Dulla 3586
Maigudo 3100
Tula Wallel 3301

**Ethiopia**

Jebel Marra, Sudan (AC Collection)

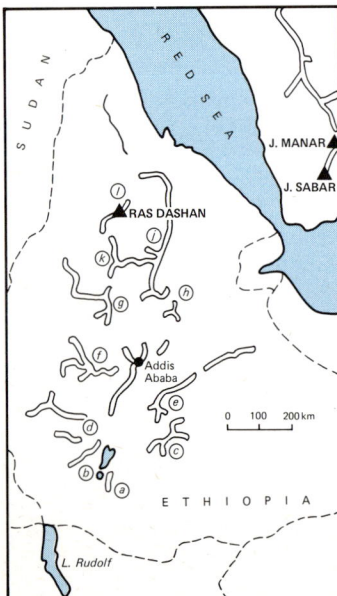

**Mount Kenya**, **the finest of Africa's mountains**, is the remains of an extinct volcano, the crater of which has eroded away leaving a steep core of hard rock which formerly filled the vent. Conspicuous by a permanent cover of snow, it was **first sighted by European explorers in 1849**. Dr Ludwig Krapf's report of 'snow on the Equator' was treated with scorn. The mountain was not approached closely until 1899 when a party led by Sir Halford Mackinder, supported by more than 160 Africans, took a month to travel from Nairobi. *En route* they were attacked by rhinos and by hostile natives firing poisoned arrows; two of their African porters were killed. They eventually established a base camp at the foot of the mountain, from which **Mackinder and his two Swiss guides, César Ollier and Joseph Brocherel, reached the highest summit** (called Batian after a local chieftain) at their second attempt. A nearby secondary peak (Nelion) was not climbed for another 30 years until Eric Shipton and P. Wyn Harris traversed both summits in the course of one expedition in January 1929.

There are twelve small glaciers, with exotic vegetation such as giant groundsel up to 4m high on the lower slopes. Batian and Nelion will always be for mountaineers only, but Point Lenana is readily accessible and presents no very special problems. The whole area is now a National Park; the Park authorities control the huts and in general set conditions for access. In addition there is plenty of hard climbing on the mountain on buttresses and faces, as well as the famous Diamond Couloir which runs up the south face to the col between the twin summits.

An interesting ascent was made during World War II when some Italian prisoners-of-war escaped from a nearby prison camp and made the climb with very primitive home-made equipment. After thus enjoying mountain adventure far from their homeland, they gave themselves up on descending from the summit snows.

Nairobi-based climbers practise at **Hell's Gate** in the Mau

**e   Kakka-Chilalo group**
Kakka 4190
Badda 4133
Enkuolo 3806
Mt Gugu 3623

**f   W of Addis Ababa**
Dendi 3298
Goroken 3276

**g   Chokké highlands**
Birhan 4154
Tala 4100
Chokké 4070
Amedamit 3619

**h   High lands between the Blue Nile and Dessie**
Abuia Mieda 4000

**j   Pks above Lalibala**
Abuna Yosef 4189
Sarenga 3938
Alaji 3437

**k   Guna uplands**
Guna 4231, has a road across the summit.

**l   Semien (Simen) Mtns**
Ras Dashan (Rasdajan) 4620 (fa 1841 Ferret, Galinier), **the highest point in Ethiopia.**
Buahit 4510
Selki 4502
Abba Jared 4459
Emiet Gogo 4015

**The 'Horn of Africa'**
Surud Ad 2408, **the highest point of Somalia.**

**Miscellaneous volcanoes**
In Eritrea on the Red Sea coast is Mussa Ali (V) 2063; there are others, also a depression of −116m with volcanic cones round and about. In Ethiopia are Dubbi (V-le 1863) 1590, Afderà (V-le 1915) 1200, Abidà (V-le 1928) 1308 and Erta-ale (V-le 1960).

**WEST AFRICA**

**Mauretania** – 915

**Senegal** – Gounou Mtn 1515

**Mali** – Adrar des Iforas 1000+, SW end of the Hoggar massif.

**Sierra Leone** – Loma Mtns 1948

**Liberia/Guinea frontier** – Mt Nimba 1768

**Ivory Coast** – Mts de Droupole 2103

**Upper Volta** – Mt Tema 749

**Ghana** – 908

**Togo** – 919

Range and at **Lukenya ridge** (40 km south-east). There is a record at the latter site of a climbing party being attacked and driven off by bees.

**Longido** (2629) mountain in Tanzania was the scene of an Anglo-German skirmish during World War I.

In Tanzania close to the Kenya frontier is Kibo Peak of **Kilimanjaro** (5895). A dormant volcano and **the highest point of the African continent**, it has a permanent snow and ice cover. **First sighted by** a German missionery, **Johann Rebmann**, **in 1848**, it was **not climbed until October 1889 by another German**, **H. Mayer**, and his guide L. Purtscheller. It now lies within Kilimanjaro Game Reserve.

The crater is ice-filled, with glaciers spilling over lower points of the rim and tumbling down the outer walls. These glaciers – Great Barranco, Heim, Penck, Kersten and so on – provide modern ice climbs of reasonably high standards, and there are also some great rock walls like the Great Breach Wall.

From this summit to that of Mount Kenya, 325 km away, is **one of the longest confirmed lines of sight on the surface of the earth** under normal visibility conditions; the air is often clear, while the prominent snow cover of the distant peak attracts the eye. These summits have been linked in an expedition of 23 hours – down 4250 m, followed by a long car drive and finally up 3650 m.

The climb is an easy one from Moshi on the Tanzanian side; it is a popular tourist mountain and is climbed for example by schoolboys on Outward Bound courses. The rim of the crater is reached at Gilmans Point and then traversed clockwise to Uhuru Peak (formerly King Wilhelm Spitze), the highest point. Ascents from the Kenyan side are forbidden.

Like other outstanding mountains, it has attracted the usual spate of freak records – two one-legged Austrians on crutches in 1961, eight blind African boys in 1969 and so on. Three French parachutists equipped with oxygen dropped on to the summit in 1962, then walked down to the valley therefrom. In 1978 a descent was made by hang-glider.

Nearby, Mawenzi Peak, also part of the Kilimanjaro massif, is a much more impressive rock mountain. It is the remains of the plug of an ancient volcano, the crater rim of which has disappeared. There is a straightforward way to the summit, the Hans Mayer Spitze, but it is hardly a tourist route. **The first ascent was made in 1912 by F. Klute and E. Oelher**. The impressive east face, 1800 m high, has been climbed in recent years.

The **Ngorongoro Crater**, a former volcanic feature now a famous game reserve, lies close to the Great Rift Valley in northern Tanzania. It is 600 m deep and with an area of 260 km² is completely enclosed. Around is a conservation area of 8000 km².

In 1960, an American, J. Graham, became **the first man to stand on all the peaks of East Africa over 16000 ft** (4875 m), when he completed an ascent of Mount Stanley in the Ruwenzori.

**Benin** – 635

**Nigeria**
Mt Dimlang 2042
Bauchi Plateau 1781

**Cameroun**
Mt Cameroun (Mongo-ma-Loba) (V-le 1959) 4070 (fa 1861 G. Mann, R. F. Burton), **the highest point in Cameroun.**
Pt 2740
Tchabel Gangdaba 1960
Mbang Mtns 1800

**Fernando Poo**
Pico de Sta Isabel 3007
Clarence Pk 2850 (fa 1860 J. P y Rodriguez)
The chain of islands, a line of volcanoes part-submerged, continues through Principe and Sao Tomé 2024.

**Central African Republic** – Mt Gaou 1420

**Equatorial Guinea** – Pico de Moca 2850

**Gabon** – Mt Iboundji 1580

**Congo** – 1040

**Angola** – Serra Môco 2610

**KENYA AND UGANDA**

**Elgon-Karamoja** (from N to S)
Mt Elgon (Oldoinyo Ilgoon, Masawa) (V-ext) 4321 (fa 17/2/1890 Jackson, Gedge, Martin), crater 600 m deep, 6·5 km dia.
Kadam (Debasien) (V-ext) 3068 (fa 1931 Preston, though possibly by the Krenkel exped of 1911–12)
Napak (V-ext) 2537
Moroto 3084
Morungole 2750

**Acholi Province**
Rwot, inselberg 600 m above the plain (fa 1957 Austrian party)
Amiel, inselberg 300 m above the plain (fa 1959)

**Turkana & Karasuk** (from N to S)
Murua Ngithigerr 2149 (fa 1921 E. von Otter)
Chemongerit Hills ∼ 1800
Kapitugen ∼ 2750
Kachagalan 2791
Lorosuk ∼ 2750
Tenus 2548
Kapcholio ∼ 2750
Tarakit 2517
Kapchok 2124
Sepich ∼ 2450

**Pokot & Cherangani Hills**
Sekerr 3325 (fa 2/1923 J. G. Hamilton Ross)
Sondhang 3206

The **Ruwenzori**, otherwise 'the Mountains of the Moon', a mountain range on the Uganda/Zaire frontier, occupies an area of $120 \times 65$ km. Often obscured by tropical rain clouds, **the range was first sighted definitely by H. M. Stanley's expedition in 1888**, though there is some possibility that he had in fact glimpsed it twelve years earlier in 1876. In 1889 Lieut. Stair of Stanley's expedition reached 3000 m and he was followed by several other explorers around the turn of the century. The first climbing expedition in 1905 was unsuccessful, but the following year the eminent Italian mountaineer, the Duke of the Abruzzi, with a massive Italian expedition, including four guides, and carried in by 150 porters, made a clean sweep of all the major summits. In terms of first ascents this was **one of the most successful mountaineering expeditions of all time**. Thirty years were to elapse before these mountains were climbed for a second time.

The range, which has a folded structure, offers mountaineering essentially Alpine in nature. There are, however, serious problems of access through the tropical rain forest. The huts, approached from Mubuku, are said to be in poor condition at the present time, while frontier mountains, such as these, often present political problems to the would-be traveller.

North of **Kazinga Channel** in Uganda between Lakes Edward and George is an 'explosion area' with more than 80 extinct volcanoes.

A portion of the landscape of the **Virunga Volcanoes**, Zaire–Ruanda frontier, is preserved in the Albert National Park. Since 1930 there has been a sanctuary ($45 \mathrm{km}^2$) on the volcano Sabinya for the rare Mountain Gorilla.

The hinterland of **Mozambique** and the adjacent areas of Africa provide the **best examples of inselberg scenery** (*see* p. 11) **in the world**. This land-form was recognised and its origins elucidated by Holmes and Wray in 1913. There are numerous steep-sided rock peaks varying from tens to hundreds of metres high, presenting quite considerable problems of access. There is no record yet of climbing developments in Mozambique, but Rhodesian mountaineers have made full use of their inselbergs – Bari 300, Hurungwe, Zhombwe Musungwa, and so on.

The **Blyde River Canyon** is a striking feature of the **Transvaal Drakensberg**; 600 m deep in places, its walls are of quartzite and dolomite. Blyde Towers are two peaks isolated by erosion, not climbed until 1959. Nearby is Swadeni, not climbed until 1961.

**Mont aux Sources, the best known peak of the Lesotho Plateau**, is the source of the Rivers Tugela, Eland and Khubedu. It rises on the plateau immediately behind the Amphitheatre, a wall of rocks and peaks 4 km long and 1500 m high which is a striking feature of the Royal Natal National Park. The vertical face of the Amphitheatre is for rock climbers only; from the foot of the escarpment the easiest route to the plateau and the peak follows

Tavach 3296
Koh ~ 2750
Chemnirot 3350
Chesugo ~ 3050
Kamelogon 3520
Kalelekelat (Kalelaigelat) 3350
Longoswa ~ 3350
Kipkunurr 3063
Kaipos 2360
Kaibwibich 2687
Kaisungur 3167
Chemurkoi 2910
Kapsiliat 2603

**Mtns of Samburu** (from N to S)
Kulal 2293
Teleki (V-le 1922) 646
Nyiru (Oldoinyo Ng'iro) 2829
Oldoinyo Ndoto (Bukkol) 2534
Alimision 2637
Oldoinyo Lengeyo 2288
Mathews Pk 2375
Mathews South Pk 2285
Warges (Wamba Mtn) 2688
Ololokwe 1853
Losiolo 2470
Poror 2583
Paua 2254
Leshingeita 2206

**Tuken (Kamasia) Hills** (from N to S)
Tiati 2351
Saimo (Ketingwan) 2501
Marop 2306
Kibimchor 2347
Kapkut 2800
Gaisamu 2726

**Aberdare Ra** (formerly Sattima Ra)
  (Aberdare NP) (from N to S)
Oldoinyo la Satima 3999
Kipipiri 3349
The Kinangop 3906
The Elephant 3590

**Mau Ra** (from NW to SE)
Summits on the Mau escarpment – the
W wall of the Great Rift Valley.
Tinderet 2640
Loldian 3011
Mau 3050
Eburu 2854
Melili 3098

**Mount Kenya**
Batian 5199 (fa 9/1899 H. J. Mackinder;
  C. Ollier, J. Brocherel), **the highest
  point of Kenya.**
Nelion 5188 (fa 1/1929 E. E. Shipton,
  P. Wyn Harris)
Pt Lenana 4985
Pt Pigott 4958
Tereri 4714
Sendeyo 4704
Midget 4700

**Mtns of Maasailand** (in Kenya) (from
N to S)
Mostly isolated volcànic peaks in or
near the Great Rift Valley.

Longonot (V-steam jets) 2777 (fa 1893
  J. W. Gregory), crater 1·5 km dia.

two chain ladders of 100 rungs installed in 1930 at the head of
the next embayment to the north.

The **Drakensberg**, which stretch for some 1100 km from the Tropic
of Capricorn in northern Transvaal to eastern Cape Province,
were **first sighted by Europeans in 1593**. The survivors of the wreck
of a Portuguese ship on what is now the coast of Natal were
forced to attempt a trek to Lourenço Marques. They succeeded
in crossing 1000 km of unmapped country; their route, which was
directed somewhat inland to avoid large embayments in the
coastline, brought them within sight of the high Drakensberg.
    The most spectacular portion forms the edge of the Lesotho
Plateau along the western frontier of Natal. The plateau consists
of a layer of basaltic lava some 1400 m thick, which forms the
scarp face and the uppermost rocks of the plains at the foot.
Below it is a layer of sandstone containing many caves with wall
paintings by the prehistoric inhabitants. The highest points hereabouts are in fact on the plateau, and are easy of access. **Thadent-**

**Central Africa**

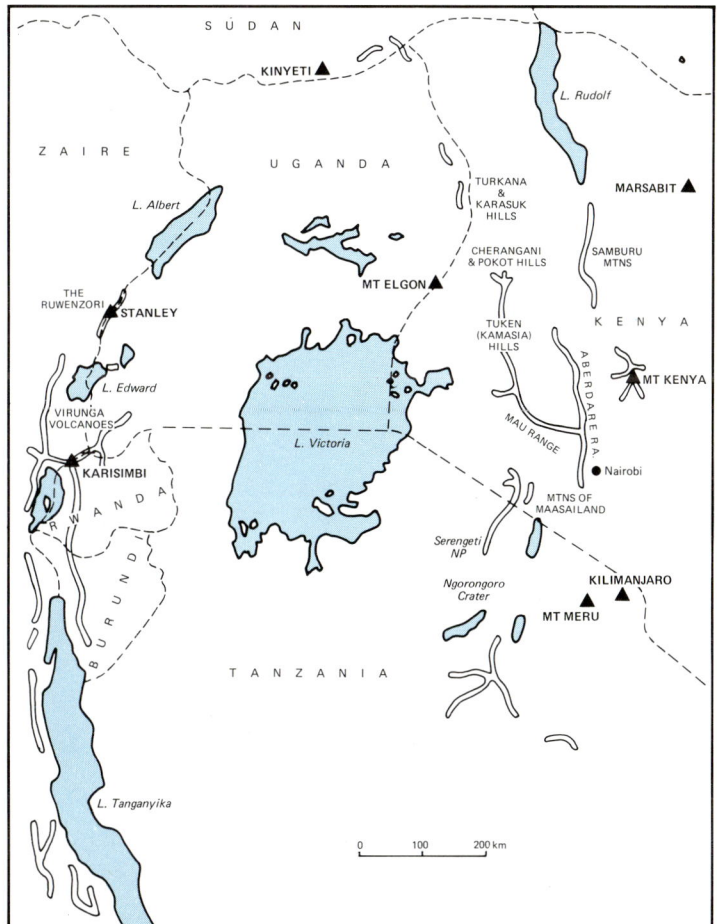

Suswa 2357 (fa 10/1897 Hall, Blackett, Welby), caldera 8 km dia. with inner caldera of 5 km dia.
Lamwia 2461
Loita 2683 (fa 1894 O. Neumann)
Shombole 1564 (fa 1904 G. E. Smith)
Lemileblu 2149
Oldoinyo Orok 2548 (fa 1902–4 G. E. Smith's party)
Chyulu 2174
Menengai, caldera 11 km dia.

## TANZANIA

**Mtns of Maasailand** (in Tanzania) (from N to S)
Gelai (V-ext) 2941
Longido 2629
Oldoinyo Lengai ('Mountain of God') (V-le 1966) 2900 (fa 9/1904 Uhlig, Jaeger, Gunzert)
Ketumbaine 2942
Kilimanjaro – Kibo Pk (V-dorm) 5895 (fa 6/10/1889 H. Mayer, L. Purtscheller), **the highest point of Tanzania and of the African continent,** formerly called King Wilhelm Spitze, renamed Uhuru Pt in the mid 1960s.
Kilimanjaro – Mawenzi Pk (Hans Mayer Spitze) 5148 (fa 6/1912 F. Klute, E. Oelher)
Shira Pk 4005
Loolmalasin 3648
Meru (Socialist Pk) (V-le 1910) 4565 (fa Jaeger or Uhlig)
Oldeani 3188
Hanang 3418

**Kipengere Ra** (at the head of Lake Nyassa)
Rungwe Mtn (V-ext) 2959

**Livingstone Mtns** (E of Lake Tanganyika)
Kungwe Mtn 2685
Mt Jamimbi (Chamembe) 2399
Nearby Lake Tanganyika reaches down to −700 m.

**Mtns W of Lake Albert**
Aburo 2437

## ZAIRE/RWANDA/BURUNDI

**The Ruwenzori** (Zaire-Uganda frontier) (from N to S)
Mt Emin 4798 (fa 28/6/1906 Duke of the Abruzzi; J. & L. Petigax, C. Ollier)
Subsid pk 4791
Mt Gessi 4715 (fa 16/6/1906 Duke of the Abruzzi; J. Petigax, C. Ollier)
Subsid pk 4699
Mt Speke 4896 (fa 23/6/1906 as for Mt Gessi)
Subsid pks 4865, 4834
Mt Stanley – Margherita Pk 5109 (fa 18/6/1906 Duke of the Abruzzi; J. Petigax, C. Ollier, J. Brocherel), **the highest point of Uganda and of Zaire.**
Subsid pks 5087, 5091, 4977, 4918
Mt Baker 4843 (fa 10/6/1906 as for Mt Stanley)

Kilimanjaro – Mawenzi in front, Kibo behind (Aerofilms)

sonyane (Thabana Ntlenyana), the highest of all at 3482 m was not discovered and climbed until 1951. The striking craggy walls of the plateau edge, called 'the Barrier of Spears' by the Zulus, are said to be eroding at a rate of 30 cm per 200 years.

Mont aux Sources was climbed in 1836 by two French missionaries; Champagne Castle may have fallen in 1861 by a route which cannot be traced and Giant's Castle in 1864. Real mountaineering began in the 1880s when an alpinist, the Rev A. H. Stocker from Britain, visited his brother who farmed locally, and Champagne Castle and Sterkhorn were climbed. Some of the peaks and pinnacles present quite serious problems – the Monk's Cowl was not climbed until 1942, Inner Mnweni Pinnacle until 1949 and North Ifidi Pinnacle until 1959.

Ruwenzori and giant groundsel (J G R Harding)

Subsid pks 4795, 4624, 4623
Mt Luigi di Savoia 4627 (fa 4/7/1906
    V. Sella; J. Brocherel, Botta)
Subsid pks 4620, 4545
Humphreys Pk 4578
Portal Pks 4391 (& 4370, 4267)
Cagni 4487
Kinyangoma 4431
Catafalque 4450

**Mtns W of Lake Edward**

**Virunga (Birunga, Mfumbiro) Volca-
noes** (Zaire-Ruanda frontier) (from
W to E)
Nyamuragira (Nyamlagira) (V-le 1971)
    3060
Nyiragongo (V-active lava lake) 3469
    (fa 1894 Count Goetzen exped)
Mikeno (V-dorm) 4437 (fa 1927)
Karisimbi (V-dorm) 4506 (fa 1903
    Berthelmy), **the highest point of
    Ruanda.**
Vishoke (V-dorm) 3711
Sabinya (V-dorm) 3634
Gahinga (V-dorm) 3474
Muhavura (Muharura) (V-dorm) 4126
    (fa 1900 Betke)

**Matumba & associated ras** (from N to S)
The main range runs from the Virunga
area down the W sides of Lakes Kivu
and Tanganyika, turning finally SW
between Lakes Mweru and Upemba.
All this is Zaire.

Mt Kahusi 3292, group of extinct vol-
    canoes, W of Lake Kivu.
Pt 2987, W of foot of Lake Tanganyika.
Marungo 2460, W of head of Lake
    Tanganyika.
Upemba NP 1889

**Mtns of Rwanda-Burundi** (from N to S)
Mt Karonje
Pt 2685, **the highest point in Burundi.**

**ZAMBIA**

Pt 2067, close to Tanzania frontier, **the
    highest point in Zambia.**
Muchinga escarpment – Chimbwin-
    gombi 1788

**MALAWI
North** – Nyika Plateau 2606

**South** – Mt Mulanje-Sapitwa Pk 3000,
    **the highest point in Malawi.**

**ZIMBABWE-RHODESIA**

**Melsetter NP** (Chimanimani Mtns)
Binga Mtn (Kweza) 2436, on frontier
    with Mozambique.
Turret Towers 2367
Ben Nevis 2179

**Inyanga NP**
Inyangani 2595, **the highest point in
Zimbabwe-Rhodesia.**

**Umvukwe Ra** – Nyamanji 1500+

The summit of **Toverkop** in the Kleinswartberge is a rock tower
of 150m. This was climbed in early days (1885) by a local man,
Gustav Nefdt, solo; he returned to his companions minus one
sock which he claimed to have left on the summit. Because of
some apparent inconsistencies his story of the ascent was not
believed, so he was forced to repeat the climb in front of witnesses,
taking up a rope which enabled others to follow. The discovery of
his abandoned sock on the top was a triumphant vindication.

The best part of the story is that the tower was not climbed
again for 21 years and then by another route. In spite of various
attempts Nefdt's route was not repeated until 1947.

The very familiar **Table Mountain** (1086), above Cape Town and
close to the southern tip of the east coast of Africa, was **first
sighted** by Bartholomew Diaz. It was **climbed for the first time** by
Antonio de Soldanha in 1503 to discover whether he had yet
rounded the Cape of Good Hope. In the ensuing years other
navigators did likewise. Jan van Riebeck, who commanded the
Dutch settlement from 1652 to 1662 climbed the mountain
several times. Lions and tigers were then plentiful on the summit
plateau.

**The first recorded ascent by a lady was made in 1797** by Lady
Anne Barnard, her luggage being carried by twelve slaves. The
large party which accompanied her spent the night on the summit.

In 1799 a press-ganged sailor fled his ship at Cape Town and
spent one-and-a-half years in a cave on the mountain, only to
find on descending that the ship and her crew had all been lost in
a storm immediately after sailing without him.

The Mountain Club of South Africa was founded in 1891 and
serious climbing began on the Table Mountain. An outstanding
climber of this early period was G. F. Travers-Jackson, who was
responsible for many first ascents. One of them, Fernwood Gully,
had to be finished up a steep bulge of earth, vegetation and loose
rock, so before starting to climb he dropped a rope down from
above to assist his exit. Near the top of the climb he reached the
lower end of the rope and found he must climb it without other
assistance. Halfway up he cramped and had to rest by seizing the
rope with his teeth; then after massaging his arms he continued
to the top. This particular climb was not done clean until 1974.

The main flat-topped central mountain is buttressed on either
hand by the outliers of Lion's Head and Devil's Peak. Cape
Town nestles between them; climbers from the city are indeed
lucky to have such fine climbing so close at hand. Since 1902
Table Mountain has been preserved as a National Monument.

A notable cloud, the Tablecloth, which often shrouds the peak,
is formed from the moisture-laden south-east winds gusting
sometimes at 120km/hr. The flora and fauna of the summit
plateau are of outstanding richness; the latter includes the
Himalayan Tahr (a goat-antelope), which escaped from local
captivity in the 1930s, thrived and multiplied.

The line of that mighty tectonic feature, the **Mid-Atlantic Ridge**,

**Matopo Hills NP** – 1553

## MOZAMBIQUE

**S of R Zambesi**
Binga Mtn 2436, **the highest point in Mozambique,** on Zimbabwe-Rhodesia frontier.
Gorongoza 1856
Mt Dedza 2035

**N of R Zambesi**
Chiperone 2054
Namuli 2419
Pt 2095
Jeci 1836

## SWAZILAND

Emlembe 1863
Mtambama Mtn 1000+

## SOUTH AFRICA

**Transvaal – miscellaneous ras**
Soutpansberg Ra 1718
Blaauberg 2047, rises 1200 m above the plain, a labyrinth of rock corridors.
Waterberg Ra 2085
Pilanesberg/Magaliesberg Ras 1687

**Transvaal Drakensberg Ras**
De Berg 2332
Mt Anderson 2286
Pt 2128
Pt 2031
Thabakgolo 1898

**Natal/Orange Free State border**
Platberg 2396
Skeurlip 2363
Witkoppies 2338
Babangiboni 2329
Ntyele 2294
Thabanakulu 2277

**Lesotho Plateau**
Thadentsonyane (Thabana Ntlenyana) 3482 (fa 1951 D. Watkins, B. Anderson, R. Goodwin *et al*, previously it had not been realised how high it was), **the highest point of Lesotho.**
Makheke 3461 (fa 1938 J. v. Heyningen, Col Park Grey)
Mafadi 3450
Mont aux Sources 3282 (fa 1836 T. Arbousset, F. Daumas)
Pelatseou 3276
Thaba Putsao 3096

**Drakensberg Plateau edge** (Khuahlamba, the 'Barrier of Spears')
Injasuti 3459 (fa no record), **the highest point of South Africa.**
Champagne Castle 3377 (fa 1888 A. H. Stocker, F. R. Stocker)
Popple Pk 3325 (fa no record)
Giant's Castle 3316 (fa 1864 R. Spiers, A Bovill, H., E. & F. Bucknoll)
Redi 3298
Cleft Pk 3281 (fa 1936 D. P. Liebenberg, L. P. Ripley & party)
Hondson's Pk 3257

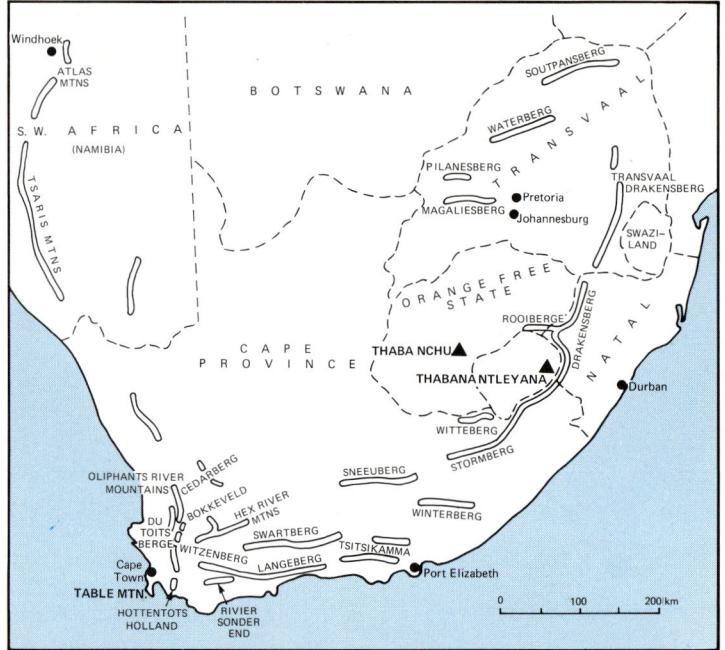

South Africa

**Blyde River Canyon, S Africa (South African Tourist Corporation)**

is marked by the islands of Ascension, St Helena and Tristan da Cunha. All show signs of volcanic activity, either past or present. Further north the Azores and Iceland mark further points of

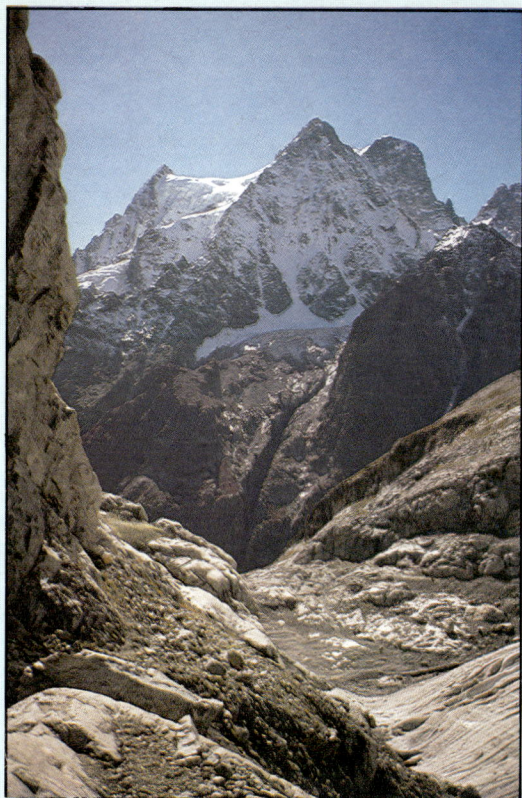

## THE ALPS

*Above:* Tremola road, St Gotthard Pass, Switzerland. *Above right:* Mont Pelvoux, French Alps (T Connor). *Below left:* Kleine Scheidegg with its view of the North wall of the Eiger, Switzerland. *Below right:* Bouquetins in Haute Maurienne, France (French Government Tourist Office)

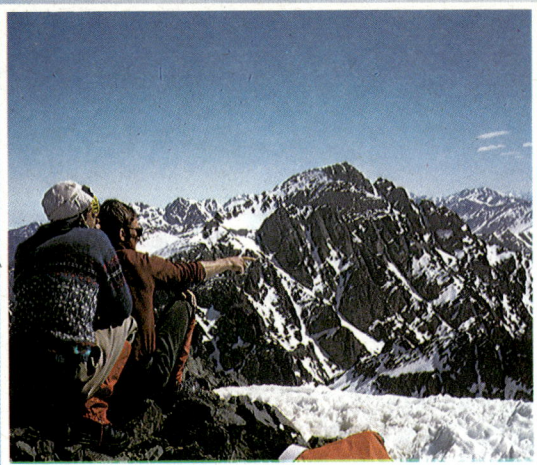

Toubkal, the highest point of the Atlas Mountains, N Africa (T Connor).

Peaks of the Ahaggar, North Africa (G Flanagan).

Filming a trek scene from the film The Third Day, Champagne Castle area, Drakensberg, South Africa (South African Tourist Corporation).

ASIA

*Above:* Eruption of volcanoes, Kamchatka, USSR (Novosti Press Agency). *Below:* Kinabalu, the highest point of Malaysia (Travel and Development Corporation of Malaysia)

Indumeni Dome 3255 (fa 1925 H.G. Botha-Reid & party)
Monk's Cowl 3234 (fa 1942 J. Botha, E. Ruhle, A.S. Hooper, H. Wong)
Western Triplet 3187 (fa 1951 R.F. Davies, D. Bell)
Subsid pks 3170, 3155
Sentinel 3165 (fa 1910 W.J. Wybergh, N.M. McLeod)
North Saddle 3153 (fa 1924 O.K. Williamson, D.W. Bassett-Smith)
Cathkin Pk 3149 (fa 1912 G.T. Amphlett, W.C. West, A.D. Kelly, T. Casement *et al*)
Mt Amery 3143 (fa 1920 D.W. Bassett-Smith, R.G. Kingdon)
Mnweni Pinnacles 3100 (fa 1949 J. de V. Graaff, G.R. de Carle, R. Buckland, P. Goodwin)
Mponjwana 3085 (fa 1946 G. Thomson, K. Snelson)
Rockeries Pk 3027 (fa 1953 R.F. Davies, J. de V. Graaff, J. Slinger, D. Williamson)
Cathedral Pk 3004 (fa 1917 D.W. Bassett-Smith, R.G. Kingdon)
Old Woman Grinding Corn 2986 (fa 1937 O.B. Godbold, C.E. Axelson, N. Hodson), a splendid descriptive name.
Sterkhorn 2973 (fa 1888 A.H. & F.R. Stocker)
Devil's Tooth 2941 (fa E.H. Scholes, D. Bell, P. Campbell)

**Southern Drakensberg** (NE Cape Province)
Ben Macdhui (Ben Maduda) 3002
Tsatsana 2952
Witteberge 2771
Draken's Rock 2726
Kopshorn 2677

**Other ras of NE Cape Province**
Stormberg – Vaalkop 2168
Bamboesberg – 2208
Winterberg – Gr Winterberg 2371
Bankberg – 2013
Sneeuberg – Kompasberg 2504, Toorberg 2280
Tandjiesberg – 2431

**S & SE Cape Province**
Grootwinterhoek – Cockscomb 1759
Baviannskloofberge – Scholtzberg 1627
Tsitsikamma Mtns – Formosa Pk 1715 (fa 1860 survey party)
Kougaberge – 1765
Grootswartberge – Spitskop (Tierberg) 2152
Kleinswartberge – 2326, also the striking Toverkop (Witches Head) 2202 (fa 1885 G. Nedft), the summit a 120m rock tower.
Langeberg – Keeromsberg 2075
Hex River Mtns – Matroosberg 2250, Roodeberg 2164, Zonklip Berg 2126, Buffelshoek Pk 2062, Groot Hoek Pk 2091, Buffels Dome 1448.
Du Toits Berge – Du Toits Pk 1997
Wittenberge 1623

emergence, while in the far south Bouvet Island occupies a similar position athwart the ridge.

To the east of the African Continent the ocean ridges are more complex. In fact most of the Indian Ocean islands also mark points of emergence of ridge systems above sea-level. A ridge runs into the Gulf of Aden joining the Great Rift Valley in the curious volcanic area known as the Danakil Depression.

Drakensberg skyline (South African Tourist Corporation)

Stanley's Expedition towards Ruwenzori (Barnaby's)

Cape Town and Table Mountain (The Argus, Cape Town). *Inset:* Climber on Table Mountain above Cape Town (M Scott)

Bokkeveld – Sneeukop 2027
Oliphant's River Mtns – Gr Winterhoek 2077, Coxscomb 1759
Cedarberg – Sneeuberg 2027, Tafelberg 1970 (fa 1896)
Table Mountain 1086 (fa 1503 Antonio de Saldanha, who climbed to see if he had yet rounded the Cape of Good Hope)
Kuruman Hills – Gakarosa 1855

**Orange Free State**
Rooiberge 2477
Witteberge 2409
Thaba Nchu 2227 (mean)
Golden Gates Ra (a NP) – Rhebokkop 2830, Sneeukop 2803, Generalskop 2734

Soutkop, inselberg on a conical hill near Ficksburg (fa 1960 no signs of a previous ascent)
Aasvogelberg 2124, 150m climbing crags with a 9m hole, the 'Eye of Zastron'.

**SOUTH-WEST AFRICA (NAMIBIA)**
Two small mountain areas – around Windhoek (Atlas Mtns 2484) and W of Grootfontein – exceed 2000m. The Karas Mtns are horsts with a graben between.

Groot Spitzkopf (fa 1945, using holds cut with hammer and chisel), the 'Matterhorn' of Namibia.

**ATLANTIC ISLANDS**

**Ascension**
Green Mtn Pk (V-not active) 859, 40 cones can still be seen.

**St Helena**
Diana's Pk 823, islands of volcanic origin.

**Tristan da Cunha**
Queen Mary's Pk (V-le 1961–2) 2060

**Cape Verde Islands**
Fogo Island – Fogo Pk (V-le 1951) 2829, **the highest point of Cape Verde Republic.**
Santo Antão Island – Topo de Coroa 1979

# Asia

Fuji-san, the highest point of Japan (Japan Information Centre)

## Western Asia

### THE CAUCASUS

**Western Caucasus** (from the Black Sea coast to El'brus) (from W to E)
Gora Pshish 3789
Dombay Ulgen 4047 (fa 1914 Schuster, Fischer)
El'brus (V-extinct) – W Pk 5633 (fa 28/7/1874 F. Gardiner, F. Crauford Grove, H. Walker; P. Knubel; fa on skis 28/7/1914 C. Egger), **the highest point in the Caucasus.**
– E Pk 5595 (fa 31/7/1868 D.W. Freshfield, C.C. Tucker, A.W. Moore; F. Dévouassoud)

**Central Caucasus** (El'brus to the Georgia Military Highway) (from W to E)
Schscheldi (Shkhelda) 4360
Ushba – S Pk 4710 (fa 26/7/1903 A. Schulze & party)
N Pk 4693 (fa 1888 J.G. Cockin; C. Almer Jnr), the 'Matterhorn' of the Caucasus.
Lyalver 4350
Gestola 4860 (fa 1886 C.T. Dent, W.F. Donkin; A. Burgener, F. Andenmatten)
Tetnuld 4853 (fa 1887 D.W. Freshfield, Déchy; M. Dévouassoud)
Katyn Tau 4985 (fa 1888 Holder, H. Woolley; U. Almer, C. Roth)

The Caucasus

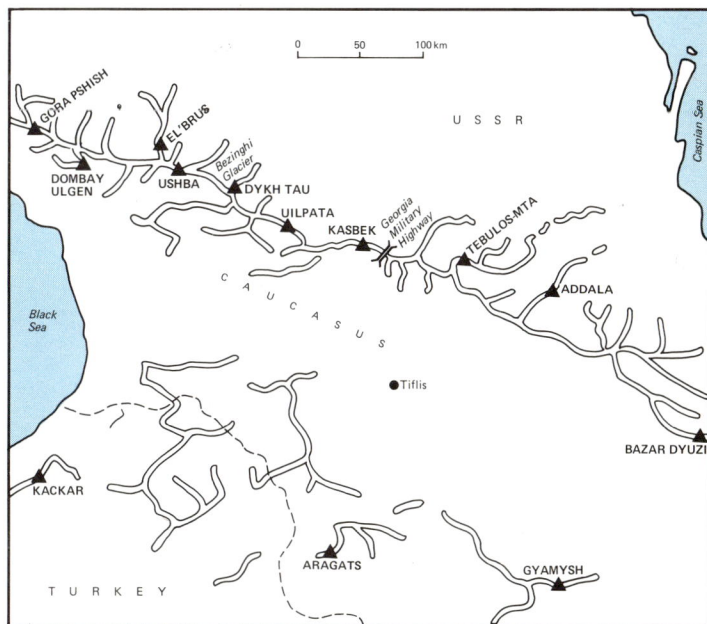

Jangi Tau 5051 (fa 1888 J.G. Cockin;
  U. Almer, C. Roth)
Pik Rustaveli 4960
Shkhara 5201 (fa 7/9/1888 J.G. Cockin;
  U. Almer, C. Roth)
Dykh Tau 5203 (fa 24/7/1888 A.F.
  Mummery; H. Zurfluh)
Mishirgi 4922 (fa 1934 L. Saladin, H.
  Graf, W. Frei)
Kunjum Mishirgi 4880
Ailama 4525
Koshtantau 5144 (fa 9/8/1889 H.
  Woolley; C. Jossi)
Uilpata 4638
Gimaraikhokh 4778
Kazbek 5047 (fa 1/7/1868 D.W. Fresh-
  field, C.C. Tucker, A.W. Moore;
  F. Dévouassoud)
 Krestovyy Pereval (Dariel Pass, Geor-
  gia Military Highway, 'Caspian' or
  'Iberian Gates' of classical times,
  gorge of the R Terek) 2388.

**Eastern Caucasus** (Georgia Military
  Highway to the Caspian Sea coast)
  (from W to E)
Arzi-choch-kort 4134
Shan Tavi 4430
Kidenais-magali 4219
Tebulos-mta 4494
Komitotavi 4272
Donos-mta 4135
Diklos-mta 4276
Bogosskiy Khrebet – Addala 4151
Bazar Dyuzi 4480 (fa 1873 Alexandrov)
Shalbuz-dag 4153
Shagh-dag 4250

**WESTERN USSR**

**Crimea**
Krymskiye Gory – Roman Kosh 1545
Chatyr Dagh – 1519
Ai Petri – 1316, dolomitic peaks

**Ural Mtns**
A 1750km range formed at an ancient
plate boundary no longer active. The
Trans-Siberian Railway crosses at a
pass of 410m W of Sverdlovsk.

G Nadodnaya 1894
Yoman Tau 1639
G Telpos Iz 1617
Yeremal 1595
G Konzhakovskiy Kamen 1569
G Denezhkin Kamen 1493

**Russian Lappland** (Kola Peninsula)
Gora Chesnachorr (Takhtawumchorr,
  Khibiny) 1216 (mean figure).

**TURKEY**

**Pontic Mtns**

*Coast ras* (from W to E)
Köroğlu Tepesi 2376
Ilgaz Daglari 2565
Yedigöz Yaylà 3095
Artabil Gediği Gumusane Mtns 3305
Cakirgöl Daği 3063

Kasbek, Caucasus Mountains (Novosti Press Agency)

The first ascent of **El'brus** (5633) has been claimed for the Cir-
cassian, Kilar Haschirow, who accompanied a scientific expedition to the mountain in 1829; this issue remains unresolved. There
was a mass ascent by 400 climbers in 1956. Oxen carrying scientific equipment have been taken to the top. The mountain was
climbed in 1963 by Tschokka Zalichanov, aged 107, his 208th
ascent. In August 1966 A. Derberashivili drove a motorcycle
equipped with skis, oxygen and a high-altitude engine up the
East Peak, abandoning the machine on the summit.

It is a vast mountain carrying 22 glaciers covering more than
140km$^2$. Local legend claims that here was the prison of Prometheus, also that the Ark grounded hereabouts before settling
finally on Mount Ararat.

Once the bulk of first ascents in the Alps had been accomplished,
the **Caucasus** became, during the 1880s, the goal of many explora-
tion-minded British mountaineers. They were able to make
substantial contributions of first ascents in an area where conditions were much as they had been in the Alps 40 years previously,
among mountains offering very similar problems. Most of these
visitors brought Alpine guides with them, their first forays into
mountains away from their native lands.

The traverse of the main ridge from **Shkhara to Lyalver**, known as
the Bezingi Horseshoe since it rings the head of the Bezingi
glacier, was first done 23–28 August 1931 by K. Moldan, J.
Schintlmeister and K. Poppinger from Austria. Up to 1965 it
had only been repeated nine times; nowadays it is frequently
done by Soviet mountaineers.

**Kazbek** (5047) is another claimant for the site where Prometheus
was bound to his rock.

Kirklar Daği 3320
Vercenik Tepe 3711
Kaçkar Daği 3931
Kaçkar Tepeleri 3502
Kükurt Tepe 3348
Krenek 3760

*Parallel ras inland* (from W to E)
Kizil Dağ 3017
Kesis Dağlari 3557
Murit Dağlari 3600
Yesercol Daği 3300
Mescit Dağ 3255
Kargapazari Dağ 3288
Allahückber Dağlari 3114
Kisir Daği 3192

**Central plain**
Hüseyin Gazi 1615, 11 km SE of Ankara, provides practice climbing.

**Taurus & Anti-Taurus Mtns** (from W to E)
Ak Dağ 3073
Bey Dağlari 3086
Barla Dağ 2734
Dedegöl Dağ 2980
Geyik Dağ 2890
Bolkar Dağlari 3240
Aydos Daği 3488
Toros Daği (Medetsiz Pk) 3585
  Cilician Gates 1158 (rd & rail pass, only 15 m wide at the summit)
Karanfil Daği 3095
Kaldi Daği 3734
Guzella 3461
Cebel Basi 3300
Kizilkaya 3723
Kizilyar 3654
Vay Vay Dağ 3563
Boruklu 3548
Demir Kazik ('Iron Post') 3756
Kurupagat 3623
Kayacik Basi 3620
Hasan Daği (V) 3253
Erciyas Daği (V) 3916
Bey Daği 3054
Besit Daği 3014
Nurhak Dağ 3090
Munzur Dağ 3138
Mercan Dağ 3449
Bagirpasa Daği 3283
Palandokan Daği 3124, a notable skiers mountain.
Bingöl Dağlari 3650
Kökse Dağ 3455
Asagi Dağ 3268

**Eastern Highlands** (from S to N)

*Cilo Daği*
  With 15 summits over 3500 m.
Gelyasin (Resko Daği) 4170
Suppa Durek 4060
Keskin Tepe 4050
Tepe Sirt 3950
Esmertepe 3900
Maunsell Pk 3860
Geniskaya 3770
Karatepe 3750
Mirhamza 3670

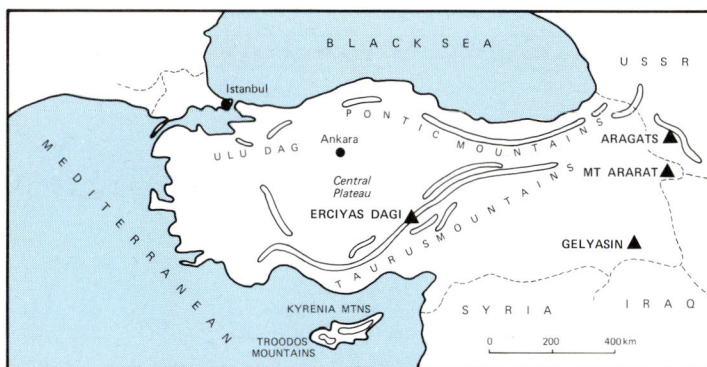

**Turkey**

Erciyas Dag – Turkey (B R Goodfellow – by courtesy of F Solari)

**Shalbuz-dag** (4153) was the home of the roc of *A Thousand and One Nights*.

The **Cilician Gates** has been an important pass throughout the course of history, having been crossed by the armies of Alexander the Great and subsequently by Romans, Byzantines, Crusaders, Saracens and Turks.

**Erciyas Daği** (3916), Mons Argerus of the Romans, is now a popular ski resort. Nearby in the valley of Gheureme earth pillars have been eroded in volcanic tuffs erupted in the past from the mountain. These have been excavated to produce underground dwellings and monasteries.

Eastern Turkey is dominated by **Mount Ararat** (5165), **the highest point in the country**. It is an extinct volcano, rising 4400 m above the surrounding plains and carrying one small glacier but no crater. Christian mythology assigns the first ascent to Noah, but misguided attempts at corroboration have failed (not surprisingly) to turn up the wreck of the Ark.

Mt Ararat from USSR side (Novosti Press Agency)

*Sat Dağ* (Mor Dağ)
   With eight summits over 3500m.
Cia e Hendevade 3810
Cia e Mazan
Knoten 3550
Bobek Tepsi 3460 (fa 1966)

This range extends S along the Turkey/
Iran and Iraq/Iran frontiers – Chalachor 3521; further S it reaches 3600m
at **the highest point in Iraq** (no name).

*Mtn groups on northwards*
Kara Dağ 3630
Ispiriz Dağ 3537
Mengene Dağ 3610
Baset Dağ 3749
Hirabit Dağ 3550
Tendürük Daği 3313
Bü Ağri Daği (Mt Ararat) (V-ext) 5165
   (fa 1829 F. Parrot, Fjodorow, the
   first ascent claim for Noah is dis-
   puted)
Küc Ağri Daği (Little Mt Ararat) 3925

*Mtns around Van Gölü* (Lake Van)
Murathasi 3519
Boz Dağ, groups of volcanic cones up
   to 3520m.
Suphan Daği (V) 4434, a perfect cone.
Nemrut Dağ (V-le 15th century) 2895
Artos Dağ 3475

**Western Turkey**
Ulu Dağ 2493, a well known skiing
   site, the 'Mount Olympus of Asia
   Minor'.
Kaz Daği (Ida Mtns) – Mt Gargarus
   1767, dominates the ruins of Troy.

Lake Van has no outlet. Christian mythology places the Garden of Eden hereabouts.

Nemrut Dağ is one of the largest calderas in the world, 8 km diameter and 600m deep, with pinnacles and lava flows. Earthquake shocks were felt here in 1881, but there has been no recent eruption.

**Mount Hermon** (2814), the highest point in Syria, is a holy mountain often featured in the Bible. It was revered in their turn by Romans and by Druses.

The **Jordan Rift Valley** is a continuation of the Great Rift Valley of Africa. It varies between 3 and 24 km wide and is 400 km long. The floor of the Dead Sea reaches down to −792m.

**Mount Carmel** (546) has long been venerated by successive races. Here the prophets of Baal were worsted by Elijah, so it is said, and here Pythagoras had a retreat.

The **Mount of Olives** (617) is another important Biblical site.

The **Elburz Mountains** exhibit a big climatic contrast between the north side – high rainfall and copious vegetation – and the south – near desert.

**Demavend** (5670), a slightly active volcano and **the highest point in the country**, is a place of pilgrimage figuring prominently

**Turkey/USSR & USSR/Iran frontiers**
(SSRs of Georgia & Armenia) (from
N to S)
B Abul 3304
Troitskaye 3157
Aragats (Alagez) (V-ext) 4090
Gegamskiy Khrebet
Ginaldağ 3373
Gyamysh 3724
Dalidağ 3661
Ishikhly 3552
Dastakert 3381
Kapydzhik 3904

**Turkey/Greece borderlands**

*Cyprus*
Troodos Mtns – Mt Troodos (Olym-
pus) 1951, **the highest point of Cyprus,**
Adelphi 1617, Papoutsa 1562.
Kyrenia Mtns – 1023
Buffavento 956

**SYRIA**

**J esh Sharqi (Anti-Lebanon)** (lining the
Lebanon frontier)
Mt Hermon (J esh Sheikh) 2814, **the
highest point of Syria.**
Talat Mūsa 2659
Halimet el Qabu 2464

**J el Drūz** – 1735

**JORDAN**

**Esh Shara**
J Ram 1754, **the highest point in Jordan.**

in Iranian legend. The summit is another alternative site for the
Ark.

   **Alam Kuh** (4826) has a great rock wall used by climbers. In the
central **Zagros Mountains** beneath Kūh e Parau, the deepest cave

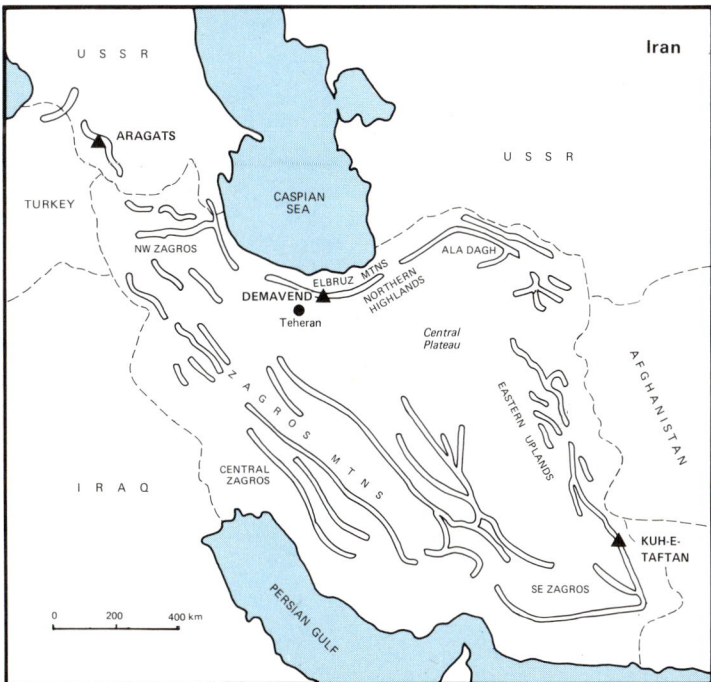

J Mubrak 1719
J el Ataïta 1641

**J Um ed Daraj** – 1234

**W bank of R Jordan** (disputed territory)
Samarian Hills – Mt Gerizim (J el Tar) 881

## LEBANON

**J Liban**
Qornet es Saouda (Qurnat as Saw̄da) 3086, **the highest point of Lebanon.**
Harf el Mreffi 2628
Mt Sannin 2481
  Col du Beidar 1520 behind Beirut is the only pass, carrying rd & rail through the range.

## ISRAEL

Har Meron (Mt Atzmon) 1208, **the highest point in Israel** (discounting Sinai).
Mt of Olives 617
Mt Carmel 546

## IRAN

**Northern Highlands (Elburz Mtns)** (stretch for 1000km in a crescent round the S side of the Caspian Sea. The Firuz Kuh pass carries road and rail across the range)
Demavend (V-slight activity) 5670 (fa 1837: W.T. Thomson), **the highest point in Iran.**
Alam Kuh 4826 (fa 1902 Bornmuller brothers)
Takht-i-Suleiman 4619 (fa 1936 L. Steinauer & party)
Lanāor Hazārchal 4500+
Sīāh Sang 4500+
Sīāh Kaman 4500+
Gämaurā 4401
Āvidar 4288
Mastechāl

**Northern Highlands (Ala Dagh)** (the Iran/USSR frontier, E of the Caspian Sea)
Kūh e Bīnālūd 3416
Kūh e Hazar 3147
  extending across the frontier into Turkmeniya SSR.
Khrebet Kopet Dagh 3050
Khrebet Bol Bolkhan 1880

**Eastern Highlands**
Shāh Kūh 2729
Kūh e Kalat 2605
Kūh e Bārān 2561
Kūh e Estand 2488

**Zagros Mtns**

*NW Zagros Mtns* (from NW to SE)
Kūh e Zakī 3367
Kūh e Saltān 3255
Kūh e Savalan

The mountains of Jebel Ram, Jordan (Aerofilms)

Demavend, the highest point of Iran (Aerofilms)

in Asia, Ghar Parau – 751m, was located and explored by British parties in 1971–2.

Kūh e Kukālār alone comprises a 50km ridge over 4000m high.

Near Shiraz are the palaces of Darius the Great and Xerxes at Perseopolis and of Cyrus the Great at Pasargadac.

Sārī Dāsh 3085
Kūh e Sahand 3710
Chalachor 3521
Kirk Bulāg Dāgh 3217 (mean)

*Central Zagros* (western ridges) (from
  NW to SE)
Kūh e Parau 3393
Kūh e Alvand (Orontes) 3572
Kūh e Garas 3638
Oshtoran (Ushtarinan) Kūh 4070
Kūh e Garbosh 4294
Zard (Zardeh) Kūh 4200
Kūh e Kukālār 4298
Kūh e Alījūq 3724
Kūh e Dinār 4432
Kūh e Būl 3661
Kūh e Tābask 3218
Kūh e Fūrgun 3279

*Central Zagros* (eastern ridge) (from
  NW to SE)
Kūh e Aliābād 3356
Kūh e Karkas 3899
Khār Kūh 3511
Shir Kūh 4074
Kūh e Masāhūn 3472
Kūhellazārān (Kūh e Hazarah) 4420
Kūh e Lāleh Zār 4374

*South-east Zagros*
Kūh e Taftan (V-slight activity) 4042
Kūh e Chihilta 3973
Kūh e Bazmān 3489

## ARABIAN PENINSULA

**West (Red Sea) coast** (a 2000-km
  range)
J al Lawz 2579
J al Magla 2426
J Dibbagh 2349
J al Hijaz 3133
J Razikh 3658, **the highest point in
  Saudi Arabia.**
J Hadur 3760, **the highest point in
  Yemen.**

A mountain range 2000 km long lines the west (Red Sea) coast of the **Arabian peninsula**. The mountains are of sandstone protected from erosion by overlying lavas. A National Park of 4000 km² is being set up in the Asir Province of Saudi Arabia with mountains up to 3000 m.

Jebel Sabir (3006), above the city of Ta'izz, is elaborately terraced nearly to the summit.

**Pieter Botte** (820) has been described as 'a ball on a finger tip', a massive boulder on top of a rock pillar with overhangs all round. It was **first climbed in 1790 by shooting a line over the top with bow and arrow**. In 1832 a British naval party used ladders and slept out on the summit. Now there are ladders and chains for all to use.

Pt 3090
Pt 3268
J Sabir 3006
J Thamar 2512, **the highest point of S
  Yemen.**

**East (Gulf of Oman) coast**
J Akhdar (J Ash Shan) 3170, **the
  highest point of Oman.**

**Socotra**
J Haggier (Hajr) 1503

**Bahrain** – J ad Dukhan 134

**Kuwait** 289

**Qatar** 73

**United Arab Emirates** – J Hafit 1189

## INDIAN OCEAN ISLANDS

**Malagasy (Madagascar)** (from N to S)
Maromokotro (Tsaratanana massif)
  2876, **the highest point on the island.**

Ankaratara 2643
Pic Boby 2658

**Comoro Islands**
Gde Comore 2361, **the highest point of
  Comoran State.**
Kartala (V-le 1972)

**Mauritius**
Piton de la Petite R Noire 827
Pieter Botte 820 (fa 1790, by a French-
  man with the aid of bow and arrow),
  the summit is a very steep pinnacle.
Le Pouce 811 (fa 1836 Chas Darwin
  from the *Beagle*)
Mtgne du Rempart 780, a 'vest pocket
  Matterhorn'.

**Réunion**
Piton des Neiges (V) 3069, **the highest
  point on Indian Ocean islands.**
Le Gros Morne 2992
Piton de la Fournaise (V-active) 2361

**Mahé** (Seychelles)
Morne Seychellois 905

# *Central and Eastern Asia*

## TIBETAN HIMALAYA

Gosainthan (Shishma Pangma) 8046
  (fa 2/5/1964 Chinese exped)
Ngojumba Ri I 7805
Gurla Mandhata 7728
Phola Gangchen (Mt Molhamongjim)
  7661
Ngojumba Ri II 7646 (fa 24/4/1965
  Jap exped)
Ngojumba Ri III 7610
Changtse 7552 (fa 1935 E. E. Shipton's
  party)
Tsunga Ri 7285
Khartaphu 7206 (fa 8/7/1935 E. E.
  Shipton's party)
Kharta Changri 7032 (fa 29/8/1935
  E. E. Shipton's party)
Nyonno Ri 6750

**Gosainthan** (Shishma Pangma) 8046 is the only 8000 m peak entirely within the Peoples' Republic of China and is thus not accessible at the present time to western mountaineers. Even in the days of an independent Tibet it was very remote and, though distantly noticed during the Everest Reconnaissance of 1921, was never seriously approached by climbers.

**1963**  A Chinese expedition reconnoitred to 7160 m.

**1964  First ascent** by some members of a Chinese expedition, 195 strong (the accounts are said to be inconclusive – presumably there are no summit photographs).

During an attempt on **Gurla Mandhata** (7728) in 1905, Dr Longstaff and his two Swiss guides were **carried down 1000 m in two minutes in an avalanche, yet survived**. There has been no subsequent

## INTERNAL RANGES OF TIBET

Shahkangsham 7660
Aling Shan (Aling Kangri) (Nain Singh
Ra) 7315
Kuhanbokang (Nain Singh Ra) 7216
Lungmari 7100
Nien ch'ing t'ang ku la Shan (Nyen-
chen Tanglha Ra) 7088
Lombo Kangra 7060
Kan Lan Shan 7000
Kailas (Kang Rimpoche) 6714
Hlako Kangri 6482
Lhari 6406
Mawang Kangri 6400
Thachap Kangri 6392
Meda Kun (Tanglha Ra) 6200
Basudan Ula 6096

## AGHIL RA (KASHGAR RA)

Qungur II (Qungur Tagh) 7719
Qungur I (Qungur Tjube Tagh) 7595
(fa 16/8/1956 Russian/Chinese exped
led by K. Kuzmin)
Muztagh Ata 7546 (fa 31/7/1956 Rus-
sian/Chinese exped)
Chakragil 6727

## KUNLUN SHAN

Ulagh Muztagh 7723
Subsid pks 7361, 7350
Mo-no-ma-la Shan (Bokalith Tagh)
7720
Muztagh 7281
Cholpanglik 7102
Chung Muztagh 6920
Zogputaran 6900 (fa 1865 W. Johnson
(disputed))

## TIEN SHAN

P Pobeda 7439 (fa 1938 claimed for
Gotman, Sidorenko, Ivanov (dis-
puted); fa (certain) 30/8/1956 V.
Abolakov & party)
P Dostok 7003 (fa 1958)
Khan Tengri 6995 (fa 1929 M Pogre-
bezky & party)
P 20 Jahre Komsomor 6930
P Vascha Pchaveli 6918 (fa 1961)
P Vojennich Topografov 6873 (fa 1965)
P Drushba 6800
P Thorez 6725 (fa 1964)
P Shater 6630 (fa 1964)
P Tchapayev 6371
P Saladin 6280
Aktau 6181
P Gorki 6050
Mt Lingsnan 6000 (fa claimed by Chi-
nese for a monk, Hsuan Chuang, in
AD 628)
P Nansen 6547
Bogdo Ula (Po-ko-to Shan) 5445, only
50km N of the Turfan Depression
−154m.
Karlik Tagh 4925
Khr Kokshaal-tau 5300
Khr Terskey Alatau 5280

Khan Tengri in the Tien Shan (Novosti Press Agency)

attack on the mountain by western mountaineers, of course, for political reasons.

**Tibet** is **the biggest and highest plateau in the world**. It lies between the Kuen Lun and the Himalaya and has an area of $1\frac{1}{4}$ million km$^3$.

**Kailas** (6714) is a very important place of pilgrimage, the throne of Siva, and a Buddhist heaven. Some pilgrims make a 40-km circuit of the mountain, measuring their length on the ground, ie a minimum of 20000 prostrations. It has never been climbed by western mountaineers. Within 60 km rise the great rivers of Asia – Indus, Sutlej, Ganges and Brahmaputra (Tsangpo) – which take widely diverging routes to the sea. The mountain was probably first seen by westerners in 1626; certainly it figures in an atlas of 1735.

The **Aghil Range** was **first discovered by Sir Clarmont Skrine in the 1920s**. An attempt on Muztagh Ata (7546) by E. E. Shipton and H. W. Tilman in 1947 reached within a few tens of metres of the top. This mountain was climbed by a joint Russian/Chinese expedition in 1956, while in 1959 eight women and 25 men reached the top.

Qungur II (7719), **the highest point of the range**, has probably not yet been climbed, but the lower Qungur I (7595) was climbed by a Russian/Chinese expedition in 1956 and by two women from an all-woman expedition in 1961.

Notable early explorers of the **Kunlun** were Sven Hedin and Sir Auriol Stein. The range runs east from the knot of the Pamirs for well over 2000km, lining the south side of the Takla Makan Desert. East of 90°E the Kunlun divides into north and south ranges enclosing the salt marsh depression of Tsaidam (600 ×

Pik Kommunizma, the highest point of USSR
(J P Zuanon)

On the summit of Pik Lenin, Pamirs (R Richards)

## PAMIR/TRANS ALAI

P Kommunizma 7495 (fa 3/9/1933 E. Abolakov), once Mt Stalin and before that Mt Garmo, **the highest point in the USSR.**
P Lenin 7134 (fa 25/9/1928 Allwein, E. Schneider, K. Wien; this is disputed by Soviet climbers but without justification), once Mt Kaufmann.
P Eugenia Korzhenevskaya 7105 (fa 1953 USSR climbers)
P Revolyutsii 6974
P Rossija 6878
P Babouchkine 6859
P Baku-Kommissare 6854
P Izvestij 6840 (fa 1964)
P Mosckwa 6785
P Oktobrskyi 6780
P Karla Marksa 6726
P Dschorsinkoyo 6713 (fa 1936)
P Fikker 6708
P Kgzyl-Agyn 6678
P Edunstwa 6673 (fa 1935 Russo-Chinese party)
P Akhmadi-Donish 6667
P Voroshilov 6603
P Garmo 6595 (fa 1948)
P Tadzhikistan 6565 (fa 1962)
P Pionerskaya Pravda 6550 (fa 1961)
P Alexznder Grin 6525 (fa 1960)
P Engels 6510
P Leningrad 6507

250km at over 2700m), where the drainage is all internal. The pass of Animbar Ulai into Sinkiang pierces the north arm, known as the Astin (Altyn) Tagh. This continues east as the Nan Shan (Humboldt Range) (*see* p. 150). The south arm, Arka Tagh, runs on to the Amne Machin Range and the Pa-yen Ko-la Shan of China.

The first ascent of Zogputaran (6900) was claimed by W.H. Johnson of the Survey of India in 1865. This was disputed and is now discounted.

Running some 1900km east from the knot of the Pamirs the **Tien Shan** encloses the Takla Makan Desert on the north side. **Pik Pobeda** (7439) **is the highest point of the range**. There are some lengthy glaciers (up to 70km). Further east Bogdo Ula (5445) rises only 100km north of the Turfan Depression (−154).

The **Pamirs** are the highest mountain group in the USSR, **rising to Pik Kommunizma** (7495). They are very much climbed upon by Soviet mountaineers and snow and ice climbing has reached a very high standard, including long traverses over numbers of peaks involving many bivouacs.

**Some of the longest mountain glaciers in the world** are to be found in the group – Fedtchenko 77km, Inylchek 70km, Koi-kaf 50km, etc.

P Abolakov 6447
P Pravda 6406

**RANGES N AND NW OF THE PAMIR/TRANS ALAI**

**Alai**
Mintaje 5500 (fa 1935)
P Granit 5308 (fa 1935)
Breithorn 5249 (fa 1935)

**Alayskiy Khr** – 5880

**Turkestanskiy Khr** – 5621

**Keravshansky Khr** – Chimtarga 5487

**Gissarskiy Khr** – 4424

**USSR, W OF ALMA ATA AND OZ ISSYK-KUL**

**Khr Zailiyskiy Alatau Talgar** – 4951

**Khr Kungey Alatau** – 4213

**Kirgizskiy Khr** – 4447

**Khr Talasskiy Alatau** – 4488

**Chatkalskiy Khr** – 4080

**USSR, W OF OZ BAYKAL (LAKE BAIKAL) AND THE LENA R, E OF ALMA ATA**

**Khr Dzhungarskiy Alatau** – 4463

**Khr Saur** – 3816

**Narimskiy Khr** – 3370

**Khr Yuzh Altay** – 3483

**Katunskiy Khr**
G Belukha 4506 (fa 1924 B. & M. Tronow), 15 glaciers hereabouts.

**G Kuytun (Taban Bogdo Ula)** 4356, the meeting place of USSR, China and Mongolian PR.

**Khr Sev Chuyskiy** – 4173

**Khr Yuzh Chuyskiy** – 3942

**Zapadnyy Sayan** – 2930

**Tannu Ola Ra** – Munkhu Khayrhan Ula 3976, G Ak-Oyuk 3615

**Vostochnyy Sayan** – Munku Sardyk 3491

**CENTRAL USSR**

**Gory Putorana** – Gora Kassen 2037

**MONGOLIAN PEOPLE'S REPUBLIC**

**Altai Ra**
Pt 4231, **the highest point in Mongolian PR**

From **'Sogdian Rock'** near the modern city of Derbent in Uzbekistan comes the first record (327 BC) of the use of pitons and ropes in climbing – in a successful assault by Alexander the Great's troops on a crag-top stronghold.

East of the **Khrebet Dzhungarskiy Alatau** (4463) is the Dzungarian Gate used as a route from Russia to China from earliest times by such as Attila and Genghiz Khan.

The **Kamchatka peninsula** and the **Kuril Islands** (which stretch on towards Japan) both have large numbers of active volcanoes. Klyuchevskaya Sopka (4850) is **the highest 'fire mountain' in Asia**. The 1956 eruption of Bezymianny ejected $2.8 km^3$ of material and removed the top 185m of the mountain. Moving initially at nearly twice the velocity of sound, finer particles reached a height of 45km. Samples of volcanic dust from here were detected in Britain within three days.

East of the island arc, the **Kuril Trench** reaches down to $-10542m$ in Vityaz Deep.

There are two outstanding mountain groups within the original bounds of China. **Minya Konka** (7590) in Szechwan Province was **first climbed** (surprisingly) **as long ago as 1932** by an American expedition (a remarkable ascent for the period in terms both of height and remoteness). The **Amne Machin Range** further north was spotted by American fliers air-lifting supplies to China in World War II. At first it was said that there was a mountain here

**Central Asia**

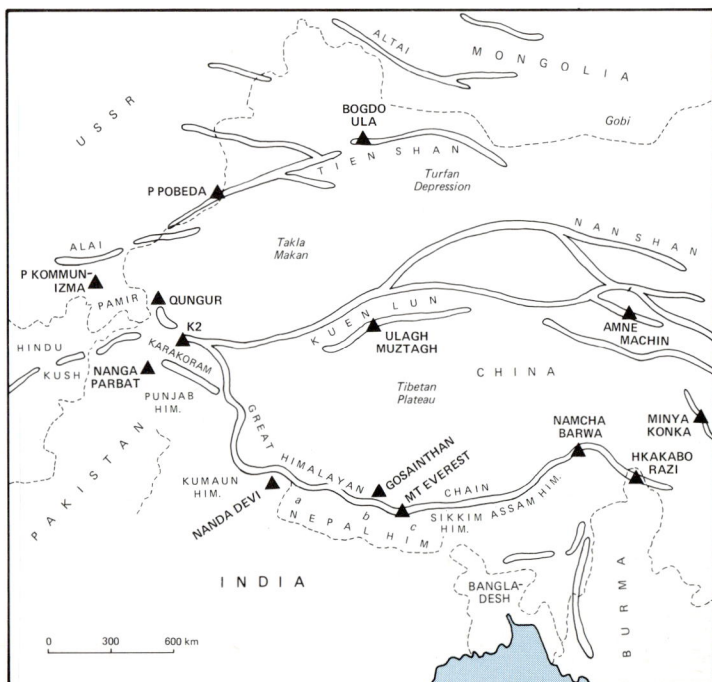

Kharkhira massif – Turgen Uul 4116
Ih Bogd Uul 4000
Gichgeniyn Nuruu Ra – 3772

**Hangayn Nuruu Ra**
Otkhon Tengri 4031
G Turgeni Ekhni Ula 4029

**Hentiyn Nuruu Ra** – 2800

**In the SW** (along the China frontier)
Pieta Shan (Baytag Bogdo) 3187

**MOUNTAINS E OF OZ BAYKAL
(LAKE BAIKAL) AND W OF OL-
EKMA R** (from SW to NE)

**Barguzinskiy Khr** – 2840

**Ikatskiy Khr** – 2573

**Khr Udokan** – G Golets-Skalistyy 2647

**Yuzhno Muyskiy Khr** – 2618

**Khr Synnyr** – G Golets-Inyaptuk 2579

**Daurskiy Khr** – G Bystrinsky Golets
2523

**EASTERN USSR**

**Stanovoy Khr** – G Golets Skalistyy 2482

**Sakhalin Island** – G Lopatina 1609

**Sikhote Alin Gory** – G Tardoki Yani
2078

**Khr Dzhugdzhur** – Topko 1906

**Verkhoyanskiy Khr**
Khr Suntar Khayata – G Mus Khaya
2959
Mt Tohen 3114

**Khr Cherskogo** – G Pobeda 3147

**Khr Kolymskiy** – 2222

**Chukotskiy Khr** – 1843

**Koryakskiy Khr** – G Ledyanaya 2562

**KAMCHATKA** (from S to N)

**Srediniy Khr**
Koryakskaya Sopka (V) 3456
Arachinskaya Sopka (V) 2741
Sopka Zhupanova 2929
Kronotskaya Sopka (V) 3528
Tolbachik (V) 3682
Ichinskaya Sopka (V) 3621
Klyuchevskaya Sopka (V-le 1966) 4850
 (fa 1829 Erman), perfect cone, crater
 200m dia.
Sopka Shiveluch (V) 3283
Shishel 2531
Bezymianny (V-le 1956) 2800

higher than Everest, but this myth is now exploded. **The highest
point** (7612) **was first climbed by Chinese mountaineers in 1960.**

**Omei Shan** (3035) has 35 monasteries and 56 pagodas on its
flanks. Swarms of fireflies account for an impressive local phenomenon known as 'tongues of flame'; hereabouts the Spectre of
the Brocken, often seen, is called 'Buddha's Glory'. Minya
Konka is impressive in the summit view.

**Tai Shan** (1551) was climbed by Confucius before 500 BC. It is a
place of pilgrimage which sometimes attracts 10000 people in a
day, who endeavour to climb the 6700 steps to the Temple of the
Jade Emperor on the summit.

**Faik tu San** (2743) (Hukutozan, Paektu), the 'Great White
Mountain', **the highest point of Korea**, is a shield volcano with a
large crater lake. Legend says that it was climbed 4000 years ago
by Tang um, founder of the nation. It was certainly ascended in
1886 by F. Younghusband in the course of a long Asiatic journey.

**Fuji-san** (Huzi-san, Fuji) (3776), one of the best known mountains
of the world, is a national symbol for Japan. A volcano, which
occasionally exhibits slight activity, it is a near perfect cone. **The
first ascent was made in AD 663** (1100 years before Mont Blanc);
there was a Buddhist temple on the top before 1150. **The first
European ascent** was accomplished by Sir Rutherford Alcock in
1860; since 1558 the mountain had been forbidden to women
until Lady Parker made the climb in 1867.

Now there is a road to 2168m and a weather observatory; it is
a frequent goal of pilgrimage.

There are some other outstanding mountains and volcanoes in
Japan:
  **Ontakesan** (3063), an extinct volcano on Honshu, is also much
visited by pilgrims. On top is a 14th-century shrine and several
lakes.
  **Tateyama** (3015) in the same area also has a summit shrine
visited by pilgrims.
  **Asama-yama** (2542) in northern Honshu is **the highest active
volcano in the country**. Its frequent eruptions are checked by a
geophysical observatory on the slopes of the mountain.
  **Asozan** (1592) in Kyūshū has a crater floor 24 × 16km, **one of
the largest in the world**. There is a group of five cones, of which
only Nakadake (1323) is active – last eruption in 1970.
  **Sakurajima** (1069), also in Kyūshū, last erupted in 1972. In
1914, two km³ of material was ejected in a two-day eruption.
Several weeks later the temperature of the sea was still 60°C. The
countryside settled up to two metres over a distance of 120km.
Some subsidence was detected as far away as Taiwan (1500km).
Both this and Asozan are under constant surveillance from geophysical observatories.

**KURILSKIYE OSTROVA (KURIL ISLANDS)** (from S to N)

**Kunashir** – Tyatya (V) 1822
**O Atlasova** (V) – 2399
 Many other volcanoes over 1000m.

**CHINA** (excluding Tibet)

**Minya Konka Ra**
Minya Konka 7590 (fa 28/10/1932 R. Burdsall, T. Moore)
Ru-dshe Konka 7100
E Konka 6900
Reddomain 6700

**Amne Machin Ra**
Amne Machin 7612 (fa 2/6/1960 Chinese exped)
Drandel Rung Shukh
Shenrezig

**Other mainland mountains**
Joma 6800
Medu Kun (Atak Hapchiga) 6200
Humboldt Ra – Nan Shan (Shu le Nan Shan) – 6100
Pa-yen-k'o-la Shan – 5500　'
Nun Shan (Yunnun Prov) 4800
Mao-mao Shan (Kansu Prov) 4070
Tai-pai Shan (Shensi Prov) 3666
Wu-liang Shan 3500
Omei Shan (Szechwan Prov) 3035, specially sacred.
 *There are five more specially sacred mountains, with well-trodden paths to summits and temples:*
Hengshan (Shansi Prov)
Tai Shan 1551 (Shantung Prov)
Heng Shan (Hunan Prov)
Hua Shan (Shensi Prov)
Sung Shan (Honan Prov)

**Hainan**
Red Mist Mtns 1879

**HONG KONG**

Tai Mo Shan 954
Kowloon Pk 601 (mainland)
Victoria Pk 551 (island)
 Rock climbing.

**KOREA**

**North**
Faik tu San (Hukutozan, Paektu) (V) 2743 (fa 1886 F. Younghusband), **the highest point of North Korea.**
Bugpotae 2435
Duryu 2369
Nangrim 2014

**Taeback Ra (Diamond Mtns)** (North/ South frontier) (from N to S)
Kumgang 1638
Shusenho 1356
Sorak San (Mt Seorak) 1708
Mt Odae 1563
Mt Taeback 1549

**Myozin-syo** was an island formed in the sea, 420km south of Tokyo, by volcanic action on 17 September 1952; 30m above sea level at its highest, it was eroded away by 21 September 1952. A ship sent to investigate, having on board 22 crew and 7 scientists, was directly above the vent when the next explosion occurred, and it vanished without trace.

Island arcs mark the tectonic plate boundaries where both plates are submerged beneath ocean waters. These, and their associated trenches, are plentiful in the western Pacific Ocean. From north to south come – the **Kuril Islands** (Atlasova Island 2399; Vityaz Deep − 10542), the **Japanese Islands** (Fuji-san 3776; Ramapo Deep − 10374), the **Mariana Islands** (Farallon do Pajaros 322; Challenger Deep − 11033, **the deepest ocean sounding in the world,** and the **Philippine Islands** (Apo 2954; Philippine Trench − 10497). All the islands are more or less volcanic, some very actively so.

There is a caldera at the crest of the Barisan highlands in **Sumatra**, the flows from which are distributed over 19 000 km$^2$ of the island and have even been traced into Malaysia.

The island of **Krakatau** in the Sunda Straits was the scene of **the most violent volcanic eruption of historical times.** On 26 August 1883 clouds of dust and gas reached a height of 28 km. The next day huge explosions, which were heard no less than 5000 km

Hong Kong – mountains and skyscrapers (Barnaby's)

away, hurled clouds of ash and pumice to 80km. This was dispersed around the world and coloured the sunsets vividly for months afterwards. Destructive sea waves (tsunami) were set up, killing 36 000 people on Sumatra and Java. Two-thirds of the island disappeared, a deep submarine hollow replacing 20 km$^2$ of land, with an estimated 18 km$^3$ of material ejected. In 1927, with the sea in the caldera 300m deep, a new cone, called Anak Krakatau, began to grow and to erupt.

**Mtns near Seoul**
Around the capital are some striking rock peaks, probably inselbergs, offering considerable climbing prospects: Hakuundai 884 (has a staircase cut in living rock with an iron handrail), Insupong 850, Bankeidai 850, Dobongsan 741, Manjoho 600, Goho, Paegundae, Obong-San.

**Sobaek Mtns & other S ras**
Mt Jiri 1915
Mt Gaya 1430
Mt Sobaek 1421

**Cheju-do (Quelpart Island)**
Halla San (V) 1950, **highest point in S Korea.**

# JAPAN

## Hokkaido

*Ranges* – Kitami, Ishikari, Hidaka, Yubari and Teshio.

*Peaks*
Taisetsuzan (Asahi-Dake) (V) 2290
Tomuranshi-Yama 2141
Tokachi-dake (V) 2077
Horoshiri-dake 2052
Ishikari-dake 1980
Meakandake 1503
Yoichi-dake 1488
Oakandake 1371
Yokutsu-dake 1167
Komagatake (V-le 1942) 1146
Tarumae-yama (V-le 1917) 1030
Usu-dake (V-le 1945) 725

## N Honshu

*Ranges* – Ohu, Dewa, Kitagami and Abukuma.

*Peaks*
Shirane-san 2578
Asama-yama (V-frequent eruptions) 2542
Nantaisan 2484
Hiuchidake 2346
Iwasuge-yama 2295
Chokaisan 2230
Komage-take 2132
Iidesan 2105
Iwatesan 2041
Azuma-yama 2024
Gassan 1980
Tanigawadake 1963
Nasudake 1918
Hayachine-san 1914
Asahi-dake 1870
Zaosan 1841
Bantaisan (V-le 1954) 1819, very violent eruption in 1888.
Iwakisan 1625

## Central Honshu

*Ranges* – Hida (N Japanese Alps), Kiso (Central Japanese Alps) and Akarshi.

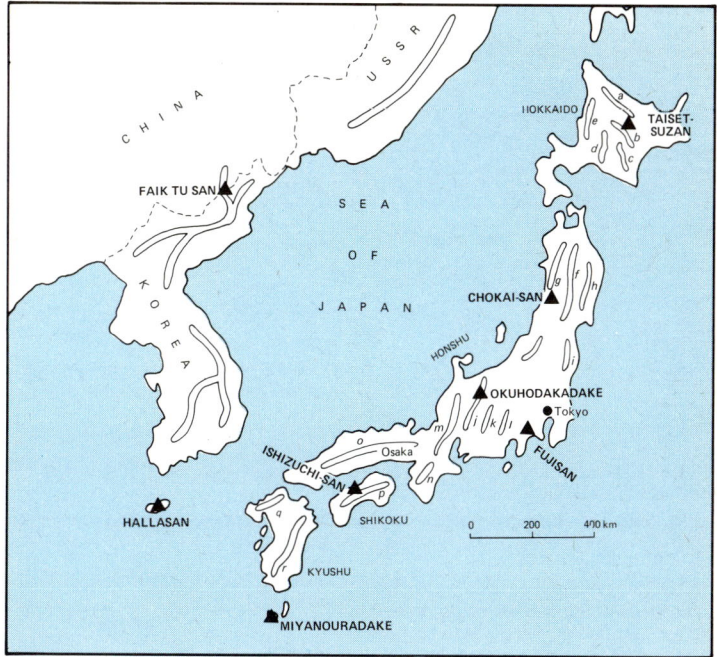

Japan

*Peaks*
Fuji-san (Huzi-san, Fuji) (V-le 1707, occasional slight activity) 3776 (fa 663, rd to 2168m, **the highest point of Japan.**
Kitadake (Shirane-san) 3192
Okuhodakadake 3190
Ainotake 3189
Yarigadake 3180
Arakawadake 3146
Akaishi-dake 3120
Ohbamidake 3120
Karasawadake 3103
Kitahodakadake 3100
Ontakesan (V) 3063
Norikuradake (V) 3026
Tateyama 3015
Tsurigedake 2998
Kiso-komagatake 2956
Shiroutnadake 2933
Yatsugatake 2899
Myokosan 2446
Amagi-san 1407

## Western Honshu

*Ranges* – Suzuka, Kii and Chugoko.

*Peaks*
Hakusan 2702
Ominesan 1915
Daisen 1731
Dainichige-take 1709
Odaigaharayama 1695
Nogohaku-san 1617
Mimuroyama 1358
Kammuriyama 1339

## Shikoku

*Range* – Shikoku

*Peaks*
Tsurugi-San 1995
Ishizuchi-San 1981

## Kyūshū & Yakushima

*Ranges* – Chikushi and Kyūshū.

*Peaks*
Miyanouradake 1935
Kujusan 1788
Sobo-san 1759
Kunimidake 1739
Kirishimayama 1700
Asozan (V-le 1970) 1592, only Nanadake cone 1323 is active.
Unzendake 1360
Sakurajima (V-le 1972) 1069

## Islands S of Japan

A belt of volcanic islands stretches from O Shimi near Tokyo to the Marianas by way of Miyake-jima, Hachijō-jima, Aogo-shima, Tori-shima, Kita-io-jima and Minami-io-jima, alongside to the E is Ramapo Deep (−10374m).

Another belt of volcanic islands stretches from Kyūshū towards Taiwan (Formosa) by way of Taki-shima, Kuchinoerabu-shima, Nakano-shima, Saiwanose-jima (le 1970) 803 and Tori-shima. Beyond is Okinawa (Yonaha-dake 498).

## MARIANA ISLANDS

Farallon do Pajaros (Uracas) (V-le
1969) 322
  To the S Challenger Deep in the
  Marianas Trench reaches −11 033
  **the deepest ocean sounding in the
  world.**

## TAIWAN (FORMOSA)

Yu Shan (Niitaka, Mt Morrison) 3997
  (fa 1896 S. Honda & Forestry
  officials), **the highest point of Taiwan.**
Tsugitaka (Mt Sylvia) 3931
Hseuh Shan 3884
Shukoran 3834
Maborasu 3806
Nankotai 3798
Hapanorau 3734 (fa 1930 Murray
  Walton)
Chusoenzan 3716
Koshi 3673
Daihasenzan (Ta Pa Chien Shan) 3572
  (fa 1928 T. Ikoma, T. Numai)
Kiranshu 3542
Momoyama 3389

As well as being very mountainous, the
Island is distinguished by very high sea-
cliffs at the N end of the E side.

## PHILIPPINES

**Luzon**
Mt Pulog 2929
Mt Tabayoo 2842
Mayon (V-le 1969) 2451, a perfect cone,
  crater 500 m dia.
Mt Sapocoy 2216
Taal (V-le 1969) 302, 1400 people
  killed by the eruption of 1911.

**Mindoro**
Mt Halcon 2585
Mt Baco 2488
San Jose (V)

**Negros**
Canlaon (V) 2497, other volcanoes.

**Mindanao**
Mt Apo (V) 2954, **the highest point of
  the Philippines.**
Mt Ragang (V-le 1915-6) 2815
Mt Dapiak 2560
Mt Malindong 2425

**Palawan**
Mt Mantalingajon 2054
The Teeth 1798
Cleopatra Needle 1593

**Camiguin Island**
Catarman (V-le 1948–52) 1332

## INDO-CHINA

**Thailand** – Doi Inthanon 2595, P
  Miang 2316

**Laos** – Phou Bia 2820

**Cambodia** – Phnom Aural 1813, Mt
  Ka-Kup 1563, Phnom Tumpor 1563

**N Vietnam** – Fan Si Pan 3143
  In Along Bay perpendicular limestone peaks rise as islands.

**S Vietnam** – Ngoc Linh 2598

## MALAYSIA

**Mainland**
G Tahan 2189
G Korbu 2182
G Chamah 2170

**Sarawak/Sabah (Borneo)**
G Kinabalu 4101 (fa 1858 H. Low,
  Spencer) **the highest point of Malay-
  sia.**
G Trus Madi (Suniatan Besar) 2597
Mt Meutapak 2000
Mt Melta 2000

## SINGAPORE

Bukit Timah 177

## INDONESIA

**Sumatra**
G Kerinci (Kerentji, Indrapura Pk)
  (V) 3800 (fa 1877 von Hasselt, Veth)
G Leuser (Lauser) 3423 (mean)
G Dempo (V) 3159
G Bandahara 3012
G Abongabong (V) 3000
G Lembu 2983
Bt Masurai 2933
G Talakmau (V) 2912
G Marapi (V-le 1957) 2891
Pt 2877 (V)
G Guereudong (V) 2855
G Patah 2817

G Peuetsagu (V) 2780
G Bepagut 2732
G Sulasih (V) 2597
G Resaru 2585
G Sumbing (V) 2508
Bt Duen (V) 2467
Sinabung (Sinaboeng) (V) 2383 (mean)
G Gedang 2383
G Malintang (V) 2262
G Resag 2232
Sibajak (V) 2225
G Bateemucica 2140
Sekincau 1718
Pematang Bata lava field (V-le 1930s)

**Krakatau** (V) an island in the Sunda
  Straits, **the most violent volcanic
  eruption of historical times** (1883).

**Java**
The island has a chain of around 50
volcanoes, of which the principal ones
are:

Semeru (V-very active) 3676 (fa 1838
  G. Clignett)
G Slamet (V-le 1967) 3428
G Sumbing (V) 3371
G Ariüna (V) 3343
G Raung (V-le 1945) 3332
G Lawu (V) 3265
Merbabu (V) 3142
G Sundoro (V) 3135
G Argopura (V) 3089
G Cereme (V) 3078
Pangrango (V) 3019
Gedeh (V-le 1949) 2975
G Merapi (V-le 1972) 2911
G Butak 2874
Tasikmalaya (V) 2821
G Merapi (V) 2800
Papandayan (V-le 1925) 2680
G Kendang (V) 2608
Bromo (V-le 1950) 2581

**Indonesia and Malaysia**

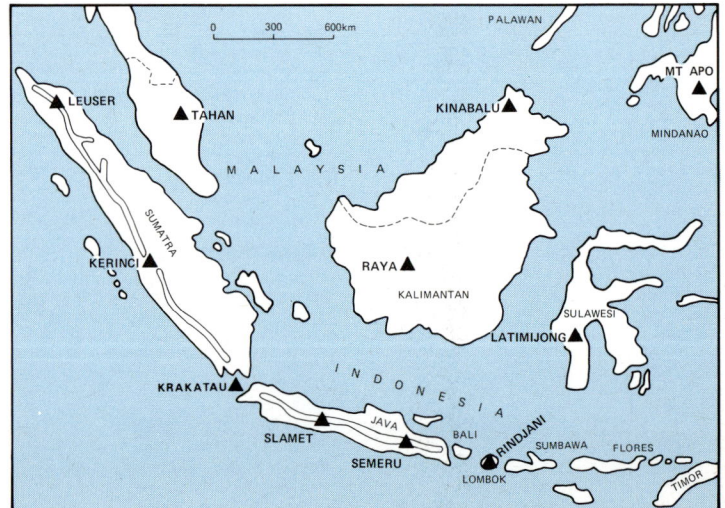

G Prau (V) 2565
G Liman (V) 2563
G Patuha (V) 2434
Kawah Idjen (V-le 1936)
G Malabar 2321
Guntur (V-le 1847) 2261
G Salak (V) 2211
Galunggung (V-le 1918) 2180
Kelud (V-le 1967) 1740
Lamongan (V-le 1898) 1678

**Bali**
Agung (V-le 1963–4) 3142, the 'Navel of the World'.
Batukau 2276
Abang 2152
Batur (V-le 1968) 1727

**Lombak**
G Rinjani (V-le 1966) 3726

**Sumbawa**
G Tambora (V) 2850

**Flores**
P Ranakah 2400
Mandasawa (V) 2382
Lewotobi Lakilaki (V-le 1968) 1593

**Kepulauen** (Solor and Alor)
Ili Boleng (V-le 1950) 1669, many other volcanoes.

**Timor**
Pt 2960
Pt 2472
Mutis 2427
Pt 2315

**Banda Sea**
Banda Api (V), the summit is a small island which rises 5000m from the sea floor and is 20km across the base.

**Kalimantan** (Borneo)
Pt 2988
Raya 2278

**Sulawesi** (Celebes)
Latimijong (Bk Rantekombola, Mt Rantemario) 3440
G Lokilalaki 3314
G Waukara 3127
Bk Gandadiwata 3074
Pt 3016
Bk Kambuno 2950
G Ogoomas 2913
G Lompobottang 2871
G Mekongga 2799
G Tokala 2630
Soputan (V-le 1968) 1794
   Many other volcanoes.

**Sangihe**
Api Siau (V-le 1967) 1794
Awu (V-le 1968) 1327
Ruang (V-le 1949) 718

**Halmahera**
Mondioli 2111
Ternate (V-le 1963) 1715
Gamkunoro (V-slight activity) 1635

**Java** is a chain of more than 50 volcanoes:
   **Gunung Raung** (3332) claimed thousands of victims in the 1638 eruption.
   **Papandayan** (2680) killed 3000 in an explosive eruption in 1776.
   **Gunung Merapi** (2911) and **Kelud** (1740), where over 5000 were killed in 1919, are very active and constant surveillance is maintained.
   **Galunggung** (2180) erupted with 4000 fatalities in 1822.

**Gunung Tambora** (2850) on Sumbawa erupted in 1815 at 15-minute intervals over a period of four months; 100km³ of material was ejected leaving a caldera 6km across in place of a cone 3650m high; 15cm blocks were hurled 40km; 90000 people perished.

The **Owen Stanley Range** is named for the skipper of the *Rattlesnake* which took a scientific party to the area in the mid-19th century.

**Adams Peak** (2243) in Sri Lanka was climbed between 1325 and 1355 by Ibn Battuta, a Moslem explorer from Tangier, during his world travels. Even then, as now, there were chains to aid the ascent. On the summit platform is a hollow in the shape of a human foot, 160 × 75cm, the footmark of Buddha, or of Adam, or of Siva – a place of pilgrimage for everyone.

**Mount Agung, Bali, Indonesia (Indonesian Embassy)**

Dukono (V-long periods of activity) 1093

**Buru** – Pt 2114

**Seram** – Mt Pina-ia 3019

**West Irian**
The nomenclature in this area is extremely confused.

Jayakusumu (Carstensz Pyramid, apparently the Indonesians call it Ngga

Pulu) 4883 (fa 13/2/1962 Harrer, Temple, Kippax, Huizenga)
Ngga Pulu 4860
Sunday Pk 4860
Oost Carstensz top 4840
P Tricora (formerly Sukarno, formerly Wilhelmina) 4730
Enggea (Idenburg top) 4717
P Mandala (formerly Juliana) 4640
P Wisnumurti 4595
Pt 4350
G Leonard Darwin 4234
Pt 4040
Peg Weyland 3897
Z Keten 3750
Angemuk 3741
Anggi Gig 3100
Kwoka 3000
Peg Arfak 2939


## PAPUA

### Central & Bismarck Ras
Mt Wilhelm 4600
Mt Kubor 4300 (fa 1965 P. Hardie and a native), the locals believe the peak to be haunted.
Mt Herbert 4267
Mt Giluwe 4088
Mt Hogen 4000
Capella 3993
Dima pks 3962
The Sugarloaf 3962
Mt Michael 3810
Mt Piora 3719
Mt Otto 3539
Aiyang 3505
Mt Kerewa 3414
Mt Yelia 3384
Mt Lolibu 3353
Crater Mtn 3231

### N coast & Huon Peninsula
Mt Bangeto 4107
Mt Sarawaket 4100
Mt Gladstone 3475
Mt Disraeli 3353

Manam Island (V-very active)
Bam Island (V-very active)

### Owen Stanley Ra
Mt Victoria 4073 (fa 1889 Sir W. MacGregor's exped)
Mt Albert Edward 3993
Mt Scratchley 3920
Mt Strong 3766
Mt Suckling 3676
Mt St Mary 3654
Mt Obree 3129
Lamington (V-le 1951, many thousands killed) 1687

## BISMARCK ARCHIPELAGO

### New Britain
Mt Sinewi 2438
Mt Utawun 2300
Mt Bomus 2249
Mt Talawe 1834
Rabaul (V-le 1969)

To the S Planet Deep in the New Britain Trench is −9140 m.

### New Ireland
Mt Konogaiang 1871
Schleinitz Ra 1480

### Admiralty Islands – 719

### INDIA/BURMA/PAKISTAN (Non-Himalayan)

### China/India (Assam) frontier
Kadusam 5108
Shalunli 4336

### Assam foothills – India (Assam)/ Burma frontier & western Burma
Daphabum 4578
Abor Hills 4500
Mishmi Hills 4500
Saramati (Patkai Ra) 3826
Moi Len 3104
Mt Victoria (Chin Hills) 3053
Japuo Mtn 3015

Kennedy Pk 2704
Blue Mtn 2164

**Burma/China frontier & eastern Burma**
Hkakabo Razi 5881, **the highest point in Burma.**
Kaolikung Shan – Pts 4100, 4036, 3404
Kumor Ra – 3410

**India** (non-Himalayan)
Cardamon Hills – Anai Mudi Pk 2695 }
Nilgiri Hills – Doda Betta 2636 } in the SW

From them a plateau edge runs N parallel to the W coast – the Western Ghats (the Gersoppa Falls 255 m sheer are noteworthy)

Eastern Ghats – Mahendragiri 1501
Aravalli Ra – Guru Sikhar 1722

The Deccan lava plateau covers 500 000 km$^2$ between the Eastern and the Western Ghats, with a total volume of 700 000 km$^3$.

**Pakistan** (non-Himalayan, non-Hindu Kush)
Zargum 3578
Khalifat 3487
Takatu 3456
Takht-i-Sulaiman 3374
Kand 3273
Sakir 3092
Ras Koh 3008
Koh-i-Patandar 2283

## SRI LANKA

Pidurutalagala 2518, **the highest point in Sri Lanka.**
Kirigalpotta 2394
Totupola 2359
Adams Peak 2243 (an ascent recorded between 1325 & 1355 by Ibn Battuta, Moslem explorer from Tangier)
World's End 2176

# Himalaya and Karakoram

Laila (A Kus)

## HINDU KUSH

**Afghanistan** (W of Salang Pass)
   *On N branch of main ridge:*
Safed Kuh ra 3084
Band-e-Baba ra 3746
Kuh-e-Hissar ra 4231
   *On the S branch of the main ridge:*
Kasa Murkh ra 3525
Band-e-Baian ra 3699
Band-e-Duakhan ra 3753
Kuh-e-Baba ra – Shah-e-Foladi 5143
   *S of main ridge:*
Koh-i-Quisar 4148
Koh-e-Sangan 3923
Paghman Ra – Takhte Turkoman 4699
Safed Koh ra – Sikaram 4761

The principal crossing of this section
of the range from Kabul to the Oxus
valley is by way of the Salang Tunnel
beneath the Salang Pass.

**Central Hindu Kush** (Salang Pass to
Dorah Pass)
Koh-e-Bandaka 6843
Koh-e-Morusg 6435
Koh-e-Chrebak 6290

The range of the Hindu Kush is 1300 km long and extends over
the whole length of Afghanistan. It starts in the east in the
Taghdumbash Pamir, where the Muztagh and Sarikol ranges
join close to the tip of the Wakhan corridor, and forms throughout its length the watershed between the basins of the Oxus and
the Indus. For 300 km it provides a frontier between Afghanistan
and Pakistan and thereafter lies entirely within the former. Inside
Afghanistan it divides into two parallel ranges – Kuh-e-Baba,
Band-e-Duakhan, Band-e-Baian and Kasa Murkh in the south
and Kuh-e-Hissar, Band-e-Baba and Safed Kuh in the north.
**Tirich Mir** (7699) **is the highest peak.**

**Noshaq** (7492) is the highest point of Afghanistan. Only a week
after the first ascent by a Japanese expedition in August 1960, a
Polish team also reached the summit. The mountain is straightforward and is often climbed. An Austrian party in 1970 used skis
on both ascent and descent. In February 1973 it was **the first high
Asian top to be climbed in winter** (by a Polish expedition). It was
climbed solo by C. Zurek in 1976 in eleven hours and the following
year by a Czech lady, D. Sterbova, also solo.

Koh-e-Mondi 6234
Sakh-e-Anjoman 6026
Danak (Koh-e-Khakestarak) 5962
Koh-e-Koran 5841
Koh-e-Marsamir (Mir Samir) 5809
Koh-e-Fergardi 5092

**High Hindu Kush** (E of Dorah Pass)
Tirich Mir-main pk 7699 (fa 21/7/1950
   Norwegian exped: P. Kvernberg,
   H. Berg, A. Naess, H. Streather)
Subsid pk – E Pk 7691 (fa 25/7/1964
   Norwegian exped: R. Hoibakk, A.
   Opdal)
Noshaq 7492 (fa 17/8/1960 Jap exped:
   G. Iwatsabe, T. Sakai), **the highest
   point in Afghanistan.**
Tirich West I 7487 (fa 20/7/1967 Czech
   exped)
Noshaq East 7480 (fa 21/8/1963 Aus-
   trian exped led by H. Gruber)
Istor-o-nal 7389 (fa 1969 Spanish ex-
   ped: an American exped, 8/6/1955,
   did not reach the highest point)
Subsid pks 7365, 7303 (fa 1969 Spanish
   exped) & 7240 (fa 29/6/1967 Aus-
   trian exped)
Saraghrar 7349 (fa 24/8/1959 Italian
   exped: F. Alleto, G. Castelli, P. Con-
   sighlio, H. Pinelli)
Tirich West IV 7338 (fa 6/8/1967 Austrian exped)
Shingeik Zom 7291 (fa 13/7/1966 Austrian exped)
Subsid pks 7170, 7150 (fa both 20/8/
   1969 Austrian exped)
Darban Zom 7220 (fa 12/9/1965 Austri-
   an exped: V. Kossler, M. Schmuch)
Noshaq West 7219 (fa 21/8/1963 Aus-
   trian exped led by H. Gruber)
Udren Zom 7131 (fa 19/8/1964 Aus-
   trian exped)
Subsid pk 7078 (fa 10/8/1977 Jap exped)
Koh-e-Skhaur (Shakhaur) 7116 (fa 17/
   8/1964 Austrian exped)
Nobaisum Zom 7070 (fa 10/7/1967
   Austrian exped)
Langar Zom 7061 (fa 8/7/1964 Austrian
   exped)
Koh-e-Urgunt (Urgend) 7039 (fa 4/9/
   1963 Austrian exped: S. Burkhardt,
   M. Eiselin, H. Ryf, A. Strickler, V.
   Wyss)
Akher Chioh (Achez Czioch) 7020
   (fa 10/8/1966 Austrian exped: H.
   Schell, R. Göschl)
Koh-e-Tez 7015 (fa 28/8/1962 Polish
   exped)
Koh-e-Shan 7010 (fa 1977 Czech exped)
Lunkho-i-Dosare 6902 (fa 4/8/1968
   Austro-Yugoslav exped)
Lunkho-i-Harvar 6895 (fa 5/8/1967
   Jap exped)
Koh-e-Hevad 6849
Languta-Barfi 6827
Koh-e-Nadir Shah 6814
Koh-e-Kishmi Khan 6800
Gumbaz-e-Safed 6800
Dirgol Zom 6778
Ghul Lasht Zom 6665
Koh-e-Mandaras 6631

The Punjab Himalaya is a roughly rectangular region, 500 km long by 250 km wide, bounded by the Indus to the north-west and the Sutlej to the south-east. The crest of the Himalaya runs from Nanga Parbat to the Nun-Kun massif and thence to the Sutlej near Shipki. Parallel to the north is the Zaskar Range; to the south are the Pir Panjal and the Dhaula Dhar. At the south-eastern end of the latter ranges are found the districts of Kulu, Lahul and Spiti, popular with climbers because of their accessibility. Between the Zaskar Range and the Karakoram is the narrow Ladakh range.

Where the Indus pierces the range close to the Nanga Parbat massif the land rises from 1000 m at the river to 8125 m at the summit in only 20 km (ie more than 1 in 3). In 1840–1 a landslide into the Indus produced a lake 6·5 km long. When the dam eventually gave way there were disastrous floods in the lower valley, the casualties including a whole Sikh army.

**Nanga Parbat** (8124), the 'Naked Mountain', is the most westerly of the 8000 m peaks of the Himalaya; it has come to be called 'the killer mountain' since 36 lives have been lost in 22 attempts to climb it. Its story mirrors all facets of the human struggle – devotion and determination, controversy and misunderstanding, tragedy and triumph:

**1895**   The famous mountaineer, A. F. Mummery, and two Gurkha companions were lost during a reconnaissance on the Diamir (north-west) face.

**1932**   A German expedition reached 6950 m on the Rakhiot (north-east) face.

**1934**   A German expedition reached 7900 m before being driven back by the onset of monsoon snowfalls with the loss of three Europeans and six Sherpas.

**1937**   A German expedition met with complete disaster when seven Europeans and nine Sherpas were buried by an avalanche at Advance Base.

**1938**   A German expedition failed at 7250 m on the Rakhiot face.

**1939**   Exploration of the Diamir face by an Austrian party.

**1953**   **The first ascent of the mountain by the Rakhiot face route** by H. Buhl of a German expedition led by K. M. Herligkoffer. Buhl was alone for the topmost 1200 m and had to spend a night in the open at 8000 m – a remarkable achievement.

**1961**   Further exploration of the Diamir face.

**1962**   **First ascent of the Diamir face** by T. Kinshofer, S. Low and A. Mannhardt. Low was killed during the descent.

**1963**   Exploration of the Rupal (south) face.

**1964 and 1968**   Failures on the Rupal face.

**1969**   A Czech expedition failed on the Rakhiot face.

**1970**   **First ascent of the Rupal face** by R. and G. Messner; F. Kuen and P. Scholtz followed the next day. The Messners traversed the mountain, descending the Diamir face, where G. Messner disappeared, believed killed in an accident.

**1971**   The second ascent by the Rakhiot face by a Czech expedition (I. Fiola and M. Orolin).

**Hawker Harts of RAF fly past Nanga Parbat, 1932 (Air Ministry Photograph)**

Asp-e-Safad 6607
Muhi Zom 6441

**Eastern Hindu Kush** (and mtns S of main ridge line)
Koyo Zom 6889 (fa 1968 Austrian exped)
Thui Zom I 6661
Buni Zom 6551
Thui Zom II 6523 (fa 1978 British exped)
Dashar Zom 6518
Pechus Zom 6514
Awi Zom 6484
Chiantar Sar 6416 (fa 1967 German exped)
Shah Dok 6320
Ghochhar Sar 6249
Chikar Zom 6110
Miangul Sar 5996
Falak Sar 5918

## PUNJAB HIMALAYA

**Nanga Parbat massif**
Nanga Parbat – main pk (Diamir) 8124 (fa 3/7/1953 German/Austrian exped: H. Buhl solo)
Nanga Parbat Vorgipfel 7910 (fa 11/7/1971 Czechoslovak exped)
Nanga Parbat North 7816
Rakhiot Pk 7070 (fa 16/7/1932 German exped)
Chongra Pk 6824 (fa 14/7/1932 German exped)

**Nun-Kun massif**
Nun 7135 (fa 28/8/1953 Franco/Swiss exped)

**1976** Fifth ascent of the mountain by an Austrian team.
**1977** An American expedition failed with two casualties.
**1978** R. Messner climbed the mountain for the second time, solo.

In June 1977 Sylvan Saudain skied down from the summit of **Nun** (7135), **the highest mountain so far descended in this way**.

**The biggest concentration of high peaks in the world**, the Karakoram is bounded by the Shyok and Indus Rivers in the south (beyond which lie the Himalaya) and the Shaksgam River in the north; the eastern limit is the 'big bend' of the Shyok River, the western limit the Taghdumbash Pamir. This last is a knot from which radiate the Karakoram (south-east), the Kuen Lun (east), the Hindu Kush (west) and the Pamir (north). Four of the world's 8000m peaks and some of the longest mountain glaciers lie within these bounds.

**K2** (8610) is **the second highest mountain in the world** and **the highest mountain in Pakistan**. A recent redetermination of the height has made it 8760m, but the exact value is by no means settled yet. A remote mountain, not readily visible from afar or from the habitations of man, it seems to be without a native name. The suggestion of Mount Godwen-Austen (by analogy with Mount Everest) seems never to have caught on and, along with a few others, it has retained its original survey designation.

**1902** A reconnaissance of the north-east ridge by an international expedition led by O. Eckenstein.
**1909** The Duke of the Abruzzi's expedition reached 6700m on the south-east (Abruzzi) spur and 6675m on the north-west ridge.
**1928–9** An Italian reconnaissance expedition led by the Duke of Spoleto.

**K2 (R D Treble)**

**High camp on Broad Peak looking across to K2 (J Ferenski)**

Kun 7087 (fa 3/8/1913 Italian exped led by Count Calciati)
Pinnacle Pk 6952 (fa 1906 Dr & Mrs Bullock Workman)
Kolahoi 5456, the Kashmir 'Matter-horn'.

**Kaskar & Pir Panjal Ras** (Kulu, Lahul and Spiti)
Leo Pargial 6791 (fa 1933 British exped led by M. Pallis)
Mulkila 6517
Shilla 6500 (fa claimed for 1860 by an Indian native surveyor, long believed to be over 7000 m)
Papsura 6451 (fa 3/6/1967 British exped led by R. Pettigrew)
Dharmsura (White Sail) 6446 (fa 1941 J.O.M. Roberts's party)
Indrasan 6221 (fa 1962 Jap exped)
Deo Tibba 6001 (fa 1952 J.de V. Graaf's party)
Pir Panjal Ra 5790

**The principal passes of the Punjab Himalaya**
Burzil Pass 4199 (E of Nanga Parbat) links Srinagar with Gilgit.
Babusar Pass 4173 (W of Nanga Parbat) carried the road from Abbotabad to Gilgit.

**1938** An American team reached 7950 m on the Abruzzi spur.
**1939** Another American party reached a similar height on the same ridge. There were four deaths when a breakdown of communications between camps left the top climbers with insufficient support.
**1953** Yet another American expedition so nearly met disaster on the Abruzzi spur. In bringing down a sick man (who was subsequently swept away in an avalanche) a slip occurred on an ice slope, but with the help of good fortune P. Schoening held the other seven single-handed.
**1954** **First ascent of the mountain by the Abruzzi spur** by A. Campagnoni and L. Lacedelli of an Italian expedition led by A. Desio.
**1960** An American/German expedition failed at 7260 m.
**1975** An American expedition failed at 6700 m on the north-west ridge.
**1976** A Polish team failed at 8400 m on the north-east ridge, while a Japanese reconnaissance reached 7160 m.
**1977** The second ascent of the mountain by a large Japanese expedition which put seven members on top in two parties with the aid of 47 Japanese, 3 Pakistanis, 125 million yen and 898 porters.
**1978** Third ascent of the mountain by an American party which started up the north-east ridge and crossed to the Abruzzi spur below the summit pyramid. In the same year a British expedition failed on the west ridge with one death.

**Gasherbrum I** (8068) has been dubbed 'Hidden Peak', since it is tucked away in a glacier basin opening out from the main stream of the Baltoro Glacier.

**1934** An international expedition led by G.O. Dyhrenfurth reached 6200 m on the south-west ridge.
**1936** A French expedition led by H. de Segogne was defeated on the south-west ridge by the early onset of the monsoon.
**1958** **The first ascent of the mountain by an American team** of P. Schoening and A. Kaufmann by the south-west ridge.
**1975** The second ascent of the mountain by a party of two, R. Messner and P. Habeler, who climbed the north-west face Alpine-style without oxygen. This was **the first time these tactics were employed on an 8000 m peak**.
The third ascent took place only a few days later – an Austrian expedition led by H. Schell.
**1976** A French two-man team failed.
**1977** Fourth ascent of the mountain by a Yugoslav expedition, with one casualty.

**Gasherbrum II** (8034) is the next peak along the cirque.

**1934** Reconnaissance by G.O. Dyhrenfurth's international expedition.
**1956** First ascent by S. Larch, F. Moravec and H. Willenpart of an Austrian expedition by the south-west spur.

Khinyang Chhish (B Jankowski)

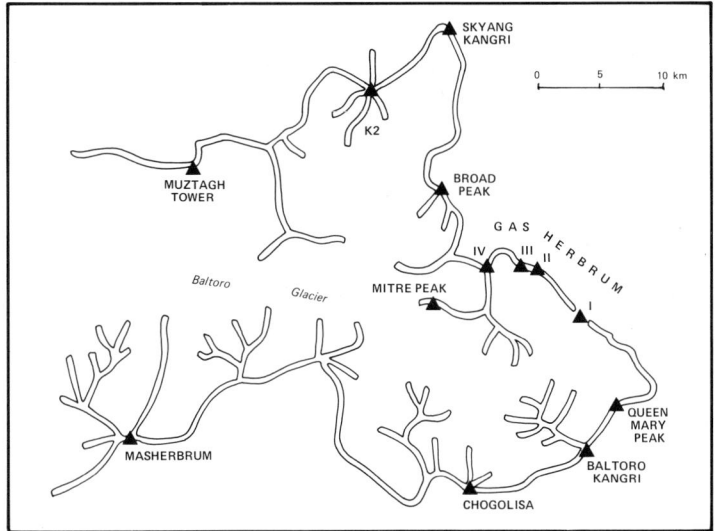

Baltoro Glacier

Rohtang Pass 3978 leads from Kulu into Spiti and Lahul.

Zoji La 3529, the lowest pass through the range, is on the caravan route from Srinagar to Leh, which continues by the Karakoram Pass into Central Asia. It was probably crossed by Genghis Khan. A road tunnel is planned.

## KARAKORAM

K 2 8610 (recently redetermined as 8760, the matter remains unresolved) (fa 31/7/1954 Italian exped: A. Campagnoni, L. Lacedelli), **the highest mountain in Pakistan, second highest in the world.**

Gasherbrum I (Hidden Pk) 8068 (fa 5/7/1958 American exped: P. Schoening, A. Kaufmann)

Broad Pk (Phalchan Ri) 8047 (fa 9/6/1957 Austrian exp: M. Schmuck, H. Buhl, F. Wintersteller, K. Diemberger: Buhl became **the first man to climb two 8000 m peaks,** both first ascents)

Gasherbrum II 8034 (fa 7/7/1956 Austrian exped: S. Larch, F. Moravec, H. Willenpart)

Broad Peak Central 8000 (fa 28/7/1975 Polish exped)

Gasherbrum III 7952 (fa 11/8/1975 Polish exped: Alison Onyskiewicz, Wanda Rutkiewicz, J. Onyskiewicz, K. Zdzitowiecki: **the highest first ascent made by women**)

Gasherbrum IV 7923 (fa 6/8/1958 Italian exped: W. Bonatti, C. Mauri)

Disteghil Sar 7884 (fa 9/6/1960 Austrian exped: G. Stärker, D. Marchart)

**1975** A French expedition made the second ascent by the same route, with one casualty.

A Polish expedition put three parties on the top (one of them all women) by the south-west spur and the north-west face.

**1976** A Japanese expedition failed with three deaths.

**Broad Peak** has a main summit of 8047 and a central summit of 8000 m, there is also a much lower north summit of 7600m. The name was given by the Conway expedition in 1892; the local name is Phalchan Ri.

**1954** A reconnaissance by a German/Austrian expedition failed at 7000m.

**1957 First ascent of the main summit** by the west spur by a four-man Austrian team; F. Wintersteller and M. Schmuck reached the summit first, H. Buhl and K. Diemberger arriving two hours later. Buhl thus became **the first man to reach two 8000 m peaks, and both were first ascents**. (Only 18 days later Buhl fell through a cornice on Chogolisa and was killed.)

**1974** A Japanese reconnaissance.

**1975** The first ascent of the central summit by a Polish expedition; severe storms during the descent led to the loss of three lives.

**1976** A French expedition failed at 7850m.

**1977** Second ascent of the main summit by a Japanese expedition.

The summit of **the Ogre** (7285) was reached for the first time by D. Scott and C. S. Bonington in 1977. There followed on the way down a remarkable *tour-de-force* by Scott, who broke both legs below the knee in an abseiling incident high up on the mountain. He completed the descent on his knees! They spent one night in the open and one trapped in their topmost camp by bad weather. Then with the aid of two members of their support party they set

Khinyang Chchish 7852 (fa 26/8/1971
  Polish exped: A. Zawada, J. Stryc-
  zynski, A. Heinrich, R. Szafirski)
Masherbrum East 7821 (fa 6/7/1960
  Pakistan/American exped: G. Bell,
  W. Unsoeld: two days later N.
  Clinch, J. Akhtar Khan)
Masherbrum West 7805
Rakaposhi 7788 (fa 25/6/1958 British
  exped: M. Banks, T. Patey)
Batura Muztagh I 7785 (fa 30/7/1976
  German exped: H. Bleicher, H.
  Oberhofer: the summit may have
  been reached by an English exped in
  1959 but all perished on the mountain)
Gasherbrum II East 7772
Kanjut Sar 7760 (fa 19/7/1959 Italian
  exped: C. Pelissier solo)
Saltoro Kangri I (K 36) 7741 (fa 24/7/
  1962 Jap/Pakistan exped: Y. Taka-
  mura, Capt Bashir, A. Saito)
Batura Muztagh II 7730 (fa 1978 Jap
  exped)
Saltoro Kangri II (K 35) 7705
Saser Kangri I (K 22) 7672 (fa 5-7/6/
  1973 Indian exped)
Chogolisa South-west 7665 (fa 2/8/1975
  Austrian exped)
Chogolisa North-east (Bride Pk) 7653
  (fa 4/8/1958 Jap exped: M. Fujihira,
  K. Hirai)
Trivor 7650 (fa 17/8/1960 British/
  American exped: W. Noyce, J. Sad-
  ler)
Shisparé (Pasu Pk) 7619 (fa 2/7/1974
  Polish/German exped)
Broad Peak North 7600
Skyang Kangri (Staircase) 7544 (fa
  11/8/1976 Jap exped led by G.
  Mitsui)
Mamostong Kangri 7526
Saser Kangri II (K 24) 7513
Yukshin Garden Pk 7498
Saser Kangri III (K 23) 7495
Kanjut II (Pumarikish) 7492
K 12 (Chumik group) 7468 (fa 30/8/
  1974 Jap exped: S. Tagachi, T. Ito,
  both killed on descent)
Teram Kangri I 7464 (fa 1975 Jap
  exped: Y. Kobayashi, K. Odaka)
Malubiting West 7452 (fa 23/8/1971
  Austrian exped)
Sia Kangri 7422 (fa 12/8/1934 Int
  exped led by G.O. Dyhrenfurth: H.
  Ertl, A. Höcht)
Skil Brum 7420 (fa 19/6/1957 Austrian
  exped: M. Schmuck, F. Winter-
  steller)
Saser Kangri IV (Cloud Pk) 7416
Teram Kangri II 7406 (fa 1975 Jap ex-
  ped: A. Ohishi, H. Kato, K. Hamaji)
Haramosh 7406 (fa 4/8/1958 Austrian
  exped: H. Roiss, F. Mandl, S.
  Pauer)
Mt Ghent 7400 (fa 4/6/1961 Austrian
  exped: W. Axt)
Rimo I 7385
Teram Kangri III 7381
Sherpi Kangri West 7380
Momhil Sar 7343 (fa 29/6/1964 Austri-
  an exped led by H. Schell)

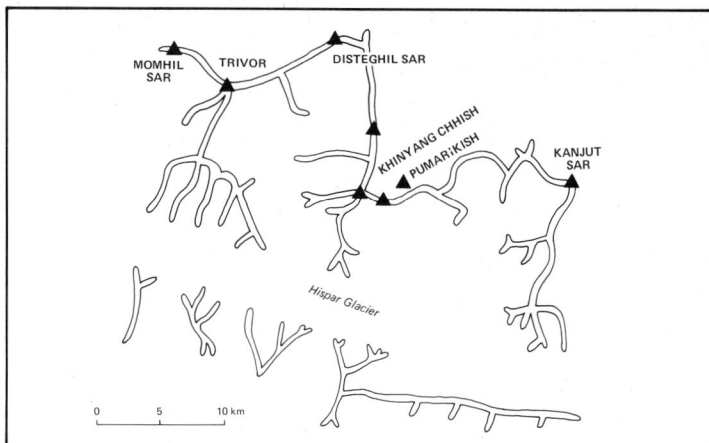

**Hispar Glacier**

off downwards. First came an ascent over the mountain's west
summit, which called for superhuman effort by all concerned;
thenceforward it was five days of arduous downhill to Base Camp.
Then came four days of waiting, three days of carrying out by
local porters and finally helicopter to hospital and eventual
recovery.

The **Muztagh Tower** (7273), from some aspects an inaccessible-
looking blunt wedge of steep rock and ice, was tackled in 1956
from different sides by British and French parties, the presence
of each unknown to the other. The British came first to the top
by a few days. It was **the first big mountain to be attacked simultaneously by two different expeditions**; it was **the first time** too
**that a man was actually seen on a high summit, other than by
members of his own expedition.**

**Boiohaghur Duanasir** (7318) is said to mean 'where only the horse
of the Devil can go' – an attractive and perhaps meaningful name
for a mountain.

The Kumaun Himalaya lies entirely within India and owes its
considerable exploration to its accessibility to European climbers
in the days of the Empire and to large numbers of Indian climbers,
who have come to the fore since World War II. It is bounded in
the north-west by the Sutlej, which rises in Tibet and breaks
through the Great Himalaya; in the south-east by the Kali River
which is the frontier with Nepal; the north-eastern boundary is
the frontier with Tibet (the Peoples' Republic of China).

**Nanda Devi** (7816), the Goddess Nanda, **is the highest mountain
in India**. It is situated within a ring of mountains known as the
Sanctuary, the only access to which is through the narrow gorge
of the River Rishi Ganga. On the rim of the Sanctuary, 2·5km
away, is the East Peak (7434), connected at high level to the main
peak by a formidable ridge.

Gasherbrum V 7321
Boiohaghur Duanasir 7318
Sia Kangri West 7315 (fa 1934 as for
  Sia Kangri)
Baltoro Kangri 7312 (fa 4/8/1963 Jap
  exped: S. Kono, C. Shima, K. Fuji-
  moto, T. Shibata)
Sherpi Kangri East 7303 (fa 8/1976 Jap
  exped led by K. Hirai)
Urdok I 7300 (fa 4/8/1975 Austrian
  exped)
Yutmaru Sar 7300
Baitha Brakk (The Ogre) 7285 (fa 1977
  British exped: D. Scott, C.S. Bon-
  ington)
Batura Muztagh III 7284 (fa 1954
  German/Austrian exped: D. Mayer,
  M. Schliessler)
K 6 7282 (fa 17/7/1970 Austrian exped)
Muztagh Tower 7273 (fa 7/7/1956
  British exped: J. Hartog, J. Brown,
  T. Patey, I. McNaught Davis)
Minapin (Diran) 7266 (fa 17/8/1968
  Austrian exped)
The Crown 7265
Savoia I 7263

**The Ogre and others from the
Baltoro glacier (Tony Riley)**

**1934**  E. E. Shipton and H. W. Tilman penetrated the gorge and entered the Sanctuary.
**1936**  A British/American expedition put H. W. Tilman and N. E. Odell **on the summit for the first time** by way of the South ridge.
**1939**  **First ascent of the East Peak** by a Polish expedition.
**1951**  In an attempted traverse from the main peak to the East Peak by a French party, the main peak was climbed successfully (second ascent), but the two climbers disappeared on the traverse between the peaks; their support party made the second ascent of the East Peak.
**1957 and 1961**  Unsuccessful Indian attempts.
**1964**  Third ascent by an Indian expedition.
**1976**  **First traverse of the two peaks** (from east to west) by a Japanese/Indian party with two bivouacs.

An American expedition climbed the main peak by way of the north ridge (a woman member, named for the mountain – Nanda Devi Unsoeld, died in the course of the action).

In 1855 the Schlagintweits reached 6785m on the slopes of **Abi-Gamin** (7355), thus at long last taking the progressive height

Baltoro Kangri East (Golden Throne)
    7260 (fa 3/8/1934 Int exped: J.
    Belaieff, P. Ghiglione, A. Roch)
Malubiting Central 7260 (fa 2/8/1975
    Jap exped)
Apsarasas I 7245 (fa 7/8/1976 Jap
    exped led by H. Misawa)
Apsarasas II 7239
Rimo II 7233
Apsarasas III 7230
Lupghan Sar 7200
Rimo III 7169
Serac Pk 7163
Masherbrum II 7163
Hachindar Chchish 7163
Savoia II 7156
Mt Depak 7150 (fa 13/8/1960 Austrian
    exped)
Latok I 7145
Kampir Dior (Karambar Sar) 7143 (fa
    14/6/1975 Jap exped)
Singhi Kangri (Mt Rose) 7141 (fa 8/8/
    1976 Jap exped led by H. Sato)
Latok II 7108 (fa 28/8/1977 Italian
    exped)
Mazero Pk 7100
Changtok 1 7092
Ghenta Pk 7090 (fa 21/7/1974 Polish/
    German exped)
Savoia III 7060
Changtok II 7045
Yengutz Har 7028 (fa 5/7/1955 German
    exped)
Spantik 7023
Malubiting East 7010 (fa 2/8/1959
    British exped)
Gasherbrum VI 7003
Paiju 6553 (fa 30/7/1976 Pakistan exped
    led by M. Hussain)

**The principal passes of the Karakoram**
The Karakoram Pass 5671 leads from
    the upper Shyok valley into Central
    Asia, to Yarkand and Kashgar. This
    is the line of the main caravan route
    from the S of the mountains, which
    came from Srinagar to Dras and Leh
    in the Indus valley and over the
    Ladakh Ra by the Khardung La to
    the Shyok valley. The first European
    to reach the Karakoram Pass was
    Dr T. Thomson in 1847-8.
The Saser Pass 5328 leads from the
    lower Shyok valley to the upper
    Shyok valley, cutting off the 'big
    bend'.
The Mintaka Pass 4709 carries the
    road from Gilgit by way of the
    Hunza valley into the Taghdumbash
    Pamir and so to Kashgar.

**The principal glaciers of the Karakoram**
The longest mountain glaciers in the
world are here – Siachen (72 km), Hispar (61 km), Biafo (59 km), Baltoro (57
km) and Batura (57 km).

## KUMAUN HIMALAYA

Nanda Devi West 7816 (fa 29/8/1936
    British/American exped: N. E. Odell,
    H. W. Tilman), **the highest point in
    India.**

record away from the Atacama Indians of the Andes (Llullaillaco). This new figure was not passed until Conway and Zurbriggen reached 6950m on Baltoro Kangri (7260) in 1892.

**Changabang** (6864), a monolithic mountain also on the rim of the
Sanctuary, **was not climbed until 1974**, by a British/Indian team
by the East arête. Yet other routes followed fast. In 1976 the
South-west ridge was climbed by a Japanese party, using 300
pitons, 20 bolts and 2500m of rope. A British party of five climbed
the South-east face in three days Alpine-style. Another two-man
British party succeeded on the extremely difficult West face in
25 days.

Many of the highest mountains in the world are in the Kingdom
of Nepal, which has only been open to foreign mountaineers for
the last three decades. The Nepal Himalaya are bounded on three
sides by the frontiers with India and Tibet (the Peoples' Republic
of China) and in the east by the Kosi valley. It divides readily into
three areas – Karnali (from Kali to Kali Gandaki), Gandaki (from
Kali Gandaki to Sun Kosi) and Kosi (from Sun Kosi to the
Sikkim frontier).

Nowadays, foreign mountaineering expeditions are closely
controlled by the Nepalese authorities. Only certain peaks are
available (the list includes most of the best), permission has to be
obtained in advance, a fee paid and Nepalese nationals included
in the expedition at assigned rates of pay. Anyone flouting these
rules is likely to be banned from future climbing activity in the
country.

**Dhaulagiri I** (8167), 'White Mountain', was one of the earliest
Himalayan peaks observed and measured from the foothills, by
Lieut W. S. Webb in 1809–10. With, probably, a certain amount
of luck on his side, he made it 8188m, which is not too inaccurate.
In the early 19th century it was listed as the highest mountain in
the world and continued thus until the publication of the figures
for Everest around 1850. However, for mountaineers it was un-
approachable until the opening up of Nepal after World War II.

**1950**    Reconnoitred by the French expedition which went on to
climb Annapurna.
**1953**    (Swiss), **1954** (Argentinian), **1955** (Swiss/German), **1958**
(Swiss) and **1959** (Austrian) attempts all failed.
**1960**    A successful assault by the North-east ridge by a Swiss/
Austrian expedition, **the first Himalayan attack supported by an
aeroplane** landing on the mountain (it crashed and had to be
abandoned there). Six Europeans and two Sherpas reached the
top in two parties, among them K. Diemberger who here became
**the second man to climb two 8000m peaks (both first ascents).**
**1969**    An American expedition was overwhelmed by an avalanche, five Americans and two Sherpas died.
**1970**    Second ascent by a Japanese expedition by the same route.
**1973**    Third ascent by an American expedition, supplied by parachute drop, again the same route.

Kamet 7756 (fa 21/6/1931 British ex-
ped: R. Holdsworth, F.S. Smythe,
E.E. Shipton, Lewa, R. Greene,
Kesar Singh)

Nanda Devi East 7434 (fa 2/7/1939
Polish exped: J. Bujak, J. Klarner,
M. Klavouer)

Abi Gamin 7355 (fa 22/8/1950 British/
Swiss exped: G. Chevalley, R. Dittert, A. Tissieres)

Mana Pk 7273 (fa 12/8/1937 British
exped led by F.S. Smythe)

Mukut Parbat 7242 (fa 11/7/1951 New
Zealand exped: H. Riddiford, E.
Cotter; Passang Dawa)

Hardeol (Tirsuli S) 7151

Chaukhamba 7138 (fa 13/6/1952
French exped)

Trisul I 7120 (fa 12/6/1907 British exped: T. Longstaff; H. & A. Brocherel, S. Karbir)

Tirsuli East 7074 (fa 9/10/1966 Indian
exped)

Dunagiri 7067 (fa 5/7/1939 Swiss exped)

North Santopanth 7062 (fa 1/8/1947
Swiss exped)

Tirsuli North-west 7035

Kedarnath 6940 (fa 1947 Swiss exped)

Kalanka 6931 (fa 1975 Jap exped)

Pt 6957 (ENE of Kalanka, Nanda Devi
basin)

Panch Chuli 6904 (fa 26/5/1973 Indian
exped)

Changabang 6864 (fa 4/6/1974 Anglo/
Indian exped)

Nanda Kot 6861

Mrigthuni 6855

Devtoli 6788 (fa 13/6/1973 Indian
exped)

Deo Damla (fa 1934 E.E. Shipton,
H.W. Tilman)

Nilkantha 6596 (fa 1961 Indian exped)

Shivling 6543 (fa 6/1974 Indian exped),
the Gangotri 'Matterhorn'.

## The principal passes of the Kumaun Himalaya

The Mana Pass 5608 and the Niti Pass
5068 lead from the Bhagirathi valley
and the Alaknanda valley respectively
to the Sutlej valley.

**Shivling, the 'Gangotri Matterhorn'
(AC Collection)**

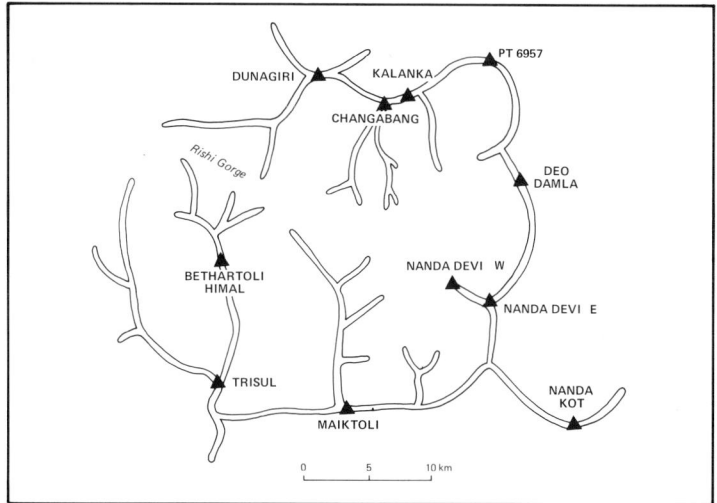

The Nanda Devi basin

Indian Himalaya

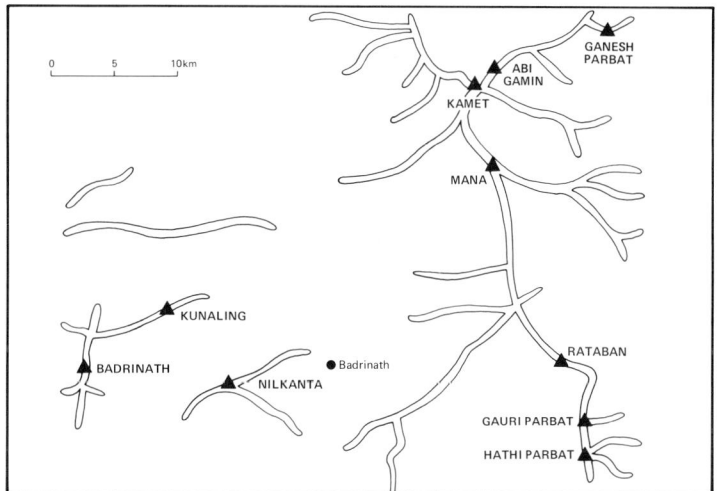

**1975** A Japanese attempt failed at 6200m with the loss of five
lives.

**1976** Fourth ascent by an Italian expedition by the same route.

**1977** R. Messner failed in an Alpine-style attempt, while a
Japanese team failed on the West ridge.

**1978** Fifth ascent by a Japanese expedition by a new route, the
South Pillar (one casualty).

Dhaulagiri is really a considerable mountain massif and there
are five other summits between 7250 and 7750m, numbered from
II to VI; all have been climbed during the 1970s. The Japanese
first ascenders of Dhaulagiri IV (7661) were both killed during
the course of the descent.

# NEPAL HIMALAYA

**Karnali section** (from the Kali R in the W to the Kali Gandaki, which penetrates to the Tibet frontier between the massifs of Dhaulagiri and Annapurna)

Dhaulagiri I 8167 (fa 13/5/1960 Swiss/Austrian exped: K. Diemberger, P. Diener, E. Forrer, A. Schelbert, Nyima Dorji, Nawang Dorji: Diemberger became **the second man to climb two 8000m peaks,** both first ascents; the mountain was climbed again ten days later by M. Vaucher, H. Weber of the same exped)

Dhaulagiri II 7751 (fa 18/5/1971 Austrian exped: A. Huber, R. Fear, A. Weissensteiner; Jangbu)

Dhaulagiri III 7715 (fa 23/10/1973 German exped)

Dhaulagiri IV 7661 (fa 9/5/1975 Jap exped: S. Kawazu, E. Yusada – both killed during the descent)

Dhaulagiri V 7618 (fa 1/5/1975 Jap exped: M. Morioka; Pemba Tsering)

Churen Himal Central & Churen Himal North both 7371 (fa claimed for Korean exped 29/4/1970, disputed by Jap exped which climbed both summits in 10/1970)

Dhaulagiri VI 7268 (fa 1970 Jap exped)

Putha Hiunchuli 7239 (fa 11/11/1954 British exped: J. O. M. Roberts; Ang Nyima)

Gurga Himal 7193 (fa 1/11/1969 Jap exped)

Sauwala 7175

False Junction Pk (Gama Pk) 7150 (fa 25/10/1970 Jap exped)

Sharpu 7100

Api 7132 (fa 10/5/1960 Jap exped)

Saipal 7035 (fa 21/10/1963 Jap exped)

Nampa South 6940 (fa 1978 Jap exped)

Kanjiroba 6882 (fa 7/11/1970 Jap exped)

Kubi Kangri 6855

Nampa 6754

**Gandaki section** (from the Kali Gandaki in the W to the Sun Kosi; the Buriganga and the Trisuli Gandaki penetrate the Himalayan crest line after rising in Tibet)

Manaslu I 8156 (fa 9/5/1956 Jap exped: T. Imanishi; Gyalzen Norbu: two days later M. Higeta, K. Kato)

Annapurna I 8091 (fa 3/6/1950 French exped: M. Herzog, L. Lachenal), **the first 8000m peak to be climbed.** Later (1970) the S face was the scene of **the first big Himalayan face climb** (fa British exped: D. Haston, D. Whillans)

Annapurna East 8010 (fa 29/4/1974 Spanish exped)

Annapurna II 7937 (fa 17/5/1960 British/Indian exped: C.S. Bonington, R. Grant; Ang Nyima)

Manaslu East 7895

Himalchuli 7864 (fa 24/5/1960 Jap

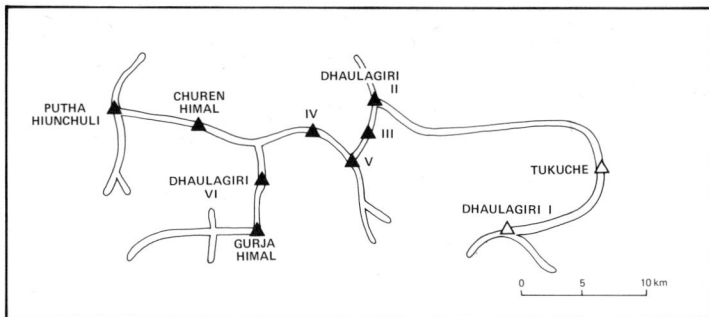

Dhaulagiri group

**Manaslu I** (8156) is hidden from view by lower fore-peaks, and in consequence had never even been photographed before 1950. As one of the giant potential goals of the new generation, it quickly attracted the attention of Japanese mountaineers who continued to mount attacks until the first ascent had been won.

**1952** Japanese reconnaissance.

**1953** Japanese expedition (15 Japanese, 15 Sherpas and 280 porters) failed at 7750m.

**1954** Because of local difficulties this year's expedition (14 Japanese, 22 Sherpas, 450 porters) was not allowed to approach the mountain; they turned (unsuccessfully) to Ganesh Himal (7406).

**1956** **First ascent of the mountain** from the north-east by a Japanese expedition, which placed four on the summit in two parties. Here the Sherpa, Gyalzen Norbu, became **the first Sherpa to reach two 8000m summits** (he had been with the French on Makalu (8481) the previous year).

**1964** First ascent of the North Peak of Manaslu (7050) by a Dutch/Austrian expedition.

**1971** Second ascent of the main mountain by the North-west wall by a Japanese expedition, two members reaching the top.

**1972** Third ascent by an Austrian team; R. Messner alone reached the top without oxygen by way of the South wall. The weather was very bad and two Europeans were killed.

A South Korean expedition failed at 6950m with 15 deaths when a camp was buried by an avalanche. This was **the worst ever mountaineering disaster**.

**1973** Fourth ascent by a German expedition by a further new route on the South face.

A Spanish attempt failed.

**1974** Fifth ascent by a Japanese all-woman expedition by the original route. This was **the first ascent of an 8000m peak by women** (M. Mori, N. Nakaseko and M. Uchida with Sherpa Jambu).

**1975** Sixth ascent by Spanish expedition.

**1976** Seventh ascent by a Japanese/Iranian expedition.

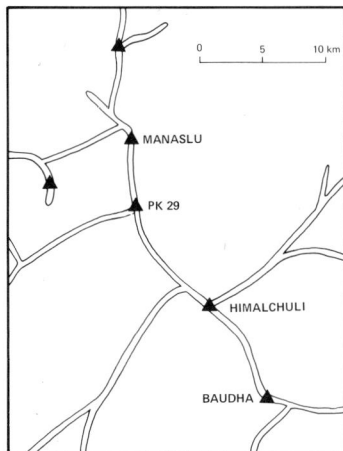

Manaslu group

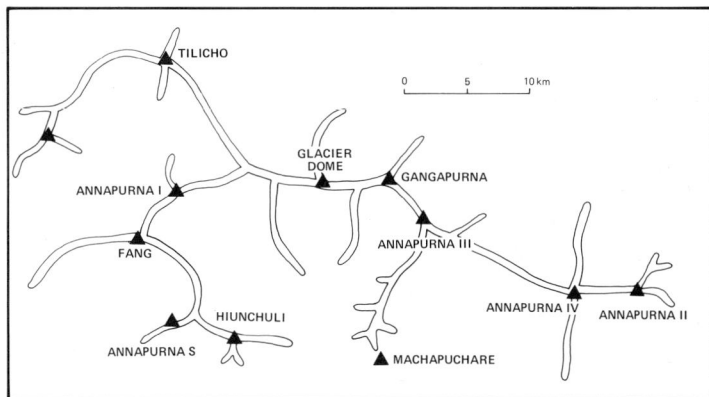

Annapurna group

exped: M. Harada, H. Tanabe: next day H. Miyashita, K. Nakazawa)
Subsid pks 7540 (fa 1977 Jap exped), 7351
Manaslu II (Dakura, Pk 29) 7835 (fa 19/10/1970 Jap exped: H. Watanabe; Lhakpa Tsering – both killed during descent)
Annapurna III 7555 (fa 6/5/1961 Indian exped: M. Kohli, Sonam Gyasto, Sonam Girmi)
Annapurna IV 7525 (fa 30/5/1955 German exped: H. Biller, H. Steinmetz, J. Wellenkamp)
Roc Noir 7485 (fa 9/5/1969 German exped)
Ganga Purna 7454 (fa 6/5/1965 German exped)
Ganesh Himal 7406 (fa 24/10/1955 French exped: E. Gauchat, C. Kogan, R. Lambert)
Glacier Dome 7255 (fa 16/10/1964 Jap exped)
Annapurna South (Modi Pk) 7218 (fa 15/10/1964 Jap exped)
Sringi Himal-Chamar 7177
Tilitso Himal 7134
Himlung Himal 7126
Jugal Himal – Lömpo Gang 7084 & Dorje Lakpa 6988
Manaslu North 7050 (fa 25/10/1964 Dutch/Austrian exped)
Nilgiri North 7032 (fa 19/10/1962 Dutch exped)
Kang Guru 7010
Macha Puchare 6998
Lamjung Himal 6983 (fa 1974 British exped)
Tukuche 6920 (fa 10/5/1969 Swiss exped)
Damodar Himal 6712
Baudha Pk 6672 (fa 1970 Dutch exped)

Kosi section (from the Sun Kosi to the Nepal/Sikkim frontier, a short distance to the W of which the R Arun penetrates the range)

Annapurna I (8091), the 'Goddess Rich in Sustenance', was the first 8000 m peak to be climbed. Though visible from the foothills it had never been approached by mountaineers, or even photographed, prior to the first climbing assault in 1950.

1950 The first attackers managed to bring off the first ascent. After a preliminary reconnaissance, a French expedition decided to tackle this mountain in preference to Dhaulagiri. By way of the North-west spur they put M. Herzog and L. Lachenal on the first 8000 m summit. These climbers were badly frostbitten during the summit push and had to be helped down in arduous circumstances and finally carried off from the mountain.

1969 A German team failed on the long East ridge.

1970 Second ascent of the mountain by the original route by a British/Nepalese Army expedition.

The first ascent of the South face, the first big Himalayan wall climb, was made without oxygen by D. Haston and D. Whillans; they arrived on the summit the day after the Army climbers from the other side. This team lost one member in an ice fall on the lower glacier.

1973 A Japanese attempt on the original route failed with the deaths of four Japanese and one Sherpa in an avalanche.

An Italian expedition lost two members in a similar avalanche accident and the attempt was abandoned.

1974 First ascent of the East summit by a Spanish expedition.

1975 An Austrian expedition to the South-east ridge failed at 6400 m with one casualty.

1977 Successful ascent by a Dutch expedition.

Annapurna is an extensive mountain massif. There are three other summits, numbered from II to IV, between 7000 and 7950 m, some of them have quite formidable mountaineering problems. All were climbed for the first time in the period between the first and second ascents of the highest Annapurna. All are to the east, Annapurna II being some 40 km away. Annapurna III (7555), climbed for the first time by Indians, received its second ascent from a Japanese ladies' expedition in 1970.

*Above:* Classic photograph taken at the highest point reached by the 1924 Everest Expedition (AC Collection). *Below:* Everest and Nuptse (Barnaby's)

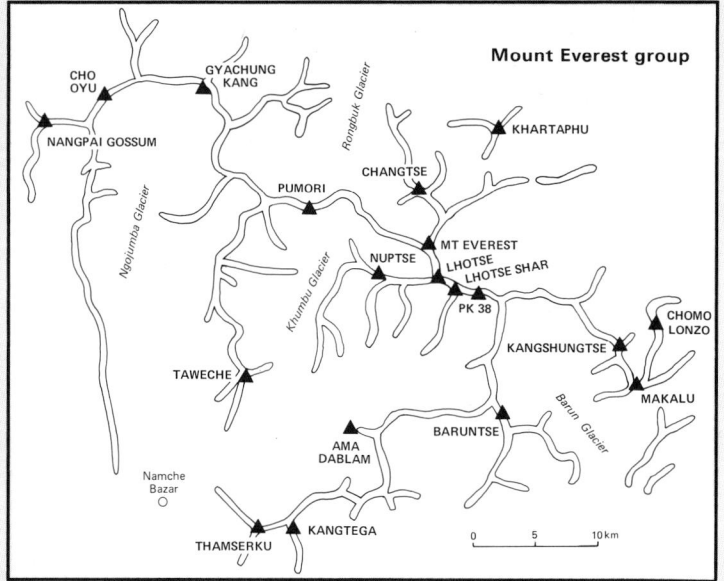

Mount Everest group

CHO OYU

GYACHUNG KANG

NANGPAI GOSSUM

*Rongbuk Glacier*

KHARTAPHU

CHANGTSE

PUMORI

*Ngojumba Glacier*

*Khumbu Glacier*

NUPTSE

MT EVEREST
LHOTSE
LHOTSE SHAR

PK 38

CHOMO LONZO

KANGSHUNGTSE

TAWECHE

*Barun Glacier*

MAKALU

AMA DABLAM

BARUNTSE

Namche Bazar

THAMSERKU

KANGTEGA

0      5      10 km

Makalu (AC Collection)

Mt Everest (Qomolangma Feng, Sagarmatha) 8848 (fa 29/5/1953 British/ New Zealand exped led by J. Hunt: E. Hillary, Tenzing Norgay), **the highest mountain in the world.**
Lhotse 8511 (fa 18/5/1956 Swiss exped: E. Reiss, F. Luchsinger)
Subsid pk 8410
Makalu 8481 (fa 15/5/1955 French exped: J. Couzy, L. Terray: on subsequent days J. Franco, G. Magnone, Gyalzen, P. Leroux, J. Bouvier, A. Vialette, S. Coupe)

Sherpa porter (M Fantin)

**Mount Everest** (8848) is **the undisputed highest mountain in the world.** The height was first determined with reasonable accuracy in 1852 and it was named in 1865 for Sir George Everest, one-time Surveyor General of India. The Nepalese call it Sagarmatha, the Chinese – Qomolongma Feng, the Tibetans – Mi-ti Gu-ti Cha-pu Long-na, but the name 'Everest' has universal acceptance among mountaineers and geographers. Reports of higher mountains have been made from time to time, notably of the Amne Machin range of China by fliers during World War II. All are now discounted.

Mount Everest lies on the border between Nepal and Tibet (Peoples' Republic of China). Until 1950 the only available approach was from the north through Tibet, reached by way of Sikkim. The opening of Nepal enabled a shorter approach to be made from the south and all attempts by western mountaineers have subsequently been made from this side. The taking over of Tibet by the Chinese led to complete closure of the north face of the mountain and only communist-bloc countries can now go that way. This was a British preserve until 1952; thereafter many nations have essayed the ascent. Two expeditions per annum are admitted, one before and one after the monsoon, and a waiting list has developed for many years ahead.

The climbing of Everest was first contemplated soon after the turn of the century, but no practical steps were taken until after World War I.

**1921** A reconnaissance led by C.K. Howard Bury discovered the North ridge route.

**1922** An expedition led by C.G. Bruce reached over 8225m.

**1924** This year's expedition was led by E.F. Norton, who with T.H. Somervell reached 8575m. G.H.L. Mallory and A.C. Irvine disappeared on a second summit attempt, after being seen by N.E. Odell below going strongly for the top. Whether or not they reached it is still an open question, unlikely to be resolved.

**1933** In an expedition led by H. Ruttledge, P. Wyn Harris, L. Wager and F.S. Smythe reached roughly the same point as Norton nine years earlier.

**1934** A solo attempt by M. Wilson failed and he died of exhaustion.

**1935** A reconnaissance led by E.E. Shipton examined the mountain thoroughly and climbed several outlying peaks.

**1936 and 1938** Expeditions led by H. Ruttledge and H.W. Tilman respectively made no further progress.

**1947** A solo attempt by E. Denman ended in failure.

**1950** The opening of Nepal enabled an Anglo/American reconnaissance of the south side to be made.

**1951** A British/New Zealand expedition, led by E.E. Shipton, reconnoitred the south side and found the route into the Western Cwm, used subsequently on all assaults.

**1952** A Swiss expedition (leader E. Wyss Dunant) made two attempts, one before and one after the monsoon. By way of the South Col and the South ridge they reached 8600m.

A Russian claim to an ascent from the north is unconfirmed.

Lhotse Shar 8383 (fa 12/5/1970 Austrian exped: S. Mayerl, R. Walter)

Cho Oyu 8153 (fa 19/10/1954 Austrian exped: H. Tichy, J. Jochter, Pasang Dawa Lama)

Makalu South-East 8010

Gyachung Kang 7921 (fa 10/4/1964 Jap exped: Y. Kato, K. Sakaizawa, Pasang Phutar: next day K. Machida, K. Yasuhida)

Nuptse 7879 (fa 16/5/1961 British exped: D. Davis, Tashi: next day C.S. Bonington, L. Brown, J. Swallow, Ang Pemba)

Subsid pks 7815, 7745

Chomo Lonzo 7815 (fa 30/10/1954 French exped: J. Couzy, L. Terray)

Makalu II (Kangshungtse) 7640

Lhotse II (Pk 38) 7589

Shartse 7502 (fa 23/5/1974 Austrian exped)

Nangpai Gosum I 7341

Chamlang 7319 (fa 31/5/1962 Jap exped: S. Anma, Pasang Phutar)

Nangpai Gosum II 7296

Langtrang (Langtang) Lirung – Gangchhen Ledrub 7245 (fa 1978 Jap exped)

Baruntse 7220 (fa 30/5/1954 New Zealand exped: G. Harrow, C. Todd: two days later W. Lowe, W. Beaven)

Menlungtse (Jobo Garu) 7181

Pumori 7145 (fa 17/5/1962 German/Swiss exped: G. Lenser, V. Hurlemann, E. Forrer)

Gaurishankar (Jomo Tseringma) 7145

Numbur 6957

Pyramid Pk 6873

Ama Dablam 6856 (fa 1961 British/New Zealand exped)

Kangtega 6809

Tamserku 6623

**Sikkim Himalaya** (from the Nepal frontier to the Chunei valley and the Tang La 4633, which is the trade route from Sikkim to Tibet)

Kangchenjunga 8597 (fa 25/5/1955 British/New Zealand exped: G.

**Kangchenjunga group**

**1953** The expedition, led by John Hunt, which tackled the mountain this year, had a scientific backing such as had never previously been seen in the Himalaya. This strong British/New Zealand team discovered and way-marked a route through the Khumbu Icefall into the Western Cwm. A series of camps was established here and supplies of food and equipment carried higher and higher by bands of Sherpas. A route was then forced up the headwall of the Cwm to the South Col (7925). The first attempt from here on 26 May by C. Evans and T. Bourdillon produced **a first ascent of the South summit** (8750). Then on 28 May a higher camp was placed at 8500 m above the South Col and E. Hillary and Sherpa Tenzing installed there. The next day they crossed the South summit and, continuing along the interconnecting ridge, **reached the top of the world for the first time**.

**1956** Second ascent by a Swiss expedition (led by A. Eggler) by the same route.

**1960** The Chinese claim an ascent from the north, but there are unresolved doubts about how high they really went.

**1960 and 1962** Failed attempts by Indian expeditions.

**1963** Third ascent by an American expedition (led by N. Dyhrenfurth). Two parties reached the summit by the South Col route. W. Unsoeld and T. Hornbein **climbed the West ridge for the first time** and, **traversing the summit**, met the second o' the South Col parties. Descending together by the original route, all four men were benighted and spent a night in the open at 8535m.

**1965** The fourth ascent was made by an Indian expedition which, in four successive parties, placed **a record total of nine climbers on the summit**. One of them was Gombu, who thus became **the first man to climb Everest twice**.

**1969** The Chinese again claim to have climbed the north side – once again there is no confirmation.

The Japanese made the first attempt on the South-west face, the steep mountainside above the Western Cwm.

**1970** The fifth ascent of the mountain by the South Col route by a Japanese expedition, which failed, however, on the South-west face.

**1971 and 1972** A series of failures on various routes.

**1973** The sixth ascent of the mountain by the South Col route by an Italian expedition which had its supplies lifted by helicopter into the Western Cwm.

The seventh ascent of the mountain, the first post-monsoon, by a Japanese expedition using the South Col route; a further failure on the South-west face.

**1974** Various failures.

**1975** The eighth ascent of the mountain by a Japanese expedition. The summit party included **the first woman to reach the top**, Junko Tabei, age 35, housewife and mother.

Ninth ascent only eleven days later by a Chinese team which, without oxygen, placed nine climbers, including a woman (Phantog, age 37, mother of three), on the summit by **the first ascent of the North ridge route**, the first authenticated ascent from this side.

*Above:* Mount Everest, South Col and Lhotse (*right*) seen over the Nuptse ridge, Nepal Himalaya (Aerofilms)

*Top right:* Macha Puchhare at sunset (Barnaby's Picture Library)

*Right:* Mount Gaudry, Adelaide Island, Antarctica (C Swithinbank)

*Left:* A Nepal village and Annapurna (Barnaby's Picture Library)

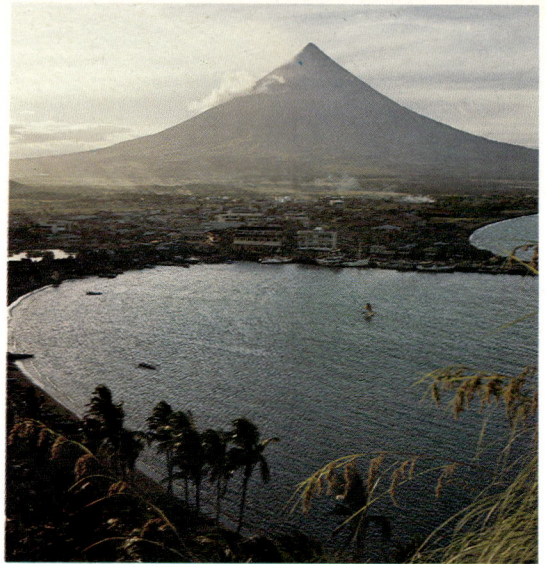

**AUSTRALASIA**

*Above:* Mount Olga, Central Australia (Australian News and Information Bureau). *Above right:* Mayon Volcano, Luzon, Philippine Islands (Barnaby's Picture Library). *Below:* Mount Cook and the Hooker glacier, New Zealand (New Zealand High Commission)

**Kangchenjunga from Darjeeling (The Times)**

Band, J. Brown: next day N. Hardie, H. R. A. Streather), **the third highest mountain in the world.**
Yalung Kang (Kangchenjunga West) 8502 (fa 14/5/1973 Jap exped: Y. Ageta, T. Matsuda – the latter was killed on the descent)

**Monks at Tengpoche Monastery (Nepal) (M Fantin)**

Tenth ascent by a British team (D. Scott and D. Haston) by **a first ascent of the South-west face**. They had to bivouac on the descent at 8750m. Two more expedition members reached the top, two days later, one other was lost on the mountain.
**1976** British/Nepalese Army and American Bicentennial expeditions each made successful ascents.
**1978** During May the **South Col route was climbed Alpine-style without oxygen** by R. Messner and T. Habeler, for the former it was **his fourth 8000 m peak**. Nine members of an Austrian expedition led by Wolfgang Nairz, and one Sherpa also reached the top.

In October several members of a Franco/German expedition made the climb, including **the first Frenchman; the first European woman;** P. Mazeaud, **the oldest man** (49) so far to climb it, and K. Diemberger, for whom it was **his fourth 8000m peak.**

**Lhotse** (8511), the fourth highest mountain, faces Mount Everest across the South Col, from which it is reasonably accessible. From the summit a ridge runs somewhat south of east to Lhotse Shar (8383), Lhotse II (7589) and Shartse (7502). Westwards is a ridge, the wall of the Western Cwm, leading to Nuptse (7879).
**1956** **First ascent of the main summit** by E. Reiss and F. Luchsinger of the Swiss Everest expedition.
**1970** **First ascent of Lhotse Shar** by an Austrian expedition from the west.
**1971** South Korean attempt on Lhotse Shar by the South-east ridge failed at 8000m.
**1973** Attempt on South wall of main peak (and later on West ridge) by a Japanese expedition failed.
**1974** A winter attempt by a Polish expedition failed.
**1975** An attempt on the South face by an Italian expedition failed.
**1976** A Japanese team failed similarly.
**1977** Second ascent of the mountain by a German expedition with one death on the descent.

**Makalu** (8481), the fifth highest mountain, is conspicuous in the view from the foothills, from which it partially hides some of its higher neighbours. It was neglected until Mount Everest had been climbed, then in 1954 a number of nations made an approach.
**1954** An American attempt on the South-east ridge failed at 7050m.

Reconnaissances by New Zealand and France (Makalu II (7640) climbed).
**1955** **First ascent of the mountain** by a French expedition from the north-west. Nine climbers reached the summit on three successive days.
**1961** An attempt without oxygen by the British/New Zealand Himalayan Scientific expedition failed at 8330m.
**1969** A Japanese reconnaissance.
**1970** Second ascent by the South-east ridge by H. Tanaka and Y. Ozaki of a Japanese expedition which comprised 20 Japanese, 32 Sherpas and 440 porters.

Kangchenjunga South 8488 (fa 19/5/ 1978 Polish exped)
Kangchenjunga Middle 8475 (fa 22/5/ 1978 Polish exped)
Kangbachen 7902 (fa 26/5/1974 Polish exped: W. Brański, W. Klaput, M. Malatynski, K. Olech, Z. Rubinowski)
Zemu Gap Pk 7780
Jannu 7709 (fa 27/4/1962 French exped: P. Keller, R. Paragot, R. Desmaison, Gyalzen Mitchu, J. Bouvier, P. Leroux, Bertrand, Pollet-Villard, L. Terray, Ravier, Wangdi)
Yalung Glacier Pk (Kangbachen South-west) 7535
Jongsong Pk 7473 (fa 3/6/1930 Int exped: F.S. Smythe, G. Wood-Johnson, H. Hoerlin, H. Schneider, G.O. Dyhrenfurth, M. Kurz, Lewa, Tsering Norbu)
Kabru 7393 (fa 18/11/1935 C.R.Cooke) An ascent was claimed by W. Graham, with E. Boss & U. Kaufmann, 1883, but exactly what they did climb is unclear.
Subsid pk 7316 (1907 C.W. Rubenson, Monrad Aas reached within 30m of summit)
The Twins 7350
Talung Pk 7349 (fa 19/5/1964 F. Lindner & a Sherpa)
Tent Pk 7342 (fa 29/5/1939 Swiss/German exped: E. Grob, L. Schmaderer, H. Paider)
Dome Khang 7260
Sharphu 7212 (fa 22/10/1963 Jap exped)
Nepal Pk South-west 7180 (fa 27/5/ 1939 Swiss/German exped: E. Grob, L. Schmaderer, H. Paider)
Subsid pks 7163 (fa 1936), 7145 (fa 1930)
The Pyramid 7065 (fa 6/6/1949 Swiss exped: R. Dittert, A. Sutter, Gyalgen, Ajeeba, Dawa Thondrup)
Pauhunri 7065 (fa 16/6/1911 A.M. Kellas)
Nupchu 7028 (fa 20/5/1962 Jap exped)
Langpo Pk 6949 (fa 1909 A.M. Kellas)
Dodang Nyima 6925 (fa 1930 Int exped: E. Schneider, H. Hoerlin)
Siniolchu 6892 (fa 1936 German exped: A. Gottner, K. Wien)
Kangchenjau 6889 (fa 1912 A.M. Kellas)
Chomiomo 6837 (fa 1910 A.M. Kellas)
Simvu 6816 (fa 1936 German exped)
Pandim 6708
Ramthang Pk 6640 (fa 1930 Int exped)

## ASSAM HIMALAYA

**Bhutan section**
Khula Kangri I 7554, **the highest point of Bhutan.**
Khula Kangri II 7541
Khula Kangri III 7532
Khula Kangri IV 7516
Gankar Punzum 7480

**1971** Third ascent by the West ridge by a French expedition.
**1972** A Yugoslav attempt on the South face reached 8000m.
**1973** A Czech attempt on the South-west Pillar failed at 8000m.
**1974** An Austrian and an international expedition both failed on the South face.
**1975** Fourth ascent by a Yugoslav expedition, which succeeded on the South face and placed seven men on the summit.
**1976** Fifth ascent by a combined Spanish/Czech expedition. Spaniards on the South-east ridge and Czechs on the South spur combined forces high up on the mountain. Two Czechs and one Spaniard reached the top, one Czech disappeared during the descent.
**1977** An international expedition had a tremendous run of bad luck and only reached 6950m.
**1978** Sixth ascent by a West German/Austrian/Swiss team by the South-east ridge. Four Europeans and three Sherpas reached the summit in three parties, one was **K. Diemberger,** for whom it was **his third 8000m peak**.

**Cho Oyu** (8153), 'Goddess of the Turquoise', rises on the Nepal/ Tibet frontier. Relatively less accessible than the other Nepal giants, it has not yet attracted as many expeditions.
**1952** British reconnaissance reached 6825m.
**1954** **First ascent** by H. Tichy, J. Jochter of an Austrian team with Pasang Dawa Lama by the West buttress. A French/Swiss party, having abandoned Gaurisankar and Melungtse, arrived at the foot of Cho Uyo at the same time as the Austrians. After some argument the latter, who had permission anyway, were allowed to make the first attempt. The Swiss/French expedition failed subsequently to make the second ascent, turning back at 7500m, but Mme Claude Kogan here set up a new high altitude record for women.
**1958** Second ascent by an Indian expedition by the same route.
**1959** An international all-woman expedition failed when two members and two Sherpas were buried by avalanches.
**1964** A German ski-ing expedition possibly made the third ascent; there is some dispute about it. Two expedition members perished.

Sikkim, formerly an independent state, is now a State of the Republic of India. The **Sikkim Himalaya** have been readily accessible for many years. During the days of the Empire this was popular holiday country for British climbers in India, while a number of expeditions came particularly from Germany during the 1930s. A number of 7000m peaks were ascended.

The ascent of Kabru (7393), claimed by W. Graham in 1883, provided an early mountaineering controversy. Some years later, in 1907, two Norwegians reached 7385m on the same mountain, then a height record. Interest in Kangchenjunga (8597) began before the turn of the century, but it is a difficult mountain and was not climbed until 1955 and then not again until 1977.

Chomo Lhari 7314 (fa 21/5/1937 F.
  Spencer Chapman, Pasang Dawa)
Tsulim Khon 7300
Teri Kang 7300
Takha Khon 7300
Jejekangphu Kang 7300
Gyu Khon 7200
Tsenda Kang 7200
Kangphu Kang 7200
Masa Gang 7165
Tserim Kang 7000
Melunghi Kang 7000
Kangchita 6840
Tsheringme Gang (Kungphu) 6789
Chum Gang (Takaphu) 6526

**Eastern section**
Namcha Barwa 7756
Gyala Peri 7151
Kangto (Kangdu) 7089
Nyegyi Kansang 7047
Takpa Shiri 6655
Gori Chen 6538

**Chomo Lhari (R D Treble)**

The trade route from India to Tibet crossed the Tang La (4633) and this was the route taken by all pre-1939 expeditions to Mount Everest.

**Kangchenjunga** (8597), which rises on the Nepal/Sikkim frontier, is **the third highest mountain in the world**. It is a difficult mountain and, in spite of prodigious efforts by German mountaineers during the 1920s and 30s, did not fall until the post-war period.

**1899**  Reconnaissance by a British expedition led by D. Freshfield.

**1905**  An international expedition failed with four deaths.

**1929**  An American, E. Farmer, disappeared on a solo attempt.

A German expedition led by P. Bauer failed at 7400m on the North-east spur from the Zemu glacier.

**1930**  An international expedition led by G.O. Dyhrenfurth failed on the North-west face with one death.

**1931**  A second P. Bauer expedition, taking the same line as in 1929, failed at 7700m with two deaths.

**1951, 1953 and 1954**  Reconnaissances by British teams.

**1955**  **First ascent of the mountain** by G. Band and J. Brown of a British expedition led by C. Evans by way of the South-west face. They stopped just short of the top having promised the Maharajah of Sikkim that it should not be trodden. Two other members reached the top on the following day.

**1977**  Second ascent by an Indian expedition by the North-east spur (one death).

The nearby summits of Kangchenjunga South (8488) and Kangchenjunga Middle (8475) were **first climbed by a Polish expedition in 1978**. Since they only had permission for the main summit, they were banned from future mountaineering in Nepal.

•

**Yalung Kang** (8502) was formerly called Kangchenjunga West.

**1973**  **First ascent** by a Japanese expedition, one of the summiters being killed on the descent, by the South-west face.

**1975**  Second ascent by an Austro/German party by the South face.

**1978**  Third ascent by a Spanish expedition.

The Assam Himalaya stretches from the Sikkim/Bhutan frontier to the 'big bend' of the River Brahmaputra, which having flowed eastwards for some 1250km on the north side of the Himalayan chain, flows south, then west, then south again for another 1150 km to join the Ganges and the sea in the Bay of Bengal.

Access to the Kingdom of Bhutan is restricted, access further east even more so, with the result that this is the least known part of the Himalaya. In the extreme west the conspicuous peak of Chomo Lhari (7314) looks down on the trade route from India into Tibet and has been climbed in 1937. Though some journeys have been made among the rest of the mountains, no serious mountaineering expeditions have yet approached the high peaks.

# Australasia

The summit of Mount Cook (JGR Harding)

## Australia

### WESTERN AUSTRALIA

**Hammersley Ra** (in NW of State)
Mt Meharry 1244, **the highest point in the state.**
Mt Bruce 1226
Mt Vigors 1145
Mt Brockman 1114
Mt Augustus 1106
Mt King 1017
Mt McRae 1014

**Stirling Ra** (near SW coast)
Bluff Knoll 1109
Ellen Pk 1042
Pyongorup 1036

**Porongorup Ra** (S of Stirling Ra)

**Kimberley Plateau** (NE corner) – Mt Wells

**Peterman Ra** (E border) – Mt Deering 1219

### NORTHERN TERRITORY

**Macdonnell Ra**
Mt Liebig 1524, **the highest point in the State.**

The **Hammersley Range** is the great iron-bearing area of the State of Western Australia. Mount Tom Price is made of iron ore.

**Ayers Rock** (867) is an inselberg, 1·6km long and 340m high, rising from an arid desert in Northern Territory some 300km south-west of Alice Springs. Named for a former premier of South Australia, it was first climbed in July 1873, then not again until 1931. Now it is a popular tourist spot, often ascended by an easy ramp which runs up on one side. There are some climbing prospects on the remainder, especially near a detached pillar, the Kangaroo's Tail. Since the air is hot and dry, exfoliation can be seen taking place on the walls.

Forty-five kilometres away is **Mount Olga** (1069), a cluster of 31 smaller inselbergs, the highest 450m above the plain. This was named for Queen Olga of Spain. Both peaks are now included in a National Park.

As serious climbing prospects these mountains are too remote to excite serious attention in a country where there are so many relatively unexplored crags much nearer to civilisation.

The **Flinders Ranges** are named for Matthew Flinders who **first sighted** them **in 1802**. His ship anchored near the present Port

Mt Ziel 1510
Mt Heughlin 1469
Mt Edward 1416
Mt Mann 1174
Mt Laughlen 1168
Mt Conway 1136
Mt Brassey 1128
Mt Riddock 1105
Mt Olga 1069, 450m above the plain,
  31 inselbergs.
Mt Harris 1067
Ayers Rock 867, inselberg 1·6km long,
  340m above the plain (fa 20/7/1873
  W.C. Gosse, Kamram; second as-
  cent not till 1931)
Mt Conner 760, an isolated plateau
  with crags all round.

## SOUTH AUSTRALIA

**Mann & Musgrave Ras** (NW of State)
Mt Woodroffe 1440, **the highest point
  in the State.**
Mt Morris 1254
Mt Whinham 1231
Mt Everard 1174
Mt Davies 1058

### Flinders Ra
St Mary's Pk 1189
Mt Hack 1128
Mt Remarkable 969
Mt Brown 967 (fa 1802 party from
  Flinders's ship)

## QUEENSLAND

### North
Mt Bartle Frere 1612, **the highest point
  in the State.**
Mt Dalrymple 1277, in Eungella NP.
Mt Elliot 1234
Mt Halifax 1063
Mt Abbot 1055
Mt Bellenden Ker, carries Australia's
  most elevated TV transmitter.

### South
Mt Roberts 1387
Mt Barney 1356
Mt Norman 1257 (Giraween NP)
Mt Lindesay 1239
Mt Warning 1156
Mt Mowbullan 1101
The Pyramids 1067
Mt Buckland 1018

## NEW SOUTH WALES

### Great Dividing Ra – N section
Round Mtn 1610
Pt Lookout 1600 (New England NP
  233km²)
Mt Barrington 1585
Chandlers Pk 1564
Capoompeta 1555
Mt Bajimba 1524
Black Sugarloaf 1494
Mt Hyland 1439
Ben Lomond 1364
Bald Rock 1342, in Bald Rock State P,
  a granite inselberg comparable with
  Ayers Rock.

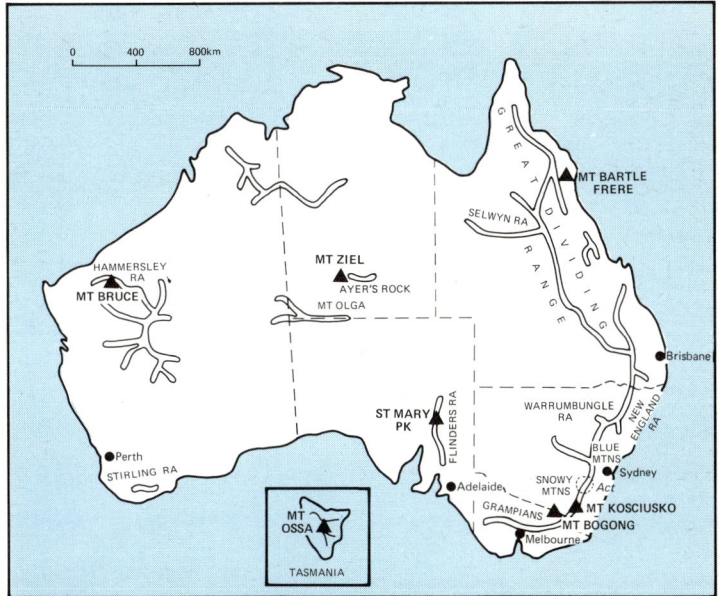

Australia

Ayers Rock, Northern Territory (Australian Tourist Commission)

Augusta and a party climbed Mount Brown (967). There are three National Parks – Gammon Ranges, Flinders Ranges and Mount Remarkable, the rock-climbing possibilities remain almost untouched.

**The highest point, Mount St Mary** (1189), is on the edge of the unusual formation of Wilpena Pound, a bowl of 83km², almost entirely enclosed by a mountain ridge. It is a massive synclinal hollow in quartzite.

In the far north a booming sound can sometimes be heard in the heart of the Gammon Ranges. This is probably due to rock

Mt Banda Banda 1263
Oxley's Pk 1241

**Mt Kaputar NP (Nandewar Ra)**
A rugged volcanic landscape.

Mt Kaputar 1527
Mt Lindesay 1442
Mt Lindsay 1440
Mt Grattai 1402
Mt Coryah 1402
Yollodunida 1280, a volcanic crater.
Ginns 1100
Ningadhun 1067

**Warrumbungle NP (Warrumbungle Ra)**
Another area of volcanic scenery, much
used by rock climbers.

Mt Exmouth 1227
Needle Mtn 1200
Siding Spring Mtn 1200

**Blue Mountains**
Hunter Ra 1274
MacQuarie Mtn 1204
McAlister 1033

**Snowy Mtns (eastern Australian Alps)
& Australian Capital Territory**
Mt Kosciusko 2230 (fa either 1834 Dr
  Lhotsky, or 1840 Count Strzelecki,
  unresolved), **the highest point in
  Australia.**
Mt Townsend 2209
Mt Twynam 2196
N Ramshead 2177
Mt Carruthers 2145
Muellers Pk 2130
Mt Northcote 2130
Mt Crackenback 1936, c-wy to summit
Mt Bimberi 1912
The Pilot 1836
Mt Currockbilly 1131

**VICTORIA**

**W Australian Alps**
Mt Bogong 1986, **the highest point in
  the State.**
Mt Feathertop 1922
Mt Hotham 1863
Mt Cobboras 1836
Mt Buller 1809
Mt Howitt 1742
Mt Buffalo 1721
Mt Tamboritha 1640
Mt Baw Baw 1560
Lake Mtn 1486
Mt Bowen 1372
Mt Ellery 1296
Mt Donna Buang 1244

**The Grampians**
An area of former volcanic activity, as
recently as 1000 BC.

Mt William 1167
Mt Macedon 1011
Mt Arapiles (fa 1836 T. Mitchell), iso-
  lated 65 km N of the Grampians,
  Victoria's Ayers Rock.

**Wilpena Pound, South Australia (Supplied by the courtesy of the South Australian Government)**

**The Three Sisters and Katoomba Scenic Skyway, Blue Mountains (New South Wales Government Offices)**

falls, but the aborigines believe, let us hope correctly, that the Great Snake is buried there.

The **Great Dividing Range** of Australia, one of the great ranges of the world, is not a structural unit but a diverse series of mountains of different forms and origins. It stretches for 3600 km along the whole length of the east coast from the Cape Yorke Peninsula in north Queensland to the New South Wales/Victoria border, then turns west through Victoria to the South Australia border. All the important ranges and mountains of these three States fall within its bounds. The area of higher than average mountains, extending from Australian Capital Territory to the far side of the New South Wales/Victoria border, is called the **Australian Alps**, while the very highest part of this is the **Snowy Mountains**.

The **Warrumbungles** in northern New South Wales present a volcanic landscape, where the cones have eroded away leaving the plugs of harder rock in the vents as vertical pillars. The Bread-

**The Warrumbungle Ranges, New South Wales (New South Wales Government Offices)**

## TASMANIA

**Ben Lomond Ra** (Ben Lomond 160 km²
  & Mt Barrow 5 km² NPs)
Legges Tor 1573
Stacks Bluff 1527
Mt Barrow 1413
Ben Nevis 1367

**Great Western Tiers**
Cumner Bluff 1452
Ironstone Mtn 1444
Western Bluff 1424

**Cradle Mtn group** (Cradle Mtn/Lake
  St Clair NP 1360 km²)
Mt Ossa 1617, **the highest point in
  Tasmania.**
Barn Bluff 1559
Mt Pelion W 1554
Cradle Mtn 1545
Mt Olympus 1447
Eldon Pk 1439

**S & SW areas** (Frenchman's Cap 102
  km² & Mt Field 163 km² NPs)
Frenchman's Cap 1444
Mt Field West 1434

knife is a wall many tens of metres high but only a few metres thick. Other noteworthy features are Boulougery Spire, Tonduron Spire, Crater Bluff, the Needle and Bluff Mountain. The rocks are strikingly reddish when lit by the sun, but blue in shadow.

Only recently did **Australia** come on to the world scene **as a rock-climbing country**. Twenty years ago there was almost nothing, now the climbing standards are as high as any in the world and climbers' crags are being exhaustively explored in almost every State. The principal crag areas are as follows: Queensland – Frog Buttress in the Glasshouse Mountains; New South Wales – Warrumbungle Range, Nandewar Range and Blue Mountains; Australian Capital Territory – Booroomba; Victoria – Mount Stapylton, Mount Rosen, Bundaleer and Mount Arapiles; Tasmania – Geryon, Frenchman's Cap, Federation Peak and on sea-cliffs and stacks.

The **Blue Mountains**, the portion of the Great Dividing Range closest to Sydney, is a dissected plateau, the valleys lined with crags and waterfalls, the ridges between providing scenic routes across the massif. There are hundreds of kilometres of sandstone

Mt Picton 1327
Mt Wellington 1269

**Tasmanian coast**
The Totem Pole is a 60m stack only
4m sq.

**LORD HOWE ISLAND**

Mt Gower 803
Ball's Pyramid 550 (fa 1965 Australian
party)

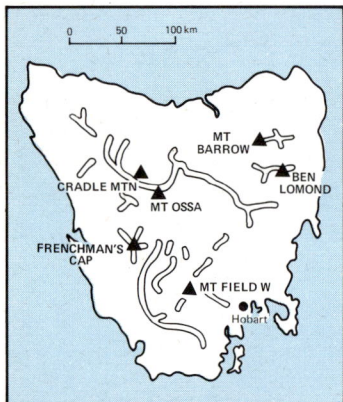

Tasmania

Ball's Pyramid (R D Treble)

# New Zealand

**NORTH ISLAND**

**Tongariro NP** (650 km²) **& extensions N**
Ruapehu (V-le 1950-1) 2797 (fa 1886

cliffs up to 300m high – pink, grey and pale gold. Tourist facilities
are well developed. At Katoomba the Scenic Railway descends
230m on a track 415m long, **one of the steepest in the world**, while
the Scenic Skyway nearby soars from cliff to cliff over a 450m
drop. The Jenolan Caves are also noteworthy.

It is not known who made the first ascent of **Mount Kosciusko**
(2230), **the highest point of Australia**; this is in dispute between Dr
Lhotsky (in 1834) and Count Strzelecki (in 1840). The former
wished to call it Mount King William IV, but the latter seems to
have won this part of the contest by asserting the name of a great
Polish/American patriot. Since the ascent is so very easy, both
the above may have been forestalled by local stockmen. Ski-ing
was pioneered hereabouts as early as the 1860s by Norwegian
goldminers, who founded a club at Kiandra in 1878. It is now
very popular ski country with a full range of amenities.

All this is now Kosciusko National Park (6134 km²). A motor-
able road from Charlotte Pass to the summit of the mountain is
kept closed to preserve the wilderness.

The huge **Snowy Mountains Hydro-electric Scheme** has been set
up since 1949. The fast east-flowing rivers of the range are
diverted westwards by tunnels into the Murray-Murrumbidgee
river system to provide annually 2·5 km³ of water for irrigation
and 4 GW of electric power. The catchment area is 5180 km²;
there are 16 large dams and many smaller, 7 power stations, 130
km of tunnels, 130 km of aqueducts and 1600 km of roads and
tracks.

The **Australian Alps** is the name given to the area of highest
mountains between Mount Bimberi (1912) (ACT) and Mount
Buller (1809) (Victoria). Winter snow is plentiful, in fact the locals
claim more snow cover than Switzerland. Mount Kosciusko
(2230) and its nearby peaks fall in this part of the range.

**Mount Bogong** (1986), the highest point of Victoria, is named for
a species of large moth caught and grilled by the aborigines when
food was short.

**Ball's Pyramid**, which rises from the sea off Lord Howe Island,
is a large sea stack with a base of 200m and a height of 550m. It
has been claimed as **the world's highest rock pinnacle**, but too much
depends here on definition. Certainly the first ascent in 1965, by
an Australian party, presented massive problems of approach,
securing lodgement, living during the ascent, retreat and re-
embarkation. This type of expedition is one of the highlights of
the sport of coasteering; the climbing challenge is the same as
that of the mountains, but every other aspect is different.

The **North Island** of New Zealand is notably volcanic. **The highest
point, Ruapehu** (2797), **first climbed in 1886**, is an active volcano
(last eruption 1950–1). In 1953 water in the crater lake was re-

Prof J. Park's party), c-wy to 2225 m.

Girdlestone Pk (Little Matterhorn) 2643

Nguarahoe (V-le 1948-9) 2291 (fa 1839 J.C Bidwell)

Tongariro (V-dormant) 1986

Pihanga (V) 1326

Tauhara (V-dormant) 1087

Maungakakaramea (V)

Tarawera (V-le 1886) 1111

Putauaki (Edgecumbe) (V-dormant) 822

White Island (V-le 1926) 328, in the Bay of Plenty.

## Mt Egmont NP

Mt Egmont (Taranaki) (V) 2518 (fa 1839 Dr E. Dieffenbach & party), discovered by Capt Cook in 1770, a symmetrical cone.

## N & NE Ras

*Rangitoto Ra* – Pureora 1166

*Hauhungaroa Ra* – Hauhungaroa 1079

*Raukamara Ra* – Mt Hikurangi 1748

*Haiarau Ra* – Pokokura 1383

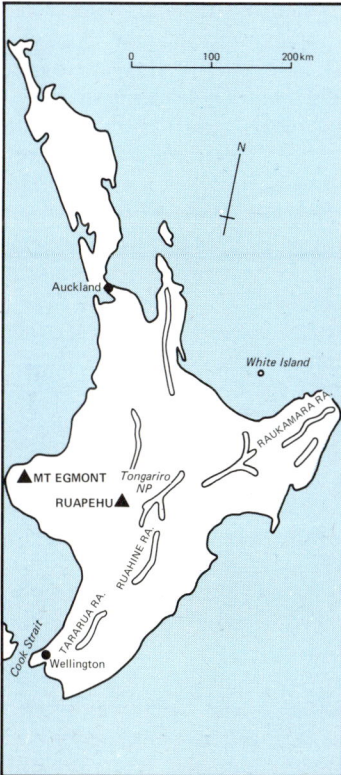
Mount Egmont, North Island (High Commissioner for New Zealand)

**New Zealand – North Island**

leased by the breaking of an ice dam; it destroyed a railway bridge 40 km away causing the Wellington–Auckland express to plunge into the Whangaehu River with the loss of 151 lives. This mountain is included with others in Tongariro National Park (650 km$^2$).

Further north the volcano of Tarawera erupted in 1886 from a fissure 14 km long. The geyser field of Waimangu, now dormant, used, around the turn of the century, to shoot boiling mud, water and stones to a record height of 450 m. Since 1959 geothermal power has been harnessed at Wairakei.

The line of volcanoes runs out into the ocean where White Island in the Bay of Plenty erupted last in 1926. A mid-ocean ridge continues in the same direction (north-north-east) for more than 2000 km through the Kermadec Islands to Tonga. Immediately east of it lie the very deep Kermadec and Tonga Trenches.

An isolated volcano, Mount Egmont (2518), rising as a symmetrical cone in the south-west corner of North Island, was first sighted by Captain Cook in 1770.

The **Rimutaka Tunnel** (8·8 km), opened in 1935, replaced the line over the Rimutaka Range, which had previously involved 5 km at gradients of up to 1 in 15 operated by the Fell System.

**Kakapo Peak** (1769) was named for a rare indigenous species of ground parrot.

**Arthur's Pass** was discovered in 1864 by Sir Arthur D. Dobson and the road through it opened in 1886. This route also provides the only reasonable rail crossing of the South Island. The 8·8 km Otira Tunnel below the Pass was started in 1907, using pneumatic drills powered by local hydro-electric generators. There were

*Kaweka Ra* – Kaweka 1724

*Kaimanawas Ra* – Makorako 1727

**Southern Ras**

*Ruahine Ra* – Mangaweka 1733

*Tararua Ra* – Mitre 1571, Hector 1529

*Rimutaka Ra* – Mt Matthews 940

## SOUTH ISLAND

### Nelson

*Arthur Ra* – Twins 1826

*Tasman Mtns* – Kakapo Pk 1769

*Lyell Ra* – Owen 1876

*Paparoa Ra* – Uriah 1501

*Robert Ra* – Travers 2326

*St Arnaud Ra* – Kehu 2218

*Spenser Mtns* – Una 2301, Faerie Queen 2235

**New Zealand – South Island**

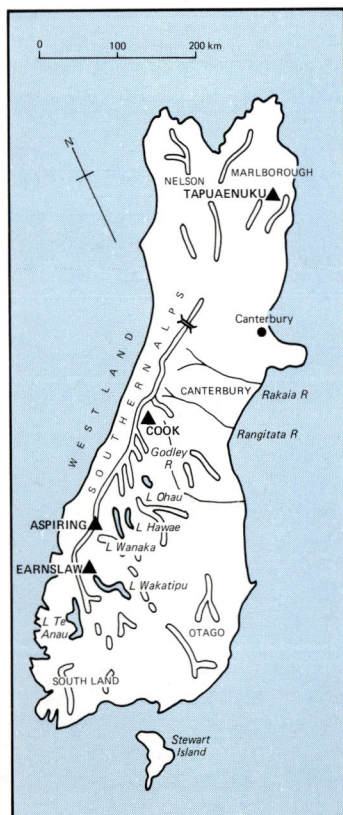

many difficulties and the headings did not meet until July 1918; the tunnel was finally completed in 1923.

The South Island is markedly mountainous, yet not in the least volcanic. The highest peaks are in the **Mount Cook** region, where the mountaineering is truly Alpine in character even though the highest point is nowhere near 4000 m. The siting of the Island directly in the path of prevailing westerlies (which have blown across thousands of kilometres of ocean) produces very heavy precipitation and violent storms leading to extensive snow and ice cover. The Tasman glacier, 30 km long, is one of the longest in the world outside Asia and the Polar regions. The Fox and Franz Josef glaciers on the steeper western side of the range fall 200 m in every km, their snouts reaching almost to sea level. Mount Cook National Park is 690 km$^2$.

**Mount Cook** (3764), named for the great navigator but called Aorangi by the natives, is **the highest point in New Zealand**. It was very nearly climbed in 1882 by Rev. W. S. Green with Swiss guides E. Boss and U. Kaufmann. In the face of very bad weather they reached a point only 60 m short of the summit; then, turning back, they were forced to spend an uncomfortable night in the open standing on a diminutive ledge before descending the next day. Bad weather prevented any further attempt. It was this climb, so nearly successful, which spurred the New Zealanders to tackle their own mountains. Twelve years later, on Christmas Day 1894, the peak finally **fell for the first time** to the locals T. C. Fyfe, G. Graham and J. Clarke. Their final inspiration came from the arrival in New Zealand of the British alpinist E. A. Fitzgerald and the Swiss guide Matthias Zurbriggen, intent on securing the same prize. Thus forestalled, the European party made first ascents of Mounts Tasman (3498), Silberhorn (3279), Sefton (3157) and Haidinger (3066), while some weeks later Zurbriggen made a solo ascent of Mount Cook by a different route.

Exploration of the high peaks proceeded apace. During **February 1907** the party of E. Teichelmann, E. Newton and A. Graham made **first ascents** of **Mounts Lendenfeld** (3185), **Torres** (3153), **Haast** (3138) and **Douglas** (3081) (in January), a remarkable *tour de force*. **An Australian woman, Freda du Faur,** with the guides, P. Graham and F. Milne, **was the first to climb Dampier** (3440) and to traverse all three peaks of Mount Cook. After the peaks came the ridges and then the faces. The great Caroline (south-east) Face of Mount Cook was finally climbed in 1970.

Other outstanding achievements include that of P. Scaife and D. MacNulty who **in 1977 traversed the main ridge from Elie de Beaumont to Harper Saddle** (34 peaks, 18 of them over 3000 m), the former carried on to Copeland Pass (42 peaks). In the same season P. Freaney and R. Brice **climbed all the 3000 m peaks in the country**, a total of 31.

Among the **lesser mountains** of **South Island**, accessibility has always presented a serious problem. A majority of the peaks had been climbed before World War I, but there were some notable

*Bryant Ra* – Red Hill 1790

**Marlborough**

*Kaikoura Ra*
Tapuaenuku 2885 (fa 1849 E.J. Eyre)
Alarm 2865
Mitre Pk 2621

*Seaward Kaikouras* – Manakau 2610

*Raglan Ra* – Scott Knob 2143

*Bounds Ra* – Bounds 2043

**Canterbury & Westland (N of Mt Cook)**

*Arthur's Pass NP* (950 km²)
Murchison 2400
Greenlaw 2286
Davie 2283
Rolleston 2272
A.P. Harper 2232, named for a pioneer
    NZ mountaineer.
Falling Mtn 1875, riven in two in the
    1929 Murchison earthquake, which
    involved land movements up to 5 m.
Arthur's Pass (rd 1886; ry below in 8·8
    km Otira tunnel 1933), discovered
    only in 1864.
*The Main Divide* – Arthur's Pass to Mt
    Cook area
Mannering 2653 (fa 1914)
Whitcombe 2638 (fa 1930)
Moffat 2636
Evans 2625 (fa 1934)
Loughnan 2589
Newton's Pk 2545
Livingstone 2540
Tyndall 2524 (fa 1911)
Malcolm 2510 (fa 1911)
McClure 2497
Blair 2495
Louper 2490
Huss 2489
Rosamund 2194 (fa 1913)
Bryce 2188 (fa 1930)

**Canterbury & Westland (The Mt Cook
    region)**
Mt Cook (Aorangi) 3764 (fa 25/12/
    1894 T.C. Fyfe; G. Graham, J.
    Clarke)
Subsid pks – Middle Peak 3710, Low
    Peak 3593
Mt Tasman 3498 (fa 6/2/1895 E.A.
    Fitzgerald; M. Zurbriggen, J. Clarke)
Mt Dampier (Hector) 3440 (fa 31/3/
    1912 Miss F. du Faur; P. Graham,
    F. Milne)
Silberhorn 3279 (fa as Tasman)
Lendenfeld 3185 (fa 26/2/1907 as
    Torres)
David's Dome (Mt Hicks) 3183 (fa
    1906 H. Newton, R. Lowe; A.
    Graham)
Malte Brun 3176 (fa 7/3/1894 M. Ross),
    another 'Matterhorn' of New Zea-
    land.
Teichelmann 3161 (fa 11/2/1929 H.E.L.
    Porter, Miss Gardiner; V. Williams)

exceptions. The summits of certain mountains at the head of the
Mathias and Rakaia Valleys, such as Whitcombe (2638), Evans
(2625) and points in the Arrowsmith and Armoury Ranges, were
not reached until the 1930s. In a remarkable expedition in 1930
J. Pascoe, R.R. Chester and A.H. Willis **traversed eleven virgin
peaks in one day – surely a record**. On the main divide they
travelled from Gerard by way of Tregear, Notman, Stout, Bal-
lance, Kensington, Shafto, Harrison, Kai-Iwi and Mystery to
Bryce.

Other peaks, such as Doris (1927) above the Wilberforce Valley, held out until after World War II.

**Aspiring** (3035), the 'Matterhorn' of New Zealand, is a pyramidal
mountain of striking beauty. Other mountains in the Haast
Range – Stargazer, Moonraker, Skyscraper, Rolling Pin and
Spike – take their names from the highest topsails of a square-
rigged ship.

**The Olivine Ice Plateau** is a very large snow-field formed between
mountain peaks; it closely resembles an ice-cap with valley
glaciers spilling down on every side. The local rainfall varies
between 5 and 7·5m per annum, so that tremendous rivers flow
in the area.

Sefton 3157 (fa 14/2/1895 as Tasman)   Wilczek Pk 3038 (fa 1936)
Torres 3153 (fa 4/2/1907 E. Teichel-   Hamilton 3022 (fa 1909)
    mann, H. Newton; A. Graham)         Glacier Pk 3007 (fa 1907)
Haast 3138 (fa 26/2/1907 as Torres)    de la Beche 2992 (fa 1894)
Elie de Beaumont 3109 (fa 1906 H.      Aiguilles Rouges 2966 (fa 1909)
    Sillem; P. Graham)                  Darwin 2961
Douglas 3081 (fa 28/1/1907 as Torres)  Chudleigh 2952 (fa 1911)
La Pérouse 3079 (fa 1/2/1906 E. Tei-   Annan 2947
    chelmann, H. Newton, R. Lowe; A.    Low 2942 (fa 1915)
    Graham)                             Haeckel Pk 2941
Haidinger 3066 (fa 8/2/1895 as Tasman) Conway 2901
Minarets 3066 (fa 9/2/1897 M. Ross;    Grey Pk 2893 (fa 1929)
    T. Fyfe)                            Footstool 2766 (fa 1894)

**Mount Cook area**

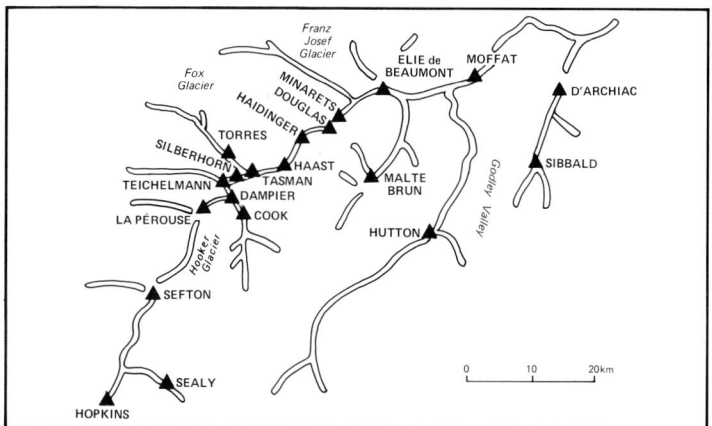

## Canterbury (E of Main Divide)

*Close to Main Divide*
Mathias valley – Kensington 2137
Rakaia valley – Arrowsmith Ra –
  Arrowsmith 2795 (fa 1917), Jagged
  Pk 2716 (fa 1931), Couloir Pk 2644
  (fa 1932), North Pk 2639 (fa 1932)
  – Armoury Ra – Warrior 2579,
  Amazon 2499
Rangitata valley – D'Archiac 2828 (fa
  1910), Forbes 2556 (fa 1912)
Godley valley – Sibbald 2798 (fa 1918),
  Wolseley 2572, Victoire 2520

*Outlying ras*
Puketeraki Ra – Chest Pk 1935
Torlesse Ra – 1820
Craigieburn Ra – 2130
Mt Hutt Ra & Old Man Ra – Taylor
  2330
Two Thumb Ra – The Thumbs 2541
  (fa 1923), Achilles 2537, Alma 2501,
  Fox Pk 2332
The Hunters Hills – Nimrod 1525
Grampian Mtns & Kirkliston Ra –
  1911

## Canterbury/Westland/Otago

*Lake Ohau region*
Burns 2738 (fa 1914)
Hopkins 2679 (fa 1914)
McKerrow 2653 (fa 1914)
Ward 2646
Brunner 2645
Sealy 2637 (fa 1895)
Thomson 2635
Glenmary (Neumann Ra) 2598
Jackson 2560
Lydia 2545
Darby 2526 (fa 1894)
Huxley 2508
St Mary 2334

Mount Aspiring – the 'Matterhorn of New Zealand' (High Commissioner for New Zealand)

*Lake Hawea/Lake Wanaka region*
Aspiring 3035 (fa 23/11/1910 B. Head;
  A. Graham, J. Clarke), the 'Matter-
  horn' of New Zealand.
Pollux 2542 (fa 1929)
Brewster 2519 (fa 1929)
Castor 2517 (fa 1929)
Ansted 2486 (fa 1914)
Tyndall 2474 (fa 1914)
Liverpool 2451 (fa 1914)
Headlong 2391 (fa 1926)
Stargazer 2380
Alba 2365
Enderby 2360
Aeolus 2349
Alta 2345
Holdsworth 2294
Awful 2202
Dreadful 2002
Doris 1927 (fa 1948)

## Otago/Southland (Lake Wakatipu region)

Earnslaw grp – Earnslaw W Pk 2819
  (fa 7/2/1914 H. Wright, J. Robert-
  son), Earnslaw E Pk 2794 (fa 16/3/
  1890 H. Birley), Sir William 2530 (fa
  1930), Pluto 2486 (fa 1931), Leary
  2438 (fa 1890)
Forbes Ra – Head 2550 (fa 1914), Ellie
  2499 (fa 1914), Moira 2499 (fa 1914)
Richardson Mtns – Centaur 2525 (fa
  1914), Clarke 2469
Humboldt Mtns – Bonpland 2470
Eyre Mtns – Jane Pk 2027
'Remarkables' – Double Cone 2343,
  Ben Nevis 2332, Ben Lomond 1752,
  Coronet Pk 1650
Olivine Ice Plateau – 1825

## Southland (Lake Te Anau & Fiordland)

Darran Ra – Tutoko 2756 (fa 4/3/1923
  S. Turner; P. Graham), Madeline
  2524, Underwood 2502, Christina
  2502 (fa 1925), Crosscut 1981 (fa
  1924)
Talbot 2225
Livingstone Ra – Moffat Pk 2085
Earl Mtns – Eglinton 1855
Mitre Pk 1695 (fa 1911)

## OFFSHORE ISLANDS

**Stewart Island** – Mt Anglem 980

**Auckland Island** – Mt Dick (Adams Island) 668, Cavern Pk 664

**Campbell Island** – Mt Honey 569

**Antipodes Island** – Mt Galloway 402

**Macquarie Island** – Mt Hamilton 433

# Pacific Ocean Islands

## HAWAIIAN ISLANDS

### Hawaii
Mauna Kea (V) 4206, rd to summit observatory; this and the next are
  shield volcanoes.
Mauna Loa (V) 4171, jeep rd to summit.
  **The bulkiest single mountain in the
  world** (42 000 km³), the base being
  5500 + m below sea level. Mokua-
  weoweo Crater is a caldera 6 km dia.
  5 km³ of lava ejected in 1955.
Hualalai 2520
Kilauea Crater (Uwekahuna) (V-le
  1971) 1247, 5 km³ of lava ejected in
  1840. The active vent is known as
  Halemaumau, 1100 m dia. and 400
  m deep.

### Maui
Haleakala Crater (Kolekole) (V) 3059,
  rd to summit.

The **Hawaiian Islands** were discovered by Captain Cook in 1778 and named the Sandwich Isles; he was killed there the following year. The volcanic peak of **Mauna Kea** (4206) **is the highest point of the islands and of the Pacific Ocean**.

Nearby **Mauna Loa** (4171), an active volcano, is the bulkiest mountain in the world (42 000 km³); the base is on the ocean-bed at a depth of 5500 + m. In Hawaiian legend Mauna Loa is the home of Pele, the goddess of fire, while Mauna Kea houses her rival, Poliahu. The earthquakes, eruptions and lava flows are signs of their frequent battles.

A large infra-red telescope for miscellaneous astronomical research was installed on Mauna Kea during 1977. The Hawaiian Volcano Observatory continuously monitors all phases of activity and makes predictions of future behaviour.

Mount Waialeale (1548) on Kauai has a recorded rainfall on the windward side of 11·9 m/annum, **one of the highest in the world**,

Lanai – Lanihale 1027

Molakai – Kamakou 1515

Oahu – Kaala 1227

Kauai
Kawaikini 1576
Mt Waialeale 1548

## MARQUESAS ISLANDS

Ua Pu – 1232

Fatu Hiva – 1200

Nuka Hiva – 1185

Hiva Oa – 1073

EASTER ISLAND – Terevaka (V) 601

## TUAMOTU AND SOCIETY ISLANDS

Tahiti – Orohena 2237, Aorai 2066, Roniu 1332

Moorea – Tohivea 1207

Uturoa – 1017

RARATONGA – 652

## WESTERN SAMOA

Upolu – 1100

Savaii – Silisili (Maungasili) 1844, Maungaafi 1720, Matavanu (V-le 1905-11) 712

AMERICAN SAMOA – Tutvila 653

NIUA FO'OU (V) 260
A circular island 8 km dia. with a lagoon 4 km dia., the upper part of a huge shield volcano.

## FIJI

Vita Levu – Tomaniivi (Mt Victoria) 1324

Vanua Levu – 1032

Kandavu – 851

## TONGA

Kao 1031
Late (V) 518
Falcon Island (V) 145, rose from the sea in 1894, periodically destroyed and reformed.

## NEW HEBRIDES

Espiritu Santo Is – Tabwemasana Pk 1811

Ambrim – Mt Marum (V-le 1967) 1334

Mount Bagona,
Bougainville,
Solomon Islands
(Aerofilms)

yet 25 km to the south-west on the leeward side it is only 0·05 m/annum.

Most of the trenches of the **Pacific Ocean** lie around its perimeter. There is a prominent mid-ocean ridge which runs from Scott Island in the Ross Sea to Easter Island and the Galapagos Islands. There are minor branch ridges, the exposure of which above sea-level accounts for the Tuamotu, the Society Islands, the Marquesas Islands and the Tubuai Islands. A further ridge running from the north-east tip of the North Island of New Zealand is exposed at the Kermadec Islands, Tonga and Niua fo'ou and has to its east the Kermadec Trench (−9476) and the Tonga Trench (−10882). Nearby is **the world's highest seamount**, rising 8690 m from the sea-bed to a summit 365 m below the surface. Beyond these trenches are American and Western Samoa.

Between here and Australia and New Guinea the configuration of the sea-bed is complex. A ridge system links Fiji with the New Hebrides, the Solomon Islands and the Bismarck Archipelago (*see* Asia). Between the New Hebrides and New Caledonia is the New Hebrides Trench (−7660), while north of the New Hebrides is the Torres Trench (−9165).

The principal areas with chains of seamounts are the Mid-Pacific Mountains, which run westwards from the Hawaiian Islands to Wake Island, and the Emperor Seamounts, which run in a line southwards from the far west tip of the Aleutian island chain. None of these breaks the surface.

Not surprisingly in an area of such obvious tectonic activity there are numerous volcanoes and volcanic manifestations.

Oba (V) 1200

Tana (V) 975

Vanua Lava (V) 951

Pentecost Is 934

Eromanga 914

Malekula – Mt Lamap 891

Mera Lava (V) 884

Aneityum 850

Epi 844

## NEW CALEDONIA

Mt Panié 1628
Mt Humboldt 1618

## SOLOMON ISLANDS

Bougainville
Mt Balbi (V-active) 3100
Takuan (V) 2251
Mt Bagana (V) 1999

Choiseul Is – 1067

Santa Isabel Is – Mt Sasari 1219

Guadalcanal
Mt Popomanaseu (Papomanasin) 2385 (mean)
Mt Kaichui 1920

Malaita Is – 1432

## GILBERT, MARSHALL AND PHOENIX ISLANDS, TUVALU
Summits of sea-mounts of 4000 to 5000 m appearing at the surface as coral reefs.

# North America

Mount Foraker, Alaska (H Adams Carter)

## *United States of America*

### WASHINGTON

**Olympic NP**
Mt Olympus 2428
Mt Constance 2370
Mt Anderson 2231
Mt Tom 2179
Mt Carrie 2140
Boulder Pk 2134

**Cascade Ra** (from S to N)
Mt Adams (V-dorm) 3751 (fa 1854
    Shaw, Aitken and Allen)
Mt St Helens (V-smoked 1841-2) 2950
    (fa 24/8/1853 T. Dryer & party)
Mt Rainier (Tacoma) (V-le at least
    1000 years ago, still steams a little)
    4392 (fa 17/8/1870 Stevens & Van

Mount Rainier, Washington
(Glaciological Society)

Trump; fa by a lady 1890 Miss
Fuller; fa, winter, 1922 J. & J.
Landry, J. Bergues), **the highest
point in the State.**
Mt Stuart 2886
Glacier Pk (V-ext) 3181
Bonanza Pk 2899
Mt Logan 2768
Jack Mtn 2765
Mt Baker (Kulshan) (V-activity in
1975, still steaming 1978) 3281 (fa
1868), 77 km² of ice-fields.
Mt Shuksan 2782

**Miscellaneous**
Stein's Pillar 125, overhangs all round,
circumference at summit $1\frac{1}{2} \times$ that
at base (fa 18/7/1950)

## OREGON

**Cascade Ra** (from N to S)
Mt Hood 3427 (fa 1854 T. Dryer &
party (disputed), certainly 1857 H.
Pittock, L. Chittenden, W. Cornell,
T. Wood), **the highest point in the
State.**
Mt Jefferson 3200 (fa 1854)
North Sister 3066
Middle Sister 3064
South Sister 3156
Mt Thielsen 2799
Mt Scott 2721
Mt McLoughlin 2894

**Wallowa Mtns**
Sacajawea Pk 3058
Matterhorn 2997, not a worthy holder
of an illustrious name.
Eagle Cap 2949

**Miscellaneous**
A large part of the interior is the desert
country of the Harney Basin. There are
summits such as Steens Mtn 2963 and
Strawberry Mtn 2755. Newberry Crater, once a 6000m peak, now has
Paulina Pk 2434 on the rim.

## CALIFORNIA

**Cascade Ra** (from N to S)
Mt Shasta (V-dorm, a few steam vents
and geysers) 4317 (fa 1854 E.D.
Pearce)
Shastine 3790
Lassen Pk (V-le 1915) 3186
Prospect Pk 3543
Mt Harkness 2450

**Sierra Nevada** (from N to S)
Matterhorn Pk 3742, another unworthy
holder.
Dunderberg Pk 3772
Mt Conness 3837
Mt Dana 3978
Mt Lyell 3998
Banner Pk 3946
Mt Ritter 4010
Red Slate Mtn 4012
Mt Morgan 4190

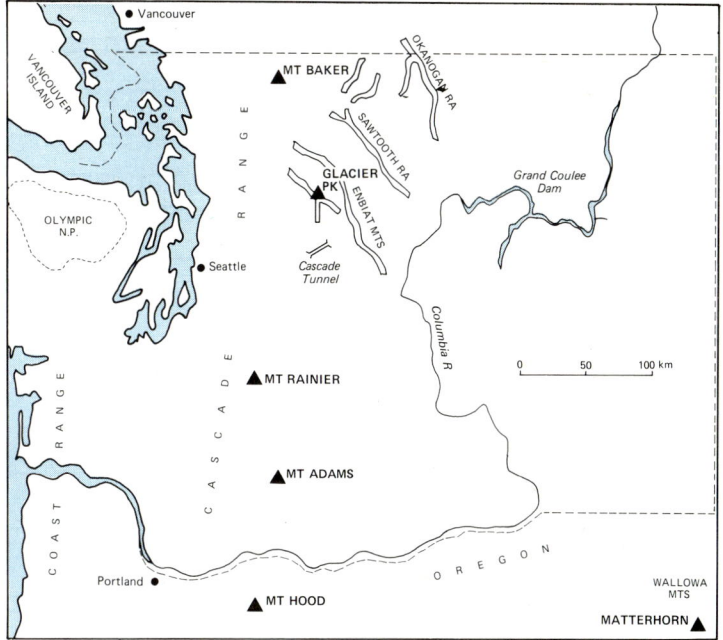

**Washington**

There are two distinct mountain systems – in the east the Appalachian Mountains, which parallel the eastern sea-board from
Georgia to the Canadian frontier, and in the west the Great
American Cordillera stretching from the Mexican to the Canadian
frontier, a formidable barrier to westward expansion a century
ago. The Cordillera is complex. Immediately west of the Plains
come the Rocky Mountains, mainly in the States of Colorado,
Wyoming and Montana, while closer to the western sea-board
are the Coast and Cascade Ranges and the Sierra Nevada. In
between these mountain systems there are a series of plateaux –
the Colorado River Plateaux (a high desert tableland, mostly in
Utah and Arizona), the Columbia Plateaux (one of the world's
largest lava fields, having an area between 500 000 and 600 000
km² with 300 000 km³ of lava, mostly in east Oregon, south Washington and Idaho) and the so-called Basin and Range Province
of Nevada (a desert area of internal drainage). The whole presents
a tremendous range of mountain scenery.

The USA pioneered the world preservation movement by
setting up the first National Park (Yellowstone) in 1871. Very
considerable areas of the country are now preserved as National
Parks (NP), National Monuments (NM), National Forests (NF)
and National Historical Sites (NHS), and at a secondary level by
similar State areas.

The USA also pioneered the long-distance trail system now
widespread in the world. The Appalachian Trail, set up in the
1920s, runs from Mount Oglethorpe (Georgia) to Mount Katahdin (Maine) – a distance of 2000 km. The Pacific Crest Trail pro-

Mt Mills 4105
Mt Abbot 4180
Mt Gabb 4179
Mt Tom 4161
Mt Humphreys 4263
Mt Darwin 4215
Mt Goddard 4136
Mt Agassiz 4235
Thunderbolt Pk 4267+
N Palisade 4345
Sill 4298
Middle Palisade 4282
Split Mtn (S Palisade) 4283
Mt Keith 4266
Mt Stanford 4262
Milestone 4158
Mt Kaweah 4207
Mt Tyndall 4275 (fa 1864)
Mt Williamson 4384 (fa 1884)
Mt Barnard 4268
Mt Russell 4325 (fa 1926)
Mt Whitney 4418 (fa 18/8/1873 A. Johnson, C. Begole, J. Lucas), **the highest point of the State and of the 48 contiguous states of USA.**
Mt Muir 4275
Mt Mallory 4228 (fa 1925)
Mt Irvine 4176 (fa 1925)
Mt Langley 4280 (fa 1871)

**Coast Ra** (from N to S)
Klamath Mtns – Mt Eddy 2755, Thompson Pk 2724 – a notable area of deep caves.
Bully Choop Mtn 2127
Mt Linn 2466
Black Butte 2270
Mt Hamilton 1299
Big Pine Pk 2081
Mt Pinos 2692
Reyes Pk 2289
Santa Lucia Ra – Junipero Serra 1781

**Sweetwater Mtns**
Wheeler Pk 3550, the range passes over into Nevada.

**White Mtns & Inyo Mtns**
White Mtn Pk 4341, jeep rd to summit.
Waucoba Mtn 3392

**Ras around Death Valley**
Panamint Ra – Telescope Pk 3367
Amargosa Ra – Grapevine Pk 2593, Funeral Pk 1946

**San Gabriel Mtns**
Mt San Antonio 3075
Mt Wilson 1881

**San Bernardino Mtns**
San Gorgonio Mtn 3498
San Bernardino Mtn 3240

**San Jacinto Mtns**
Mt San Jacinto 3301
Mt Palomar 1872

**Warner Mtns**
Eagle 3014
Warren 2960
Squaw 2635

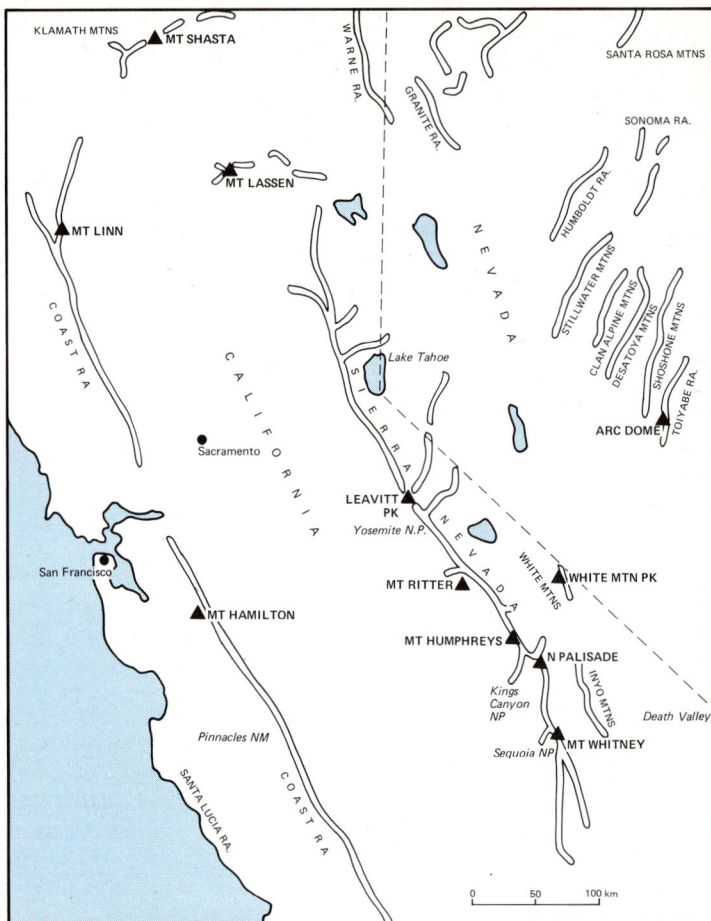

California/Nevada

vides 5000km of walkers' routes in the western States, including the Cascade Crest Trail and the John Muir Trail in the Sierra Nevada. A desert trail from Mexico to Canada has recently been proposed.

Between **25 May and 13 September 1969 F.T. Ashley reached the highest point of each of the 48 contiguous states of USA**. These vary from Florida (105m) to California (4418m) and required 30000km of motoring and 1000km of walking. The feat had been achieved previously, but not inside one year, by A.H. Marshall, Lawrence and Julia Grinnell, G. Peters and V. Hoeman. **The last, in fact, also included the highest points of the States of Hawaii and Alaska**, the latter a considerably more difficult climbing problem than any of the others.

A Spanish party consisting of V.L. de Ceballos, his wife Andrea, E.L. de Ceballos and J.P. Sangines carried out a protracted expedition between February and October 1965. Starting in Mexico with ascents of Nevado de Toluca and Popocatepetl, they worked north through Arizona and Nevada, the Sierra

*Above:* Mount Edith Cavell, Rocky Mountains of Canada
(Travel Alberta)

*Right:* Totem Pole and others, Monument Valley, Arizona,
USA (G Flanagan)

*Below:* The giant heads (18 m) of Washington, Jefferson,
Roosevelt and Lincoln (*left to right*) on Mount Rushmore,
S Dakota, USA (G Flanagan)

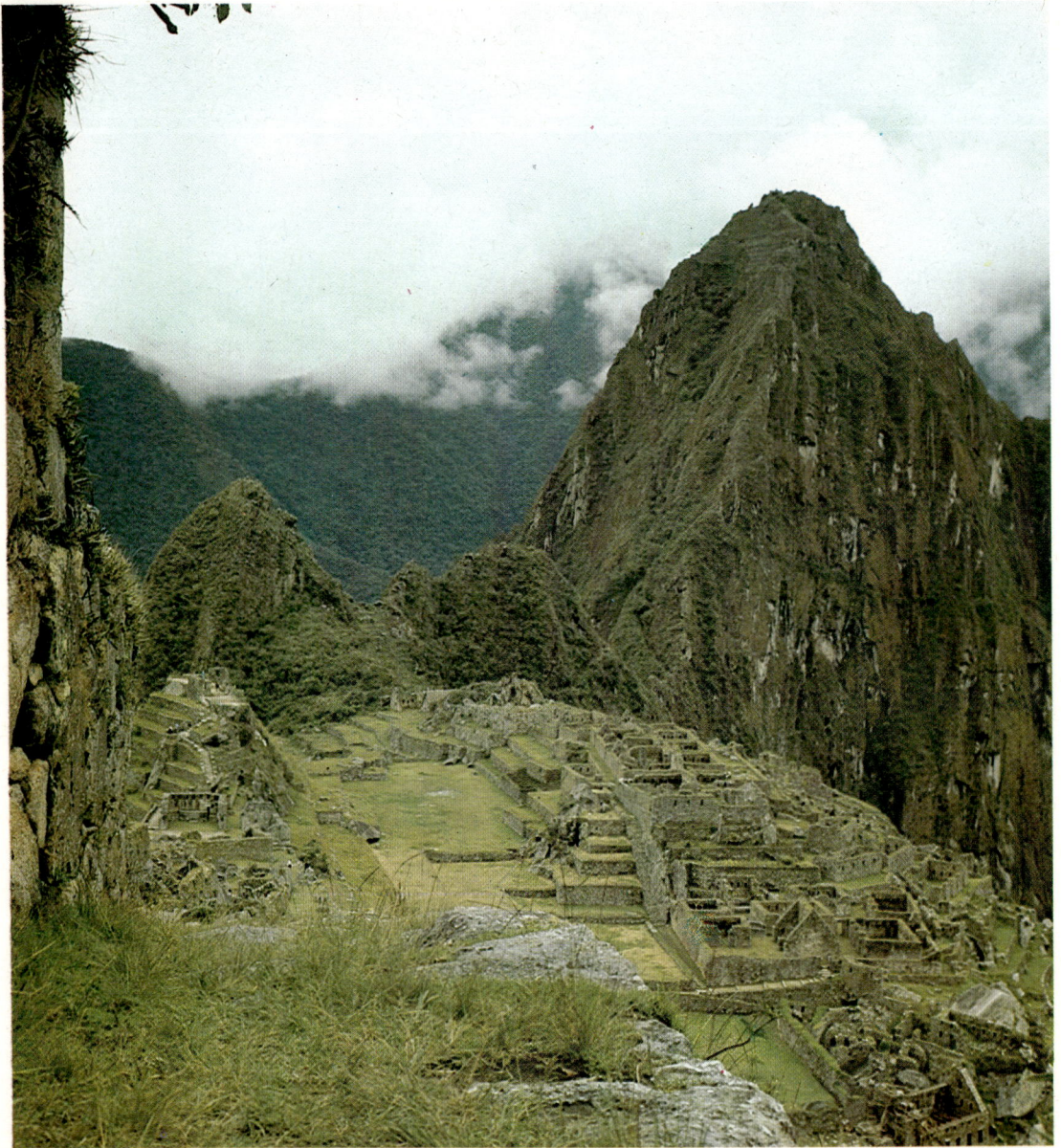

Machu Picchu, Peru (Barnaby's Picture Library)

The Observatory, Mount Palomar
(Photograph from the Hale
Observatories)

Half Dome, Yosemite (W Kirstein)

**Miscellaneous**
There is notable mountain scenery in
the Yosemite Valley (Yosemite NP),
the Tuolumne Meadows, and the Se-
quoia and King's Canyon NPs. The
following peaks rise round Yosemite:
N Dome 2295
El Capitan 2306
Basket Dome 2317
Eagle Pk 2369
Sentinel Dome 2474
Half Dome 2698 (fa 1875 G. Anderson
   by drilling a bolt ladder; 2nd ascent:
   P. Procter lassoing the bolts!)
Clouds Rest 3026

**NEVADA**

**NE corner of the State**
Matterhorn 3304, no more worthy
   than the others.

**Ruby Mtns & E Humboldt Ra**
Ruby Dome 3459
Sherman Mtn 3146

**Ras W of Great Salt Lake Desert**
Spruce Mtn 3365
Pilot Pk 3263

Nevada of California and the Cascade Range to Alaska. Forty-
six peaks were climbed, one of 2804m in Alaska, **a first ascent,**
two others by new routes. Mount Rainier (4392) was considered
the most difficult encountered in the 55000km of travelling involved. Later they carried out a similar lengthy raid in South
America (qv).

**Mount Adams** (3751) takes the form of a typical volcanic cone.
There are caverns under the summit ice-cap similar to those on
Mount Rainier (qv).

**Mount Rainier** (4392) has 28 glaciers with 120km$^2$ of ice. Beneath
the summit ice, steam has produced ice caves, first penetrated in
the 1970s. The crater can be crossed through passages large
enough for walking, with some chambers.
   Some remarkable snowfall records come from Paradise Ranger
Station on the mountain. The average annual fall is 1460cm; the
greatest twelve-month fall between February 1971 and February
1972 was 3110cm.
   The mountain was discovered by Vancouver in 1792 and named
for a British naval officer. Mount Rainier NP (960km$^2$) was
established in 1899.

Below **Stevens Pass**, south of Glacier Peak (3181), the Great
Northern Railway between Spokane and Seattle traverses the
range by the 12·5km Cascade Tunnel. This was opened in 1929
replacing a 4km tunnel of 1900. Previous to that the line had
climbed over the top by time-consuming zig-zags.

Yosemite and the Sierras (Aerofilms)

**Sierra Nevada** (extending from California)
Mt Rose 3285, with the ski resort of Lake Tahoe nearby.

**Wassuk Ra**
Mt Grant 3445
Cory Peak 3205

**White Mtns** (extending into California)
Boundary Pk 4005, **the highest in the State.**
Mustang Mtn 3144

**Clan Alpine Mtns**
Grant Pk 3428, Humboldt Sink, landmark in early overland travel, lies to the W.

**Shoshone Mtns**
N Shoshone Pk 3143
S Shoshone Pk 3070

**Toiyabe Ra**
Arc Dome 3589
Mt Callaghan 3107

**Toquima Ra & Roberts Mtns**
Mt Jefferson 3599
Wildcat Pk 3211
Summit Mtn 3189
Roberts Creek Mtn 3089

**Monitor & Antelope Ras**
Monitor 3309
Antelope Pk 3111
Sharp Pk 3079

**Diamond Mtns**
Diamond Pk 3237

**White Pine Ra**
Duckwater Pk 3410
Mt Hamilton 3275

**Grant Ra**
Troy Pk 3433
Timber Mtn 3133

**Shell Creek Ra**
Grafton 3351

**Snake Ra**
Wheeler Pk 3982
Baker Pk 3748
Mt Moriah 3673

**Spring Mtns**
Charleston Pk 3633

**Miscellaneous**
The Hoover Dam on the Colorado R is 221 m high; Mead Lake behind it is 185 km long. Sand Mtn, near Fallon, is noteworthy.

**IDAHO**

**Salmon R Mtns**
Twin Pks 3148
Bald Mtn 3132
Mt McGuire 3069

**Mount Scott** (2721) is the caldera (9 km dia.) of Mount Mazama which collapsed somewhere between 6000 and 4000 BC with an estimated disappearance of 50 km³ of material. Crater Lake, which now occupies the caldera has a surface area of 55 km² and a rim 150 to 600 m high. Now all preserved as a NP.

East of the **Wallowa Mountains** the **Grand Canyon of the Snake River is the deepest gorge in North America**. On the far side are the Seven Devils Mountains of Idaho.

**California** boasts **the highest mountain** (Mount Whitney (4418)) **and the lowest point** (Death Valley ($-85$ m)) **in the contiguous 48 states of the USA**, as well as the only active volcano (Mount Lassen (3186)). The rock climbing (Yosemite Valley, Tuolumne Meadows, Tahquitz Rock, Joshua Tree) is world famous.

The San Andreas Fault, which runs up through the State, marks the tectonic plate boundary where the Pacific plate is sliding past the North American plate. Possible movements here pose an ever-present threat. At the Geysers, volcanic activity associated with the Fault is used to generate geothermal power – the capacity will eventually reach 1 GW.

Above the smog of the cities, the mountains of the State provide attractive sites for astronomical observatories. On the summit of **Mount Hamilton** (1299) is Lick Observatory with a 0·9 m refractor telescope (1888) and a 3-metre reflector (1959). **Mount Wilson** Observatory has a 2·5-metre reflector telescope (1917). On **Mount Palomar** (1872) an observatory sited at 1697 m has a huge 5-metre reflector (1948).

**Mount Shasta** (4317), discovered in 1827, has five glaciers. There are some odd records. It was once climbed three times within a 24-hour period from Shasta Alpine Lodge at 2400 m; the record time from this point to the summit stands at 2 hr 17 min.

**Lassen Peak** (3186) was named for an immigrant of no particular distinction. It is not a volcanic cone but the upper part is a plug dome elevated 750 m above the crater of the former Mount Brokeoff. Lassen Peak NP (410 km²) contains a variety of volcanic scenery – cones and lava fields, fumaroles and mud volcanoes.

The **Sierra Nevada**, 650 km long, is a great tilted block of granite with gentle slopes west and steep fault scarps east. The land here rose 2000 m in two million years in the Pleistocene. There is a considerable amount of fine rock climbing. A high mountain route, the John Muir Trail runs, though not along the tops, from Tuolumne Meadows to Mount Whitney (4418).

The emigrant routes across the Sierra (one of the biggest obstacles to westerly expansion in the last century) followed one of three general lines, each capable of local variation. In the north the Lassen Route went from Black Rock Desert, either by Goose Lake and north of Mount Shasta, or north of Honey Lake and

**Sawtooth Ra**
Hyndman Pk 3681
Ryan Pk 3672
Castle Pk 3606
Bowery Pk 3331
Thompson Pk 3285
Mt Cramer 3268
Snowyside 3249
Williams Pk 3249

**Lost R Ra**
Borah Pk 3857, **the highest in the State.**
Leatherman Pk 3728
Dorion Pk 3662
Invisible Pk 3457
Dickey Pk 3395

**Lemhi Ra**
Portland Mtn 3298

**Seven Devils Mtns**
The Devil 2863

**Miscellaneous**
Volcanic scenery in the Craters of the
Moon NM.

## MONTANA

**Glacier NP**
Mt Cleveland 3185
Mt Stimson 3098
Kintla Pk 3082
Mt Jackson 3058
Mt Siyeh 3052
Mt Merritt 3034
Going-to-the-Sun Mtn 2927
Mt St Nicholas 2859
Chief Mtn 2763 (fa, recorded, 1892,
    but a bison skull was found on top)

**Absaroka Ra (Beartooth Mtns)**
Granite Pk 3901 (fa 8/1923 R. Ferguson, J. Whitam, E. Koch), **the highest point in the State.**
Mt Wood 3859
Tempest Mtn 3803
Mt Peal 3784
Beartooth Mtn 3773
Glacier Pk 3765
Mt Villiard 3760
Sundance Mtn 3740
Elk Mtn 3716

**Crazy Mtns**
Crazy Pk 3423

**Gallatin Mtns**
Electric Pk 3350
Hyalite Pk 3139
Mt Holmes 3139
Mt Blackmore 3138

**Madison Ra**
Koch Mtn 3440
Lone Mtn 3412
Gallatin Pk 3357

**Tobacco Root Mtns**
Granite Pk 3223

**Snowcrest Mtns & Gravelly Ra**
Hogback Mtn 3232

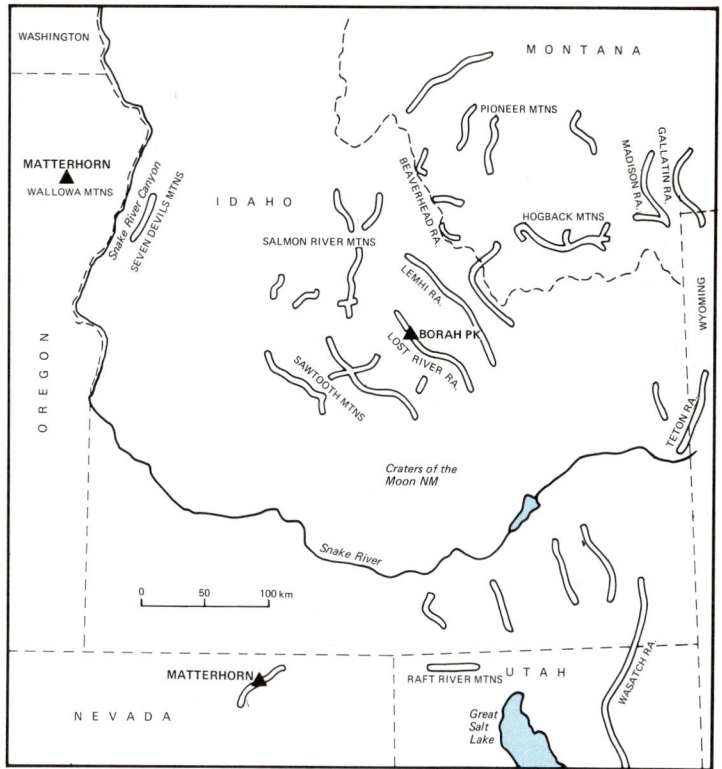

Idaho

north of Mount Lassen. The central Truckee Route went from the present site of Reno by either the Beckwith road, the Henness Pass or the Donner Pass, while the Carson Route crossed at points between Lake Tahoe and the start of the high Sierras. Only the Sonora variation from the Walker River appears to have tackled a really high pass, between Sonora Peak and Leavitt Peak.

The first transcontinental railway also hit problems in the Sierras. It was constructed by two companies – the Central Pacific driving east from Sacramento, California, and the Union Pacific striking west from Omaha, Nebraska. Payment by the mile produced bitter rivalry in which the Central Pacific was seriously disadvantaged by having to build over the Sierra Nevada very close to the start. Leaving a large work force engaged on tunnels, bridges, etc among the mountains, the engineers of the railroad took 3000 Chinese labourers over to start work on the far side. Three locomotives and 40 cars, with rails, spikes, tools and so on were dismantled and dragged over the summit (over 2100m) on sledges or on logs – the whole episode reminiscent of Hannibal or Napoleon taking their forces through the Alps. To keep the railway open to the summit tunnel, 65km of snow-sheds were built, using 65 million feet of timber and 900 tons of bolts and spikes at a cost of $2 million. The 'Great Race' between the two companies ended at Promontory Point, Utah, in May 1869.

Sunset Pk 3223
Black Butte 3214

**Pioneer Mtns**
Torrey Mtn 3427
Alder Pk 2902

**Anaconda Ra & Bitterroot Ra**
Mt Evans 3246
McDonald Pk 3139
Trapper Pk 3088

**Beaverhead Ra**
Garfield Mtn 3333
Ajax Mtn 3310

**UTAH**

**Raft R Mtns**
Raft R Pk 3027

**Deep Creek Ra** (continuing into Nevada as Snake Ra)
Haystack Pk 3688

**S of Great Salt Lake**
Deseret Pk 3362
Oquirrh Mtns 3353

**Mineral Mtns** 3414

**Pavant Ra**
Mt Catherine 3073

**Tushar Mtns**
Delano Pk 3710
Mt Belknap 3700
Circleville Mtn 3437

**Fish Lake Plateau**
Mt Marvine 3536
Mt Terrill 3514
Mt Hilgard 3513
Thousand Lake Mtn 3446

**Aquarius Plateau**
Blue Bell Knoll 3453
Barney Top 3261

**Sevier Plateau**
Signal Pk 3424
Monroe Pk 3422

**Wasatch Mtns**
Mt Nebo 3620
Mt Timpanogos 3581
South Tent 3439
Musinia Pk 3349
Spanish Fork Pk 3104

**Uinta Mtns**
King's Peak 4123, **the highest point in the State.**
Mt Emmons 4093
Gilbert Pk 4091
Mt Lovenia 4032
Tokewanna Pk 4015
Red Castle Pk 4003
Wilson Pk 3991
Hayden Pk 3801
Marsh Pk 3724

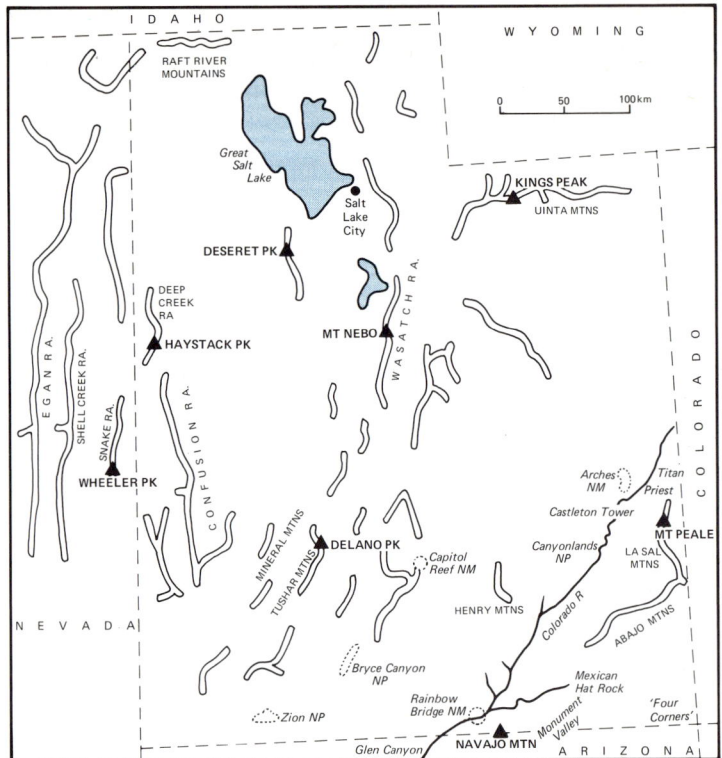

Utah

**Mount Whitney** (4418) has a road to the summit, so that this and Death Valley ($-85$ m) (**the highest and lowest points of the contiguous 48 states,** only 130 km apart) can easily be visited in a single day. Indian relics have been found on the summit which is much visited nowadays.

Palisade glacier on the slopes of the **Palisades** is **the longest in the Sierra Nevada** (2·5 km).

The **Santa Lucia Range** runs parallel to the coast, south of Monterey. The coast here, known as Big Sur, is developing as a climbing site. Inland the Pinnacles National Monument offers a further selection of volcanic scenery.

**Telescope Peak** (3367) is only 25 km from the lowest point of Death Valley $-85$ m). The Valley, named for a band of 49ers who perished there from thirst, is 250 km long and 15 to 60 km wide. The surrounding rocks are brightly coloured; temperatures as high as 57°C in the shade have been recorded.

At the east end of the **San Bernardino Mountains** there is rock climbing at Joshua Tree.

**W Tavaputs Plateau**
Bruin Pk 3135
Mt Bartles 3062

**Henry Mtns**
Mt Ellen 3540
Mt Pennell 3450
Mt Hillers 3246

**Utah/Arizona border**
Navajo Mtn 3175

**Abajo Mtns**
Abajo Pk 3463

**La Sal Mtns**
Mt Peale 3878
Mt Waas 3836
Mt Tomasaki 3740
Also the following isolated rock masses
  or pinnacles:
Fisher Towers 450
Titan 200 (fa 1962)
Moses 200 (fa 1972)
The Priest 120 (fa 1961)
Castleton Tower 120 (fa 1961)

**Miscellaneous**
Arches NM, where there are 90 natural
  arches.
Big Rock Candy Mountain, S of
  Sevier, is named for mineral out-
  croppings which glisten in the sun.
Bryce Canyon NP, outstanding ex-
  ample of Badlands topography.
Canyonlands NP, more pinnacles, not-
  ably:
  Standing Rock 100 (fa 1962) &
  North Six-shooter Peak 90 (fa 1962)
Capitol Reef NM, rock strata of multi-
  coloured hues, domed formations
  capped by white sandstone.
Dinosaur NM, prehistoric remains.
Natural Bridge NM, 3 stream-carved
  bridges.
Rainbow Bridge NM, an arch with a
  span of 75m and 95m high.
Zion NP, around Zion Canyon, rock
  climbing. The surrounding peaks
  are: W Temple 2376, Great White
  Throne 3056, Sentinel 2182, Angel's
  Landing 1763 and Three Patriarchs.

**ARIZONA**

**San Francisco Pks**
Humphreys Pk 3862, **the highest point
  in the State**, on the rim of an extinct
  volcano once 4500m high.
Agassiz Pk 3766
Fremont Pk 3648
Kendrick Pk 3175
Bill Williams Mtn 2824

**Pinaleno Mtns**
Mt Graham 3265
Hawk Pk 3231

**White Mtns**
Baldy Pk 3533
Greens Pk 3083

**Yosemite Valley** was discovered in 1851 by Major J. Savage and a party of soldiers who entered it in pursuit of Indians. It is 12km long, 1·5km wide and 600 to 1200m deep, and is now a NP and **one of the leading rock climbing sites in the world**. There are a number of famous crags.

**El Capitan**, at the west entrance, is 900m high (fa 1957). There are now many climbing routes. Typically one in 1962 involved a 42½-day period with 25 days of climbing, 3300ft (1000m) of rope, 75 pitons (some used 600 to 700 times), 75 to 80 bolts and six hammers. In recent years as techniques have developed the tendency has been to use much less in the way of artificial aids; times have also shortened.

On 1 February 1972 R. Sylvester skied down the summit slabs, then parachuted down the vertical face to land in a tree.

**Half Dome**, at the east entrance, 600m high, is only accessible to climbers (fa 1875 G. Anderson, using a ladder of eye-bolts).

Other crags are Three Brothers, Royal Arches, Washington Column, Glacier Point, Sentinel, Cathedral Spires and Leaning Tower. There is a further crag beside the Yosemite Falls, a series of waterfalls totalling 740m; Ribbon Fall (490m) is **one of the highest single falls in the world.**

**Tuolumne Meadows**, north of Yosemite and 1200m higher, are surrounded by domes of granite, which give more high-standard rock climbing. The following are worthy of note: Daff, Fair View, Harlequin, Lembert, Mariuolumne, Medlicott, Polly, Pywiack and Stately Pleasure Domes, Pennyroyal Arches and The Lamb. Mount Dana rises to the east. The granite here is orange, contrasting with the grey of Yosemite.

In **Sequoia NP** on the west side of the Sierra Nevada a fine collection of *Sequoia Gigantea* trees is preserved. These are up to 3500 years old, rise to 90m or more and with diameters sometimes exceeding 10m. Their long life is probably due to the thickness of the bark, which is as much as one metre. One, called Wawona (now fallen), had a tunnel made, 2·5m high and 3·5m wide, through it; another, General Sherman, is 11m diameter at the base and as high as a 20-storey building.

**King's Canyon NP** includes part of the highest mountains in the Sierra Nevada range. The canyon itself is 2400m deep.

To the east of **Mount Rose** (3285) is Virginia City, centre of the 'gold rush' of 1859, site of the so-called Comstock Lode, which produced $100 million of silver and gold in ten years.

Below the vast north face of **Wheeler Peak** (3982) lies the only glacier in the Great Basin region; 600 × 150m and 25m deep, it is the most southerly glacial ice in the USA.

**Sand Mountain** is a gigantic dome of white quartz sand, 5m long, 1·5 km wide and tens of metres high. It is also called the Singing

**Mtns in the extreme S**
Chiricahua Pk 2986
Miller Pk 2879
Mt Wrightson 2875

**Miscellaneous**
Barringer Crater, **one of the largest
meteor craters in the world,** (175 m
deep, 1265 m dia.) is close to Wins-
low.
Canyon de Chelly NM, notable pinnacle of Spider Rock 250 (fa 1956)
Chiricahua NM (420 km²) has three
canyons with rock monoliths.
The Grand Canyon of the Colorado
River, a colossal example of river
erosion. The surrounding peaks are
Siegfried Pyre 2412, Wotan's Throne
2327, Brahma Temple 2304, Vishnu
Temple 2294.
Monument Valley, exhibits striking
scenery of buttes, mesas and pinnacles up to 300 to 400 m. Noteworthy are: Agathla Pk 2164 (fa 29/
5/1949 L. Pedrick, H. Conn, R.
Garner) which rises 370 m above
the valley floor; Totem Pole 90 (fa
1957), which has a very small cross-
section; Castle; Stagecoach; Mitten
Buttes.
Sunset Crater NM, 400 m dia. & 120 m
deep, 400 volcanic outlets have been
traced hereabouts and here are copious signs of one-time activity, last
eruption 1900 years ago.

**WYOMING**
**Yellowstone NP & surroundings**
Overlook Mtn 3618
Pilot Pk 3569
Thorofare Butte 3480
Eagle Pk 3462
Mt Holmes 3165
Mt Washburn 3153
Mt Sheridan 3124
Mt Hancock 3048

Barringer Crater, Arizona (Aerofilms)

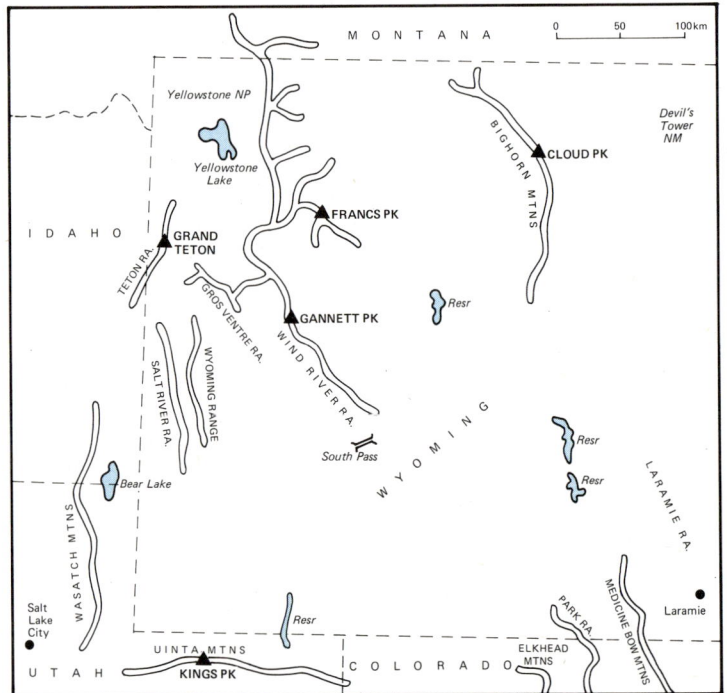

Wyoming

**The Tetons of Wyoming (AC
Collection)**

Mountain because of the rustling of the grains in the wind. Nearby
is the famous Carson Sink (a drainage hollow), a landmark on the
westerly trails of the last century.

The **Sawtooth Range** is one of the best rock climbing areas of the
country, notably on Thompson Peak (3285). Nearby, Sun Valley
is an international ski resort.

**Teton Ra**
Grand Teton 4196 (fa 29/7/1872 N.P.
  Langford, J. Stevenson (disputed)
  11/8/1898 W. Owen & party)
Mt Owen 3939
Middle Teton 3901 (fa early by Indians,
  a shelter found on top)
Mt Moran 3839 (fa 1921 Le Roy Jeffers)
South Teton 3812
Teewinot 3754
Thor Pk 3663
Rendezvous Pk, c-wy from Teton
  village.

**Salt R Ra**
Mt Wagner 3275
Man Pk 3148
Virginia Pk 3092
Indian Mtn 3008

**Wyoming Ra**
Wyoming Pk 3480
Hoback Pk 3311

**Gros Ventre Ra**
Doubletop Pk 3571
Darwin Pk 3549
Triangle Pk 3513

**Absaroka Ra**
Francs Pk 4005
Washakie Needle 3809
Trout Pk 3737
Dead Indian Pk 3735
Younts Pk 3708
Needle Mtn 3697
Fortress Mtn 3680

**Wind R Ra**
Gannett Pk 4204 (fa 1922 A. Tate, F.
  Stahlnaker), **the highest point in the
  State.**
Fremont Pk 4185 (fa 7/8/1878 J.
  Eccles & Hayden Survey party)
Mt Warren 4182
Mt Sacagawea 4147
Mt Helen 4145
Doublet Pk 4145
Mt Woodrow Wilson 4115 (fa 15/8/
  1842 J. Fremont & party)
E Sentinel 4115
Turret Pk 4115
Downs Mtn 4078
Wind River Pk 4031
Lizard Head 3914

**Bighorn Mtns**
Cloud Pk 4013 (fa 1887, recorded, but
  Indian relics found on the summit)
Black Tooth 3967
Mt Woolsey 3962

**Sierra Madre Ra** (extension of Park Ra)
Bridger Pk 3355
Vulcan Mtn 3261

**Medicine Bow Mtns**
Medicine Bow Pk 3659
Elk Mtn 3400

**Laramie Ra**
Laramie Pk 3132

The **Glacier NP** (4100 km²) is an outstandingly attractive mountain area with more than 40 glaciers and a wide selection of fauna. East and west sides are linked by the 'Going-to-the-Sun' road over Logan Pass. The 1200 m north face of Mount Cleveland (3185) is one of the great precipices of the USA. Mount Saint Nicholas (2859) is the most difficult single peak in the area. This NP area is contiguous with the Waterton Lakes NP of Canada.

**Chief Mountain** (2763), first sighted in 1792 and called the 'King', figured on Arrowsmith's map of 1795. It stands isolated east of the main mountain block. The bison skull, found on top by the first ascensionists, had presumably been carried up by Indians.

In the **Absaroka Range** there are 25 peaks over 12 000 ft (3658 m). Glacier Peak (3765) has a big rock wall on the north-east side. Grasshopper Glacier is named for millions of these insects trapped in the ice.

Towards the north end of the Idaho/Montana border is Flathead Tunnel through the **Bitterroot Mountains**. Opened in 1970, it is 12·5 km long.

The 19th-century emigrant 'California Trail' passed through the **Raft River Range**; hereabouts was the landmark known as the Cathedral Rocks.

On the north side of **Mount Timpanogos** (3581) is a permanent snow-field. There is a mass hike to the summit each year in July. Neffs Canyon Cave nearby is **the deepest in the USA**. Timpanogos Cave National Monument contains a notably large stalactite – the Great Heart of Timpanogos.

In the **Henry**, **Abajo** and **La Sal Mountains** are notable examples of laccoliths.

In **Arches National Monument** is the Landscape Arch (fa 1949). A visitor to the summit has described it as 'like standing on the wing of a big airplane taking off at a steep angle into the southern sky'.

Badlands topography, of which **Bryce Canyon** provides an excellent example, is caused by violent water erosion on sloping ground underlain by clay or soft earth; the land becomes sculptured into a pattern of gullies and ravines separated by spurs and buttresses. Here the rock colours are particularly striking, varying from light pink to deep red.

The high and graceful natural arch of **Rainbow Bridge**, which spans the waters of Bridge Creek, Utah, was, until the construction of Glen Canyon Dam, remote and little visited. It was only as late as 1909 that it was first located and approached. Now easily accessible by boat trips on Lake Powell it is seriously threatened

**Devil's Tower** 1558 (fa 4/7/1893 W. Ripley, W. Rogers using a wooden ladder pegged to the rock; fa, free, 1937), a volcanic plug monolith.

## COLORADO

### San Juan & San Miguel Mtns
Uncompahgre Pk 4354 (fa 1874 signs of previous ascents by grizzly bears)
Mt Wilson 4342 (fa 1874 The Hayden Survey)
El Diente 4328 (fa 1890 P. W. Thomas)
San Luis Pk 4312
Mt Sneffels 4311 (the grizzly bears had been here too)
Windom Pk 4295
Mt Eolus 4293
Sunlight Pk 4286
Redcloud Pk 4282
Stewart Pk 4277
Wilson Pk 4275
Sunshine Pk 4273
Wetterhorn 4272 (fa 1906)
Handies Pk 4271
Vestal Pk 4271 ⎫ the most difficult
Arrow Pk 4207 ⎭ mountains in the State.
Pole Creek Mtn 4188
Matterhorn 4142
Dolores Pk 4115
Kendall Pk 4100
Lizard Head 3997 (fa 1920), the summit is a 110m volcanic plug sheer on all sides, one of the earliest roped climbs in the country.
Chimney Pk 3591, the top is a 120m column of vertical rock.

### Elk Ra
Castle Pk (Mt Carbon) 4346 (fa 1873 H. Gannett)
Maroon Bell Pk 4306 (fa 1890s C. Willson)
N Maroon Pk 4270 (fa 1909 Hagerman, Clarke)
Snowmass 4267 (fa 1873 Hayden Survey)
Capitol Pk 4267 (fa 1909 Hagerman, Clarke)
Pyramid Pk 4247 (fa 1909 Hagerman, Clarke)
Mt Gunnison 3875, in the W Elk Ra
Crested Butte 3710

### Mosquito, Gore & Park Ras
Mt Lincoln 4354 (fa 1877 Princeton Survey), there are mines near the summit.
Quandary Pk 4345
Mt Cameron 4339
Mt Bross 4319
Mt Democrat 4311
Mt Sherman 4279
Ptarmigan Pk 4187
Mt Powell 4125 (fa 1868)
Red Pk 4018

### Sawatch Ra
Mt Elbert 4395 (fa 1874 H. Struckle), **highest point in the State and second highest in the 48 contiguous States of the USA.**

by the pollution which accompanies tourism. Moreover, when the lake is at its highest level, the existence of the Arch itself is menaced by the rising waters, not by submersion but by eroding away the supports.

**Arizona** is not a State of notable mountains, yet it is packed, nevertheless, with scenic attractions, particularly in the high plateau areas of the north and east.

Kitt Peak Observatory, near Tucson, and Lowell Observatory, near Flagstaff, take advantage of the clear skies to mount large telescopes.

**The Grand Canyon** is the result of large-scale erosion by the Colorado River, which has exposed 1500 million years of geological formations. It has a maximum depth of 1900m, the width between the rims varying from 8 to 25km. The scale of erosion is something like 14km$^3$ per km of river.

Portions of the side walls isolated by erosion by side streams have the scale and inaccessibility of mountain peaks, even though their summits are no higher than the rim of the gorge. Notable among them are: Vishnu Temple, Shiva Temple and Wotan's Throne.

The Canyon exhibits six of the seven botanical zones of the Northern Hemisphere. The climate variation travelling from top to bottom is the same as travelling from Mexico to Canada.

**The first running of the Grand Canyon** by boat in 1869, led by one-armed J.W. Powell, took three months. Nowadays, Glen Canyon Dam has modified the flow; the lake it holds back is named for the first explorer.

Colorado

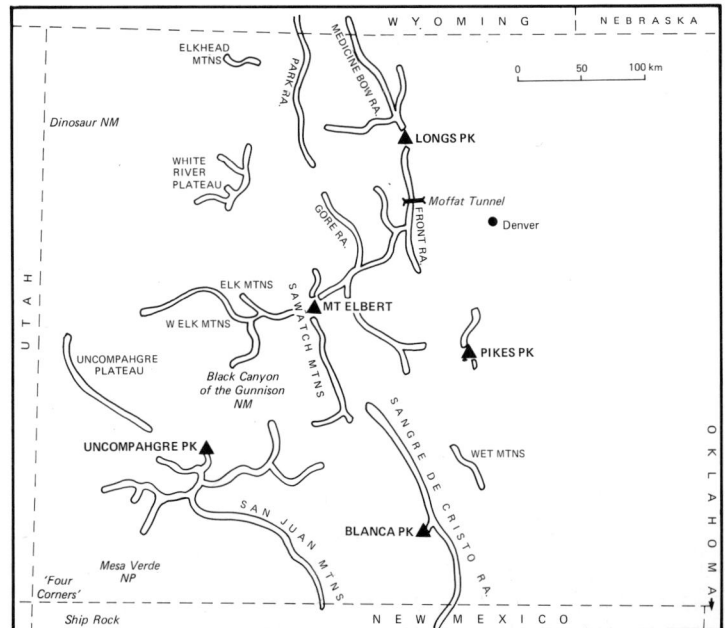

Mt Massive 4390 (fa 1874 H. Gannett)
Mt Harvard 4389 (fa 19/8/1869 W. M.
    Davis, S. Sharpless)
La Plata Pk 4371
Antero Pk 4342
Shavano Pk 4322
Mt Princeton 4321 (fa 1877 Princeton
    Survey, possibly earlier by miners)
Mt Yale 4320 (fa 18/8/1869 W.J.
    Whitney & party)
Mt Columbia 4289
Grizzly Mtn 4273
Mt Oxford 4267
Tabeguache 4267
Mt of the Holy Cross 4266 (fa 1873
    Hayden Survey party)
Mt Ouray 4253

**Medicine Bow Ra**
Clark Pk 3952
Richthofen (Never Summer Mtns)
    3948

**Front Ra**
Grays Pk 4350 (fa 1862 Parry)
Torreys Pk 4348
Mt Evans 4346, **the highest road in the
    country leads to the observatory on
    the summit.**
Longs Peak 4345 (fa 23/8/1868 Major
    Powell & party, probably climbed
    earlier by Indians in search of eagles'
    feathers)
Pikes Peak 4300 (fa 13-14/7/1820 E.
    James & party, fa by a lady 1858 Miss
    C. Archibald), rd & r-ry to summit.
Mt Bierstadt 4300
Hagues Pk 4134
Mt Ypsilon 4117
Arapahoe Pk 4116
Mt Fairchild 4115
Pagoda 4112
Mummy Mtn 4088
Kiowa 4047
Mt Audubon 4031
Lone Eagle 3627

**Sangre de Cristo Mtns**
Blanca Pk 4364 (fa 1875 F. Rhodes,
    A.J. Wilson, but Indian relics found
    on the top)
Crestone Pk 4356, Crestone Needle
    4325, Kit Carson Pk 4317 (fa 7/1916
    A. Ellingwood & party), hard rock
    climbing, probably 'the first rock
    climbs in the USA, where a conscious
    effort was made to belay'.
Old Baldy 4305
Culebra Pk 4288
Humboldt Pk 4281
Little Bear Mtn 4279 (fa 1888)
Purgatory Pk 4182
Spanish Pk 4152
Rio Alto Pk 4137

**Wet Mtns**
Greenhorn Mtn 3759

**Miscellaneous**
Black Canyon of the Gunnison NM, a
    gorge sometimes a mere 12 m wide
    in granite to a depth of 500 to 750 m.
    Some rock climbing.

The State of **Wyoming** offers the finest mountaineering of any in
the USA with two quite outstanding mountain groups – the
Tetons and the Wind River Range, which provide both rock and
snow work. The State is a lofty plateau of 1200 to 2100 m, with
mountain ranges averaging 3500 m and peaks up to 4200 m.

**Yellowstone**, discovered in 1807–8 by John Colter, was **the first**
(1871) **and is the largest** (8900 km$^2$) **NP in the USA**. It is a wildlife
sanctuary for bear, elk, buffalo, moose, etc. Scenic attractions
include: the Grand Canyon of the Yellowstone (40 km long,
360 m deep); Mammoth Hot Springs; 120 named geysers, including the world-famous 'Old Faithful' (60 m high).

The finest part of the **Teton Range** forms the Grand Teton NP
(1240 km$^2$). The range is 65 km long and another example of a
huge tilted fault block.

On 23 June 1925 there was a considerable earth slide on Sheep
Mountain in the **Gros Ventre Range**. Forty million m$^3$ of material
dammed the Gros Ventre river forming a lake; this later burst
the obstruction and flooded the valley, killing six people. The
sliding material, a 60 m thick stratum of sandstone, left a scar
1·6 km long and 0·8 km wide.

The **Wind River Range** presents a fine array of peaks and glaciers.
The mountain, climbed in 1833 by Capt B. Bonneville and claimed
to be the highest in North America, can no longer be positively
identified. The explorer John Fremont's party, once thought to
have climbed Fremont Peak (4185), are now believed to have
succeeded on the mountain now called Mount Woodrow Wilson
(4115) in 1842. J. Janisse, who accompanied Fremont, was **the
first negro to climb a major mountain in the USA**.

**Mount Sacagawea** (4147) is named for the Indian squaw who
accompanied Lewis and Clark on their early exploration journey
to the west coast. There is another mountain named for her in
Oregon.

**Wind River Peak** (4031) was visible from, and landmark on, the
California and Oregon Trails, by which settlers travelled overland
to the west in the mid-19th century. The actual route was by South
Pass, approached up the valley of the Sweetwater River, a broad
easy saddle across the Rocky Mountain ranges between the Bighorn and Wind River Ranges to the north, and the Laramie and
Park Ranges to the south. Later the Union Pacific Railway crossed
the same wide saddle some 70 km further south.

The **Devil's Tower** (1558) is a 275 m rock monolith with fluted
columnar sides. The first climbers found the top swarming with
chipmunks – it is interesting to speculate how they got there. It
became the first National Monument in the USA in 1906. In
1941 a parachutist landed on the top and had to be rescued by

Boulder Canyon, rock climbs.

Colorado NM holds the Grand Mesa, the largest flat topped mountain in the USA, 3000 m high with an area of 125 km² and sprinkled with lakes. Climbing on pinnacles – Watusi 90 etc. – and faces.

Dinosaur NM, dinosaur skeletons have been found; Steamboat Rock, in layered sandstone carved by rivers, now rises over 200 m above them.

Eldorado Springs Canyon, rock climbs.

Great Sand Dunes NM, the highest sand piles in the country, some over 240 m above the valley, these create violent thunderstorms and lightning bolts.

Mesa Verde NP, ruins of ancient cliff dwellings.

Rocky Mountain NP (1040 km²) takes in a large area of the Front Range. Grand Lake 2554 is claimed to be **the world's loftiest yacht anchorage.**

Royal Gorge has the steepest cable railway in the USA with a gradient of 1 in 1·55.

## NEW MEXICO

**Mogollon Mtns**
Whitewater Baldy 3320
Mogollon Mtn 3285

**Gallo Mtns**
Alegros Mtn 3122

**Black Ra** (and extensions to the NE)
South Baldy 3288
San Mateo Pk 3091
Mt Withington 3083
Reeds Pk 3051

**San Mateo Mtns**
Mt Taylor 3471

**San Juan Mtns** (& Valle Grande Mtns to the S)
Sta Clara Pk 3524
Pt 3430
Pt 3328
Chicoma Pk 3642

**Manzano Mtns**
Sandia Pk 3257, rd & c-wy to summit, rock climbing and ski-ing.
Manzano Pk 3078

**San Andreas Mtns** (Organ Mtns)
Organ Needle 2747

**Sangre de Cristo Mtns**
Wheeler Pk 4011, **the highest point of the State.**
Old Mike 4004
S Truchas Pk 3993
3 subsid pks, hard climbing on W Truchas Pk.
Latir Pk 3878
Gold Hill 3865
Costilla Pk 3851
Baldy Pk (A) 3847
Vallecito Mtn 3840

**Arizona/New Mexico**

climbers; 81 ascents were made during a mountaineers' week in 1952.

**Colorado** is known as the 'Mountain State', having 54 peaks over 14 000 ft (4265 m), yet none over 14 500 ft (4420 m), and no less than 1500 over 10 000 ft (3048 m). Most of the mountains are easy, and first ascents may often have been made by Indians, miners or hunters, or even by grizzly bears. Of the highest peaks, 19 have personal names (including six politicians, four explorers, a surveyor, a miner, a painter, a General, two Indians and two scientists) eleven are descriptive, seven are Spanish, five are colleges and the rest are assorted). These mountains form part of the Rocky Mountain chain which, running the length of the State, effectively blocks all east to west routes. Early travellers called the wall of the Rockies 'the Shining Mountains'.

Silver Lake, Colorado, claims **a world record for the greatest one-day snowfall**: 187 cm in April 1921.

Of **Uncompahgre Peak** (4354), one of its early climbers said – 'its striking resemblance to the Matterhorn gave us a wholesome dread of it'. The peak here actually named 'Matterhorn' is nothing like so impressive. Perhaps some confusion of names has taken place.

Cumbres Pass in the **San Juan Mountains** (3054) is **the highest**

Baldy Pk (B) 3807
Pueblo Pk 3743
Lobo Pk 3690
Cerro Vista Pk 3641

**Sacramento Mtns**
Sa Blanca 3659
Capitan Mtns 3111

**Guadelupe Mtns** continue into Texas

**Miscellaneous**
Capulin NM to the E of the Sangre de
   Cristo Mtns is a perfect volcanic cone
   reaching 2740 m with a road spiral-
   ling to the crater rim. Other cones
   also.
Carlsbad Caverns NP (185 km²) is in
   the Guadelupe Mtns.
In the deserts are:
Ship Rock 2188 (fa 1939, a three-day
   climb), rises 520 m above the desert
   floor.
Cleopatra's Needle 75 (fa 1955), pinnacle.
Venus's Needle 75 (fa 1962), pinnacle.

## SOUTH DAKOTA/NEBRASKA

**Black Hills**
Harney Pk 2207
Mt Rushmore 1943
Thunderhead Mtn

The Black Hills of South Dakota
(Aerofilms)

**point in North America reached by narrow gauge rails without a rack**, on the Cumbres and Toltec Scenic Railway.

There have been higher narrow gauge summits in the past – 3538 m on the Como-Gunnison section and 3454 m at the Fremont Pass on the Denver, South Park and Pacific Railway.

The mountains of the **Elk Range** offer some rock climbing, while the Aspen Basin between here and the Sawatch Range is one of the country's big ski-ing areas.

Leadville, to the east of the **Sawatch Range**, was the scene of a great silver strike in 1877, leading to a 'rush'. The local high level narrow gauge railways were built to support the mining activities.

The **Angel of Shavano** is a figure which is formed on Shavano Peak (4322) when all the snow has melted except for certain fissures in the peak.

**The Mountain of the Holy Cross** (4266) is named for a perpendicular gash and platform, often snow filled, which form a cross with a 425 m upright and arms of 60 m each side.

In the **Never Summer Mountains**, part of the **Front Range**, some of the peaks are named for clouds – Cumulus, Cirrus, Nimbus, etc.

**Longs Peak** (4345) was discovered by E. James of Long's expedition in 1820. Curiosities include a summit marriage (1927) and an ascent by F. Chamberlain on one leg with crutches (1928). There is hard rock climbing on the east face (the first route was made in 1927 by the Stettner brothers, who had motorcycled 1500 km from Chicago). The mountain now falls in Rocky Mountains NP.

**Pikes Peak** (4300) was discovered by Lieut Pike and party in 1806 and thought to be 18 581 ft (5663 m), only 1701 ft (518·5 m) short of Chimborazo, then considered to be the highest mountain in the world. 'Pikes Peak or bust' was the slogan for a gold rush in the 1850s which was largely a hoax. However, there was gold on the far side of the mountains which was exploited in due course. In 1876 a meteorological observatory was built on top and observers spent whole winters there. There is a road to the summit, on which hill climb races are held annually. The rack railway, which runs to the summit from 2298 m, has a maximum gradient of 1 in 4. Early locomotives were steam, now they are diesel-electric. It is the highest point reached by rail in North America.

Originally the Denver, North-western and Pacific Railroad crossed the **Front Range** over a pass of 3560 m to the north-west of Denver. Special locomotives had to be used on the gradients and average speeds were low. In 1923 work was started on the Moffat Tunnel beneath the pass, 10 km long at a maximum alti-

**Miscellaneous**
Badlands NM (435 km²), peaks, tur-
  rets, spires and steep canyons, with
  no vegetation.
Chimney Rock NHS & Scotts Bluff
  NM, once landmarks on the westerly
  trails; the former is a 150 m slender
  pinnacle, the latter a more substantial steep rock.
Wind Cave NP & Jewel Cave NM,
  caves with 40 & 80 km of passages
  respectively.

**TEXAS**

**Guadalupe Mtns NP** (continuing into
  New Mexico)
Guadalupe Pk 2667, **the highest point
  in the State.**

**Davis Mtns**
Mt Livermore 2555

**Chisos Mtns (Big Bend NP)**
Emory Pk 2388

**Chinati Mtns**
Chinati Pk 2356

**MISSOURI/ARKANSAS/
OKLAHOMA**

**Ozark Plateau**
Boston Mtns 600 (Arkansas)
St Francois Mtns – Taum Sauk 540
  (Missouri)

**Ouachita Mtns**
Rich Mtn 880 (Oklahoma)
Magazine Mtn 860 ⎱(Arkansas)
Blue Mtn 860          ⎰

**Miscellaneous**
Hot Springs NP includes some moun-
tains; the springs were discovered by
H. de Soto in 1541.

Ouachita NF.

**KENTUCKY/TENNESSEE**
Alongside Green River is the biggest
known cave system in the world; 290
km of passages honeycomb the Mammoth Cave and Flint Ridges.

**GEORGIA/the CAROLINAS**

**Blue Ridge**
Mt Mitchell 2037, **the highest point in
  the E USA,** rd to within a few min of
  top.
Balsam Cone 2025
Cattail Pk 2010
Roan Mtn 1924
Grandfather Mtn 1818
Mt Oglethorpe 1003, the SW end of the
  Appalachian Trail carries a statue of
  the founder of Georgia.

**Great Smoky Mtns** (crossing into Tennessee)
Clingman's Dome 2025, **the highest
  point in Tennessee,** rd close to summit.

tude of 2822 m. The headings met in February 1927, the first train
passing through just one year later. The cost was $18 million and
11 lives.

**Mount Taylor** (3471) is an ancient volcano. To the south-west
are extensive lava beds reaching for 60 km from the foothills of
the San Mateo Mountains, through the valley between the Zuni
Mountains and Cobolleta Mesa.

The **San Andreas Mountains** have been described as the most
impressive desert range in the USA.

The impressive **Carlsbad Caverns** include among their attractions
the 'Big Room', **one of the largest cave chambers in the world** –
100 m high, 200 m wide and 1250 m long. There is also a notable
bat colony.

**Ship Rock** (2188) is the core of an old volcano which has been
left upstanding 520 m above the desert. Narrow walls of lava
radiate like rays from the base. It was the first desert pinnacle to
be climbed, occupying a three-day period in 1939.

A large area of the **Black Hills** is set aside as Black Hills National
Forest (5000 km²). The Needles give some rock climbing on
granite. There was a 'gold rush' here in the mid 1870s centred on
Deadwood. Nowadays these hills are specially noted for large-
scale sculpting (*see* below).

In 1927 **Mount Rushmore** (1943) was selected by Gutzon Borghum
as a site for large-scale sculpture. With the support of a skilled
team he set to work on carving, on the face of the mountain,
giant heads (18 m high) of Washington, Jefferson, Lincoln and
Theodore Roosevelt. The undertaking was completed by Lincoln
Borghum in 1941 and is expected to last 100000 years (see illustra-
tion on p. 189).

A similar ambitious project is being carried out on **Thunderhead
Mountain**, where K. Ziolkowski is sculpting a giant figure of
Crazy Horse (Red Indian chief killed in 1877), 172 m high and
195 m long. The head is 27 m high, the arm 80 m long and the
head of the galloping pony 67 m high. Several million tonnes of
rock are being removed in the process.

Stone Mountain, near Atlanta, **Georgia**, said to be the world's
largest granite rock in isolation, is 250 m high. Here Gutzon
Borghum has carved a relief showing Robert E. Lee, Stonewall
Jackson and other Southern generals riding off to battle.

The **Great Smoky Mountains** have 29 peaks above 1600 m and 16
above 1830 m. The Great Smoky Mountains NP has an area of
2020 km².

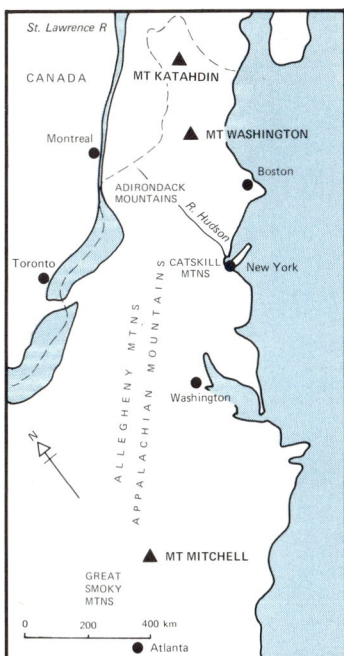

**Eastern States**

Mt Guyot 2018
Mt Le Conte 2010
Mt Collins 1886
Mt Kephart 1859

**WEST VIRGINIA**

**Allegheny Mtns**
Spruce Knob 1481
Grassy Knob 1333

**NEW YORK**

**Adirondack Mtns**
Mt Marcy 1629
Algonquin 1558

**Catskill Mtns**
Slide Mtn 1281
Hunter Mtn 1227

**Shawangunk Mtns**
An elevated plateau with precipitous sides, which provides some high standard rock climbing.

**Taconic Mtns**
Low folded ranges with monadnacks.

Equinox Mtn 1163
Mt Greylock 1068

**VERMONT**

**Green Mtns**
Mt Mansfield 1339, rd to summit.

---

Seneca Rock, near Petersburg, **West Virginia**, is a well known rock climbing site.

The **Adirondack Mountains** have 46 peaks above 4000 ft (1220 m). The Adirondack FP covers 7800 km². There is rock climbing at Wallface, Chapel Pond and Pokomoonshine.

**Mount Marcy** (1629), the source of the River Hudson, was believed by the Iroquois Indians to be the home of the 'Great Spirit'.

Dunderberg in the **Catskill Mountains** was the scene of Washington Irving's *Rip Van Winkle*.

The **Green Mountains** have five summits over 4000 ft (1220 m) and rock climbing at Smugglers Notch.

**New Hampshire** has 46 summits over 4000 ft (1220 m). Rock climbing and winter ice climbing are available at Huntington Ravine, Crawford Notch, Cathedral/White Horse and Cannon Mountain.

**The rack railway on Mount Washington** (1917) **was the first in the world** – built by Sylvester Marsh and opened in July 1869. It is standard gauge, 4·8 km long, with a maximum gradient of 1 in 3·1. The summit observatory claims **the highest recorded wind velocity in the world** – 370 km/hr. The mountain is generally noted for its grim weather. It was the scene in 1934 of a 'race' between three one-legged men.

**Mount Katahdin** (1605) marks the start of the Appalachian Trail which runs for 2000 km along all the eastern ridges to Mount Oglethorpe (1003) in Georgia. It was founded in 1921; **the first end-to-end trip was in 1948** by E. Shaffer who took 123 nights.

---

Killington Pk 1293
Mt Ellen 1260
Camel's Hump 1246
Lincoln Mtn 1243

**NEW HAMPSHIRE**

**White Mtns**
Mt Washington 1917 (fa 6/1842 D. Field and 2 Indians), rd and r-ry to summit.
Mt Adams 1767
Mt Jefferson 1745 (fa 1820 J. Weeks & survey team)

Mt Clay 1686
Mt Monroe 1641
Mt Madison 1634
Crawford Notch separates the **Presidential Ra** above from the Franconia Ra.
Mt Lafayette 1600

**MAINE**

**White Mtns**
Mt Katahdin 1605 (fa 1804 C. Turner & Survey team)
Sugarloaf Mtn 1292

# Alaska and the Yukon

**Brooks Ra**
Mt Michelson 2816
Mt Chamberlin 2749
Mt Igikpak (Schwatka Mtns) 2594 (fa
  1968 V. & G. Hoeman)
Pt 2438
Mt Doonerak 2320
Mt Greenough 2286

**Aleutian Ra & Islands**
Mt Torbert 3479
Mt Spurr 3374
Redoubt (V) 3108 (fa 1959 US party)
Iliamna (V) 3074 (fa 1959 US party)
Shishaldin (V-le 1963) (Unimak Is)
  2860 (fa 9/1872 A. Pinart got close to
  the summit)
Isanotski (V-le 1845) (Unimak Is) 2571
Veniaminoff (V) 2507
Pavlof 2504
Knife Pk 2332
Mt Denison 2301
Mt Katmai (V-le 1931) 2298
Mageik (V) 2208
Mt Chigmigak (V) 2143
Mt Vsevidof (V) (Umnak Is) 2109
Makushin (V) (Unalaska Is) 2036 (fa
  possibly 1815 O.v. Kotzbues: cer-
  tainly 9/9/1867 US coast survey
  party)
Pogromni (V) (Unimak Is) 2002
Aniachak 1357
Akutan (V-le 1948, very active) 1300

**Alaska Ra**
Mt McKinley (Denali) 6194 (fa 7/6/
  1913 H. Stuck, H. Karstens, R.
  Tatum, W. Harper: the ascent

---

**Mount McKinley (H A Gebbie)**

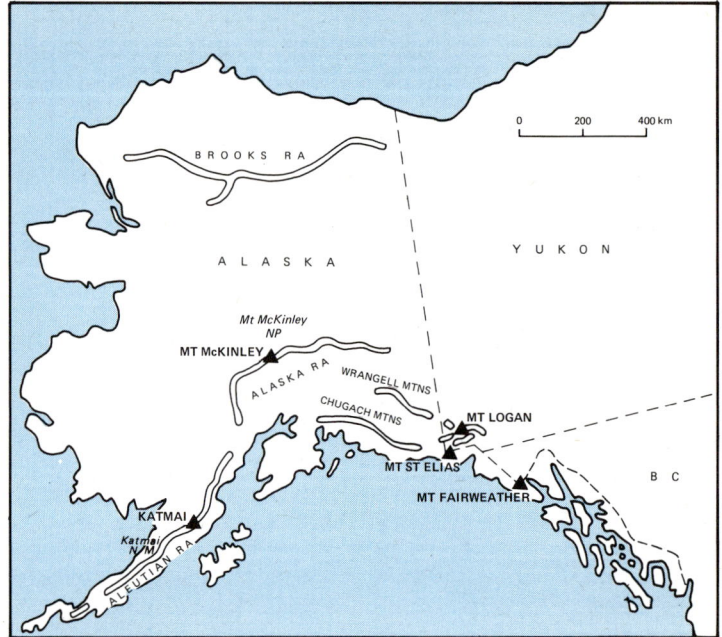

Alaska and Yukon

The mountains of **Alaska** were first sighted in July 1741 by a Russian scientific expedition commanded by Vitus Bering. Captain Cook in 1778 named the region Alaska, named Mount Fairweather (4670) and applied the name of Cape St Elias to the mountain behind it. The Comte de la Perouse in 1786 named Mount Crillon (3882) (for a French general); later on mountains were named for the leader of this expedition and for Dagalet, the astronomer of the party. Malaspina, an Italian employed by the Spanish in 1791 measured the height of Mount St Elias (5489); a 50 km glacier has been named for him. Vancouver in 1794 first sighted Mounts McKinley (6194) and Foraker (5304).

Novarupta erupted on the flanks of **Mount Katmai** (2298) in 1912 with the ejection of 20 km³ of material. Kodiak Island was buried in volcanic ash, 30 cm deep; all Alaska was darkened for 60 hours; avalanches fell from mountain peaks. A large area of fumarole activity was created, known as the Valley of Ten Thousand Smokes, now a NP of 11 700 km². Activity is dying.

**Makushin** (2036) was first climbed by US Government officials in 1867, who took up the US flag as part of the ceremonies connected with the Alaska purchase.

The caldera of **Pogromni** (2002) is 5 km diameter and 1000 m deep. The name means 'black destroying death'.

**Terminus of the Gilkey glacier, Alaska (Glaciological Society)**

claimed by F. Cook in 1906 is discounted: fa, winter, 1/1967 A. Davidson, D. Johnston, R. Genet: fa, solo, 26/8/1970 in 8 days N Uemura), **the highest point of the N American continent and of the USA.**

Subsid pk (North Pk) 5934 (fa 10/4/1910 P. Anderson, W. Taylor: C. McGonagall stopped 150 m below)

Mt Foraker 5304 (fa 6/8/1934 C. Houston, T. Graham Brown, C. Waterson)

Mt Hunter 4442 (fa 5/7/1954 F. Beckey, H. Meybohm, H. Harrer)

Mt Hayes 4188 (fa 1/8/1942 Mr & Mrs B. Washburn, S. Hendricks, B. Ferris, W. Shand)

Mt Silverthrone 4002

Mt Carpe 3825

The caldera of **Aniachak** (1357), 10 km diameter holding a lake 4 km diameter and Vent Mountain with its own crater, is not obvious on the ground. It was discovered originally during aerial mapping, and then explored. Now it is a wild life rendezvous, home of the Alaskan brown bear, **the largest land carnivore**. It is estimated that $50\,km^3$ were ejected here.

There are some other interesting volcanic features among the islands. **Bogoslof Island**, the summit of a plug dome, has appeared and disappeared several times during the last two centuries. **Buldir caldera**, between the islands of Buldir and Kiska, is $22 \times 45$ km and contains several sea-mounts.

**Mount McKinley** (6194) is named for the President of the USA in the year of its certain location (1896). The native name, Denali, is now coming back into use. An extensive massif, it is now a NP $(8000\,km^2)$.

Mt Deborah 3822 (fa 23/6/1954 as for
   Mt Hunter above)
Mt Hess 3667 (fa 24/5/1951 D. Hold-
   ron, E. Thayer, E. Huizer, H. Bow-
   man, A. Paige)
Mt Mather 3662
Mt Brooks 3639
Mt Russell 3557 (fa 1962 German/
   American party)
Mt Gerdine 3431
Mt Tatum 3395
Mt Kimball 3135
Moose's Tooth 3150 (fa 3/7/1964 Ger-
   man exp)
Revalation Mtns 2996
Kichatna (Cathedral) Spires 2739 (fa
   1966 R. Millikan, A. Davidson)

**Chugach Mtns**
Mt Marcus Baker (Mt Ste Agnes) 4014
   fa 19/6/1938 N. Dyhrenfurth, N.
   Bright, P. Gabriel, B. Washburn)
Mt Huxley 3828
Mt Thor 3816 (fa 1968)
Mt Witherspoon 3665 (fa 1955)
Mt Tom White 3545
Mt Miller 3353
Mt Goode 3231
Mt Steller 3129
Mt Gilbert 2940 (fa 1961)
Kenai Mtns – Truuli Pk 2015

**Wrangell Mtns**
Mt Blackburn 5036 (fa 19/5/1912 Miss
   D. Keen, G.W. Handy to SE sum-
   mit: true summit not until 1958)
Mt Bona 5005 (fa 1930 A. Carpe, A.
   Taylor, T. Moore)
Mt Sanford 4940 (fa 1938)
Mt Churchill 4767
Mt Bear 4520 (fa 1951)
University Pk 4410 (fa 1955)
Aello 4403
Mt Wrangell (V-steams only) 4269 (fa
   1908 R. Dunne, W. Sowle)
Mt Natazhet 4109 (fa 1913 H. Lambart
   & party)
Regal Mtn 4087
Mt Sulzer 3330

**N & E Yukon/NW Territory of Canada
(Mackenzie Mtns)**
Keele Pk 2970
Mt Sir James McBrien (Logan Mtns)
   2758
Mt Hunt 2750
Mt Savage 2744 (fa 1961)
Thunder Dome 2530 (fa 1968)
Mt Campbell 2500
Mt Joy 2237
Grey Hunter Pk 2215

**Yukon** (SW corner)
Mt Logan 6050 (fa 23/6/1925 A.H.
   MacCarthy, H. Lambart, W. Foster,
   L. Lindsay, N. Read, A. Carpe), **the
   highest point of Canada.**
Mt St Elias 5489 (fa 31/7/1897 Duke of
   the Abruzzi and large Italian/Ameri-
   can party; Italian guides)

Dr F. Cook, who later claimed the North Pole, said that he had climbed the mountain in 1906, but his 'summit pictures' were proved to have been taken elsewhere.

Four locals – T. Lloyd, C. McGonagall, W. Taylor and P. Anderson – set out in 1910, financed by leading citizens of Fairbanks. From a camp at 3350m with primitive equipment, the last two went to the North summit and back in a day, erecting there a 4·5m flag-pole which could be seen clearly from Fairbanks. Their claim to have reached the higher South summit also is discounted.

In 1912, B. Browne and H. Parker turned back in a blizzard only a few tens of metres below the top. The mountain was finally climbed in June 1913 by Archdeacon Hudson Stuck with H. Karstens, W. Harper and R. Tatum.

The use of ski-equipped aircraft to transfer men and materials to the base of the mountain was pioneered on Mount McKinley in 1932. Other expeditions in the 1930s, and since, have used similar tactics, so that the animal pack trains, dog teams and backpacking of the early explorers are now very much a thing of the past.

Ascents were infrequent until after World War II. Then in the Bicentennial Year (1976) more than 800 people made an attempt; there were 33 major accidents and ten fatalities. On 6 July 35 climbers were at or near the top (the exact date, 4 July, had been excluded by poor weather conditions).

The **Revalation Mountains** (2996) and the **Kichatna Spires** (2739) are notable rock climbing areas with impressive steep faces and pinnacles. A 1978 expedition made six first ascents, including Sunrise Spire (2400) and Cemetery Spire (2300).

The 50km Barnard glacier in the **Wrangell Mountains**, which has many tributaries, reveals a remarkable pattern of medial moraines derived from them.

G.W. Handy was the only one out of ten male members of the party to reach the summit of **Mount Blackburn** (5036) with Miss Dora Keen in May 1912. Their subsequent marriage seems a fitting conclusion to such an important joint endeavour.

A climber on **Mount Sanford** (4940) has reported seeing Mount McKinley, 370km away, and this may be one of **the longest lines of sight possible on the surface of the earth**. Unfortunately, the simple mathematical conditions for this process are seldom fulfilled and freak refractions can sometimes produce startlingly longer paths.

The summit of **Mount Wrangell** (4269) is used for research by the US Army; their buildings are continuously heated by steam from a fumarole.

The summit ridge of **Mount Logan** (6050) extends east and west for 18km above 4850m. The first ascent involved a walk of 150

Mt Lucania 5227 (fa 9/7/1937 B. Washburn, R. Bates)
King Peak 5221 (fa 1952)
Mt Steel 5011 (fa 15/8/1935 W. Wood's exped)
Mt Wood 4840 (fa 1941)
Mt Vancouver 4785 (fa 5/7/1949 R. MacCarter, A. Bruce Robertson, W. Hainsworth, N. E. Odell)
Mt Slaggard 4747
Mt McCauley 4717 (fa 1959)
Mt Hubbard 4557 (fa 1951)
Mt Walsh 4505 (fa 1941)
Mt Alverstone 4420 (fa 1951)
Mt McArthur 4420 (fa 1961)
Mt Augusta 4289
Mt Kennedy 4237 (fa 24/3/1965 R. Kennedy, J. Whittaker, B. Brather *et al*)
Mt Cook 4194 (fa 1952)
Mt Craig 4039
Mt Queen Mary 3962 (fa 1961)
Mt King George 3749
Mt Constantine 3136

**Alaska 'Panhandle'** (Alaska/British Columbia border)
Mt Fairweather 4670 (fa 8/6/1931 A. Carpe, T. Moore), only 22km from the coast.
Mt Quincey Adams 4161 (fa 1962)
Mt Root 3920
Mt Crillon 3882 (fa 9/7/1934 B. Washburn, A. Carter)
Mt Lituya 3642 (fa 1962)
La Perouse 3270
Mt Lodge 3210
Mt Ratz 3136
Mussell Pk 3127
Mt Bertha 3113
Kate's Needle 3049 (fa 6/8/1946 F. Beckey, R. Craig, C. Schmidtk)
Mt Burkett 2987
Mt Ambition 2926
Mt Gilroy 2865
Mt Aylesworth 2838
Devil's Thumb 2767 (fa 25/8/1946 same party as Kate's Needle)
Mt Pattulla 2729
Devil's Paw 2616
Cat's Eat Spire 2600 (fa 1972 P. Starr, R. Culbert, F. Douglas)
Mt Sheppard 2515
Mt Bigger 2515
Mt Barnard 2504
Mt Nesselrode 2470
Mt Edgecumbe (V) (Baranof Is) 976, **the first mountain to be ascended in Russian America.**

km up the Chitina valley, followed by a further 150km across glaciers to the top. Forty-four days were spent on snow and ice; the temperatures during a ten-day spell above 4500m varied between −18° and −33°C.

**Mount St Elias** (5489) has been described as a truly international mountain group – discovered by a Dane in the employ of the Russian government, named for a Greek saint, first mapped by an Englishman, who claimed possession for his country, again claimed in turn by official expeditions from Spain, then France, purchased by the USA from Russia. The first ascent of the highest point (which is shared by Canada) was by a royal Italian duke. It is sited only 20km from the sea.

**Mount Kennedy** (4237) was named for President John F. Kennedy, while the first ascent was made by his brother Robert Kennedy in March 1965.

Kluane NP is on the east side of the **Logan/St Elias/Lucania** massif and includes the mountains of the Icefield Range. A parallel range, the Kluanes, reaches 2400m. The mountains around **Mount Crillon** (3882) form the Glacier Bay National Monument.

During a 1949 expedition to **Devil's Paw** (2616) **and Michael's Sword**, W. Putnam was accompanied by his dog, Pinkham, which had an impressive list of first ascents. An application for membership of the American Alpine Club was submitted on his behalf and almost escaped the keen eye of the scrutineers; his record was adequate, but he stood on too many legs!

Packers ascending to the summit of Chilkoot Pass, Alaska (Public Archives Canada: C-5142)

# *Canada*

## THE COAST RANGES OF BRITISH COLUMBIA

### Northern area
Mt Monarch 3569 (fa 1936 H. Hall Jnr, H. Fuhrer)
The Queen 3050
Mt Edziza 2786
Mt Atna 2755
Weeskinisht 2755
Matterhorn Pk 2749

### West Central area
Mt Waddington ('Mystery Mountain') 3994 (fa 1936 F. Weissner, W. House)
Mt Tiedemann 3901 (fa 1939 S. Hendricks, H. Fuhrer, R. Gibson, H. Hall Jnr)
Mt Asperity 3720 (fa 1947 F. Beckey, H. King, F. Magoun, G. Matthews)
Mt Combatant 3701 (fa 1933 H. Hall Jnr, H. Fuhrer, A. Roovers, D. & P. Munday)
Serra Pks 3650 (fa 1954 D. Sowles, A. Kaufmann)
Subsid pks 3620, 3590, 2590, 3530
Mt Bell 3557 (fa 1936 B. Robinson, H. Voge, R. Bedayan, K. Austin)
Mt Tellot 3505
Remote Mtn 3353
Mt Geddes 3350
Mt Munday 3350
Mt Razorback 3251 (fa 1932)
Ottarasko 3079 (fa 1962)
Mt Blackburn 3063 (fa 1932)
Mt Dauntless 3018
Silverthrone 2960 (fa 1936)

### East Central area
Mt Queen Bess 3313 (fa 1942 D. & P. Munday, H. Hall Jnr)
Good Hope Mtn 3236 (fa 1922)
Pagoda Pk 3200
Mt Monmouth 3194
Mt Gilbert 3110
Mt Grenville 3110
Mt Raleigh 3078
Mt Tatlow 3060 (fa 1922)
Mt Taseko 3060 (fa 1922)

### Southern area
Skihist Mtn 2944
Mt Whitecap 2911
Wedge Mtn 2891
Mt Garibaldi 2678

### Vancouver Island
Golden Hinde 2200
Elkhorn Mtn 2192
Victoria Pk 2163

## THE INTERIOR RANGES OF BRITISH COLUMBIA

### Cariboo Mtns
Mt Sir Wilfrid Laurier 3520 (fa 1924 A. Carpe, R. Chamberlin, A. Withers)

Between Skagway (Alaska) and Bennett Lake on the British Columbia/Yukon border is the **Chilkoot Pass**, which was crossed in deep snow by thousands of prospectors in the Klondike Gold Rush of 1897–8. Each man had to bring a year's supplies involving him in numerous relays to reach the top. The chain of men employed the 'Chilkoot Lock Step', which made it difficult to rejoin for anyone forced to use the resting ledges. Enterprising 'climbers' cut steps up the last 50 m and then charged others for using them.

**Mount Garibaldi** (2678) falls within the Garibaldi ProvP. Close-by is a notable rock face, much used by climbers, Squamish Chief. The first ascent of the Great Wall was made by E. Cooper and J. Baldwin in 1961. They spent a month fixing ropes before commencing the ascent proper. At the week-end 12000 cars gathered below the crag as spectators turned up to watch. The climbers were given free food and lodging in the nearby town because of the tourist boom.

In 1972 a small climbers' hut was erected on a ledge on the face, a point requiring a high standard of rock climbing skill to reach.

The **Cariboo Mountains** occupy an area of 120 × 60 km inside the 'big bend' of the Fraser River, west of North Thompson River.

**Rocky Mountains of Canada and adjacent ranges**

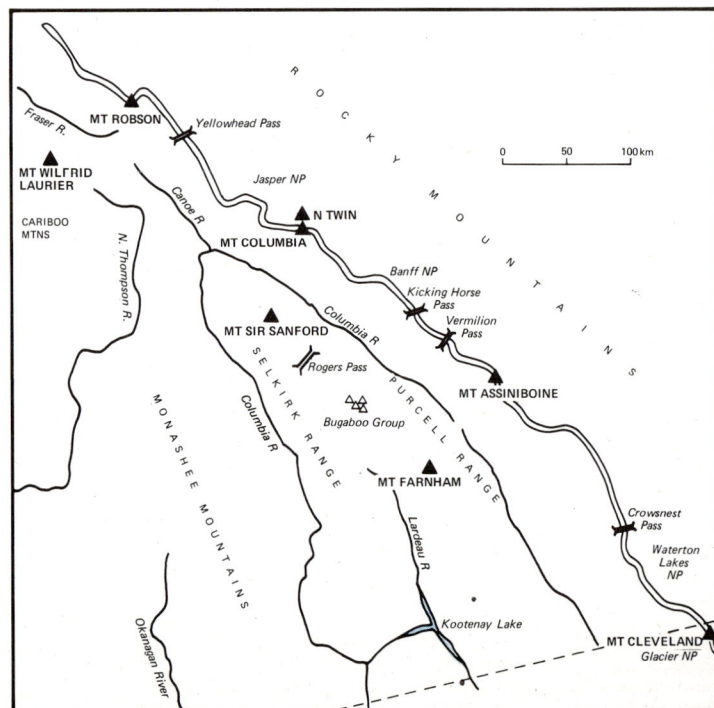

Mt Sir John Abbot 3429
Mt Sir Mackenzie Boswell 3277
Mt Stanley Baldwin 3249
Mt Sir John Thomas 3246
Mt Spranger 3024
Little Matterhorn 2850, yet another.

## Monashee Mtns
Mt Monashee 3246 (fa 1952 Hendricks,
Hubbart, Wexler)
Mt Lempriere 3207
Mt Torii 3200
Mt Milton 3185
Mt Deception 3133
Mt Hallam 3127
Dominion Mtn 3124

## Selkirk Ra
Mt Sir Sandford 3530 (fa 1912 E. Hol-
way, H. Palmer; R. Aemmer, E.
Feuz Jnr)
Mt Dawson 3390
Mt Wheeler 3386
Mt Adamant 3365
Mt Selwyn 3360
Mt Austerity 3347
Grand Mtn 3305
Mt Sir Donald 3297 (fa 1890 C. Sulzer,
E. Huber), the 'Matterhorn of the
Selkirks'.
Augustine Pk 3283
Pioneer Pk 3280
Mt Sugarloaf 3274
Cyprian Pk 3268
Blackfriar E 3255
Mt Proteus 3250
Mt Rogers 3208
Mt Bonney 3050 (fa 1888 W. Green,
H. Swanzy), among the first Euro-
pean mountaineers to come here.

## Purcell Ra
Mt Farnham 3457 (fa 1914 Mr & Mrs
A. MacCarthy; C. Kain)
Mt Karnak 3429 (fa 1916 Mr & Mrs A.
MacCarthy, Dr & Mrs W. Stone; C.
Kain)
Jumbo Mtn 3419 (fa 1915 Mr & Mrs A.
MacCarthy and party: C. Kain)
N Howser Tower 3399
Delphine Mtn 3376
Commander Mtn 3362
Mt Peter 3354
Farnham Tower 3353
Mt Hammond 3353
Eyebrow Pk 3353
Cauldron Mtn 3315
S Howser Tower 3307
Nelson Pk 3283
The Cleaver 3255
Mt Conrad 3252
Truce Mtn 3246
Mt Hamill 3243
Spearhead Pk 3230
Bugaboo Spire 3176
Pigeon Spire 3124 (fa 1930)
Snowpatch Spire 3063 (fa 1940 J.
Arnold, R. Bedayn)

**Mount Robson, highest point of the Canadian Rocky Mountains, and
Yellowhead Pass road (By courtesy of the Government of the Province of
British Columbia)**

**The far-N Rocky Mountains** (Yukon
border to Peace River)
Mt Sylvia 2952
Mt Lloyd George 2911
Great Snow Mtn 2900
Mt Stalin 2900, the only Mt Stalin not
changed after the Russian's death!
Mt Roosevelt 2900
Churchill Pk 2750
Terminus Mtn 1905, where the range
ends above the Kechika R of British
Columbia.

**The northern Rocky Mountains** (Peace
River to Yellowhead Pass, 450 km)
Mt Robson 3954 (fa 1913 W. Foster,
A. MacCarthy; C. Kain: an ascent
claimed in 1909 by D. Phillips & G.
Kinney is discounted), **the highest
point of the Canadian Rockies.**
Mt Resplendent 3426 (fa 1911 B. Har-
mon; C. Kain)
Whitehorn Mtn 3395
Mt Chown 3331
Mt Sir Alexander 3274
Mt Bess 3216
Mt Longstaff 3210
Mt Lynx 3291
Mt Resthaven 3215
Calumet Pk 2975

Mumm Pk 2969
The Ranee 2938
Whitecap Mtn 2864
Salient Mtn 2810
Sentinel Pk 2499

**The central Rocky Mountains** (Yellow-
head Pass to Kicking Horse Pass,
225 km)
Mt Columbia 3747 (fa 1902 J. Outram;
C. Kaufmann), **the highest point in
Alberta.**
North Twin 3684 (fa 1923 W. Ladd,
J. M. Thorington; C. Kain)
Twins Tower 3597 (fa 1938 C. Cranmer,
F. Weissner)
Mt Clemenceau 3658 (fa 1923 D.
Durand, H. Hall Jnr, W. Harris, H.
Schwab)
Mt Forbes 3628 (fa 1902 J. Collie, J.
Outram, H. Stutfield, G. Weed, H.
Woolley: C. & H. Kaufmann)
Mt Alberta 3619 (fa 1925 Y. Maki &
party; H. Fuhrer, H. Kohler, J.
Weber)
South Twin 3559 (fa 1924 F. & W.
Field, L. Harris; J. Biner, E. Feuz
Jnr)
Mt Lyell 3511 (fa 1902 J. Outram; C.
Kaufmann)

Mount Alberta (FS Smythe – by
courtesy of the Countess of Essex)

The Rocky Mountain Trench (By courtesy of the Government of the
Province of British Columbia)

Subsid pks 3511, 3507, 3402, 3399 (fas
  1926-7)
Mt Bryce 3507 (fa 1902 J. Outram; C.
  Kaufmann)
Subsid pks 3383, 3292
Mt Kitchener 3505 (fa 1927 A. Host-
  heimer; H. Fuhrer)
Mt King Edward 3475 (fa 1924 J.
  Hickson, H. Palmer; C. Kain)
Mt Brazeau 3470 (fa 1923 A. Carpe, W.
  Harris, H. Palmer)
Snow Dome 3456 (fa 1898 J. Collie, H.
  Stutfield, H. Woolley)
Stutfield Pk 3450 (fa 1927 A. Ost-
  heimer; H. Fuhrer)
Subsid pk 3383 (fa 1962)
Andromeda 3444 (fa 1930 W. Hains-
  worth, J. Lehmann, M. Strumia, N.
  Waffl)
Mt Tsar 3424 (fa 1927 A. Ostheimer;
  H. Fuhrer, J. Weber)
Mt Alexandra 3418 (fa 1902 J. Outram;
  C. Kaufmann)
Mt Woolley 3405 (fa 1925 Y. Maki &
  party; H. Fuhrer, H. Kohler, J.
  Weber)
Mt Hector 3394
Diadem Pk 3381
Mt Willingdon 3366
Mt Edith Cavell 3363
Mt Cline 3361
Mt Fryatt 3361
Recondite Pk 3356

It is sad to note that the peaks are named for a row of politicians,
now nearly forgotten.

The **Monashee Mountains** lie east of North Thompson and Oka-
nagan Rivers, west of Columbia River.

Road and Canadian Pacific Railway cross the range west of
Revelstoke. Further south there are road crossings at Monashee
Pass (1199) and close to the USA frontier.

In the extreme north, the Mica Dam on the Columbia River is
244 m high, 800 m wide and stores 25 km$^3$ of water.

The **Selkirk Range** falls inside the 'big bend' of the Columbia
River and west of Kootenay Lake. Glacier NP, on either side of
Rogers Pass and including many major peaks (1333 km$^2$), was
set up in 1886.

The range is crossed by Rogers Pass (1382), which carries road
and the Canadian Pacific Railway below in the Connaught
Tunnel. This, 8·1 km and **the longest in Canada**, was opened in
December 1916. It replaced an outside route over the Pass, which
was shortened by 7·25 km, had its summit lowered by 165 m and
had curves eliminated amounting to seven complete circles.

The first ascent of **Mount Bonney** (3050) by W. Green and H.
Swanzy in 1888 is regarded as **the first technical mountaineering
exploit in Canada**.

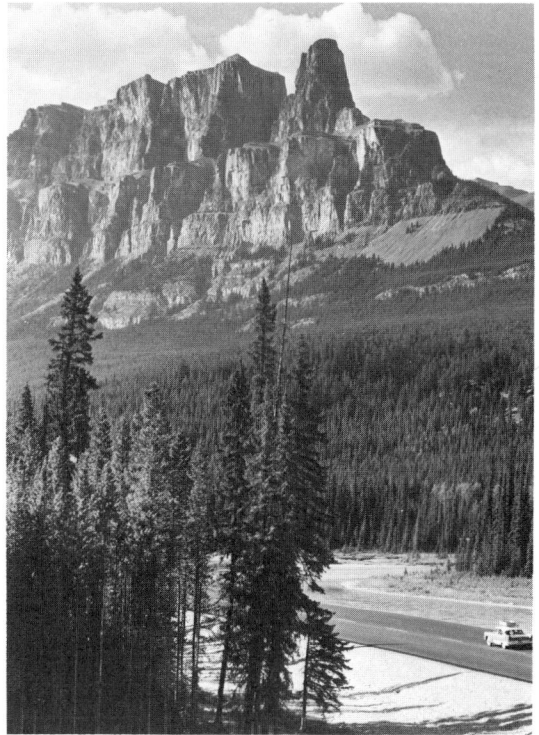

*Above:* Mount Assiniboine
(F S Smythe – by courtesy of the
Countess of Essex)

*Right:* Mount Eisenhower, Canadian
Rocky Mountains (Canadian
Government Travel Bureau)

Mt Barnard 3339
Mt Freshfield 3336
Mt Oppy 3334
Mt Amery 3329
Mt Mummery 3328
Mt St Bride 3315
Mt Stewart 3313
Mt Geikie 3308
Howse Pk 3290
Mt Nanga Parbat 3285
Rostrum Pk 3283
Mt Bulyea 3222
Hallam Pk 3219
Ghost Mtn 3204
Mt Evans 3188
Mt Brussels 3161 (fa 1948) a very diffi-
cult rock climb.
Mt Aylmer 3161
Mt Eisenhower 2766

**The southern Rocky Mountains** (Kicking Horse Pass to the US frontier,
340 km)
Mt Assiniboine 3618 (fa 1901 J. Outram; C. Bohren, C. Hasler), the
'Matterhorn of the Rockies'.

The **Purcell Range** is also inside the 'big bend' of the Columbia
River, but east of Kootenay Lake and Lardeau River and bounded
to the north by Rogers Pass.

Three 'spires' – **Bugaboo** (3176), **Pigeon** (3124) **and Snowpatch**
(3063) – are in the Bugaboo Range, a very popular climbing area
with spectacular peaks. There is a 100 km ski traverse from here
to Rogers Pass. Part of the range is Bugaboo Glacier ProvP,
while there is also a Purcell Wilderness Conservancy.

The **Interior Ranges of British Columbia** are separated from the
Rocky Mountains by the Rocky Mountain Trench, a major geo-
logical fault feature in which flow the Kootenay, Columbia and
Fraser Rivers.

The Alaska Highway, a strategic road linking Canada and the
USA with Alaska runs along the east side of the **far-north Rocky
Mountains**. From Fort Nelson it traverses the Stone Mountain
and Muncho Lake ProvPs to the Liard River and then continues
west into Yukon Territory. This road has facilitated access to
many far-away mountains, previously almost inaccessible.

**Mount Robson** (3954), the culmination of the Canadian Rocky
Mountains, lies closely north of road and railway through the
Yellowhead Pass. The Cree Indians call it Yuh-hai-has-kun, 'the
mountain of the spiral road', because of the prominent stratifica-
tion. On the lower slopes is Arctomys Pot (522 m deep).

Mt Goodsir 3563 (fa 1903 C.E. Fay,
    H.C. Parker; C. Hasler Snr, C.
    Kaufmann)
Subsid pk 3525 (fa 1909 A. Eggers,
    J.P. Forde, P. McTavish; E. Feuz)
Mt Temple 3547 (fa 1894 S. Allen, L.
    Frissell, W. Wilcox – a guideless
    American party), **the first 11000ft
    (3350m) peak to be climbed in the
    Canadian Rockies.**
Mt Hungabee 3492 (fa 1903 H.C.
    Parker; C. & H. Kaufmann)
Mt Victoria 3464 (fa 1897 J. Collie,
    C.E. Fay, A. Michael; P. Sarbach)
Subsid pk 3402 (fa 1900 J. & W. Out-
    ram, J. Scattergood; C. Clarke, H.
    Zurfluh)
Mt Deltaform 3455 (fa 1903 A. Eggers,
    H.C. Parker; C. & H. Kaufmann)
Mt Joffre 3449 (fa 1919 J. Hickson;
    E. Feuz Jnr)
Mt Lefroy 3423 (fa 1897 J. Collie &
    party; P. Sarbach)
Mt King George 3422 (fa 1919 V.
    Fynn; R. Aemmer)
Mt Sir Douglas 3406 (fa 1919 J. Hick-
    son; E. Feuz Jnr)
Lunette Pk 3399
Mt Harrison 3383
Mt Huber 3368
Mt Biddle 3319
Mt Ball 3312
Mt Eon 3310
Ringrose Pk 3278
Neptuak Mtn 3233
Mt Rae 3224
Mt Bogart 3144
Tornado Mtn 3100

## EASTERN CANADA

**New Brunswick**
Mt Carleton 820

**Labrador** (from S to N)
Kiglapait Mtns 900+
Kaumajet Mtns – Brave Mtn 1300,
    Mitre Pk 1143
Torngat Mtns – Torngarsoak Mtn
    1595, Innuit Mtn 1585, Cirque Mtn
    1569, Mt Cladonia 1555, Mt Silene
    1525, Mt Razorback 1100
Central Ra – Pt 1432, Mt Tetragona
    1345

**Nova Scotia**
Cape Breton Highlands NP 533

**Newfoundland**
Long Ra Mtns – Gros Morne 808

**Quebec**
Notre Dame Mtns (Gaspé Peninsula) –
    Mt Jacques Cartier 1268, Mt Albert
    1083
Torngat Mtns (Labrador border) – Mt
    d'Iberville 1652
Laurentides NP (N of Quebec), 10000
    km², reaches 1190m, **the oldest rocks
    in America.**
Close to Montreal there is rock climbing
    on Mont Césaire and Mont Condor.

In the **northern Rocky Mountains** there is road and rail crossing at Pine Pass in the north. The W. A. C. Bennet Dam on the Peace River holds back lakes of over 400km².

The **central Rocky Mountains** are delimited by the Yellowhead Pass (1131) in the north, carrying road and Canadian National Railway, and Kicking Horse Pass (1647) in the south, carrying road and Canadian Pacific Railway. The ends are linked by the Icefield Parkway road, which crosses Bow Pass (2068) and Sunwapta Pass (2035) and gives close-up views of peaks and glaciers. In winter there is a high-level ski route from end to end.

The main ridge is crossed by a trail through Howse Pass (1531), which provided **the very first crossing of the range early in the 19th century**. The Athabasca Pass (1748) was another historic crossing point having been traversed before 1811. It was opened up in 1811 by D. Thompson and became the route of fur-traders and other travellers between the plains and the Pacific coast. In 1827, David Douglas, Scottish botanist, climbed Mount Brown on the west side of the Pass, ascribing to it a height of 17000ft (5180m); it was figured on maps of America in 1829 at 16000ft; in fact it is 9156ft (2791m). However, this was **the first ascent above the snow-line in the Rocky Mountains of Canada and Douglas was thus the first North American mountaineer.** (At this time only a handful of snow and ice mountains had been climbed anywhere in the world.)

Notable in this part of the range is the Columbia Icefield, which sends water via the Saskatchewan, Athabasca and Columbia Rivers, respectively, to Hudson Bay, the Arctic Ocean and the Pacific Ocean. It has an area of 250km² and straddles the Divide for 25km. Castleguard Cave, beneath Castleguard Mountain, 9km long, is completely blocked at one end by the Icefield.

The first ascent of **Mount Amery** (3329) was made in 1929 by L.S. Amery, the British Conservative politician, *after* it had been named for him.

The first ascent of **Mount Alberta** (3619) marks one of the first forays into world mountaineering by the Japanese, now one of the leading mountaineering nations. A party of six crossed the ocean specially after reading of an unclimbed mountain in a climbers' guidebook. They engaged three Swiss guides in Canada. The lateness of the ascent (ie 1925) can be ascribed to the difficulty of the climbing problem and not to the inaccessibility of the mountain. The next ascent was not until 1948.

Three passes cross the **southern Rocky Mountains** – Crowsnest Pass (1396) (road); Vermilion Pass (1640) (road) and Kicking Horse Pass (1647) (road and Canadian Pacific Railway). All these were discovered by Dr Hector of the Palliser Expedition of 1857.

The original severe gradient on the railway across Kicking Horse Pass, known as 'Big Hill', was eliminated by spiral tunnels in 1909.

# Central and South America

Huascaran, Cordillera Blanca, the highest point of Peru (H Adams Carter)

## *Central America*

### MEXICO

**Baja California**
Co de la Encãntada – Picacho de Diabolo Pk 3078
Co Santa Genoveva 2406
Sa Juarez – Co Colorado 2036
   – El Gran Trono Blanco, gives climbing in an area reminiscent of the Algerian Hoggar.

**Sa Madre Occidental**
Citlaltépetl (Orizaba (V-le 1687) 5699 (fa 1846 or 1848 F. Reynolds, G. Maynard & American soldiers), rd to 4400m, **the highest point in Mexico and in Central America.**
Popocatépetl (V-le 1920) 5452 (fa 1519 Diego de Ordaz & Spanish soldiers and 1522: F. Montaño & 3 others, who brought back 140kg of sulphur)
Ixtacihuatl (V-dormant) 5286 (fa 3/ 11/1889 J. de Salis)
N de Toluca (V-extinct) 4577 (fa 1799-1804 Humboldt), rd nearly to summit.
La Malinche (Malintzin) (V-dormant) 4461

**Citlaltépetl** (Orizaba) (5699), a dormant volcano, is **the highest point in Mexico and in Central America.** The name means 'Mountain of the Star'. The mountainsides exhibit an interesting range of vegetation zoning – sugar, bananas and coffee at 600 to 1800m; maize, wheat and fruit at 1800 to 3000m; forests at 3000 to 4000m; alpine meadows at 4000 to 4400m; finally snow-cap to the summit.

**The first ascent of Popocatépetl** (5452) was made by Spanish soldiers in search of sulphur for gunpowder; in fact the mountain has been described as a 'gigantic natural sulphur manufacturing plant'. The name means 'Smoking Mountain'.

**Parícutin** began to erupt from flat fields on 20 February 1943 and then grew as follows: by 21 February 9m; by 27 February – 150m; in five months – 300m; in a year – 370m. It stopped in 1952 when 410m above the original land surface, though only 100m above its self-created platform of lavas and ashes. In the nine years, $35 \times 10^8$ tonnes of material were ejected. Parícutin village was buried in ashes and then covered by lava flow, while

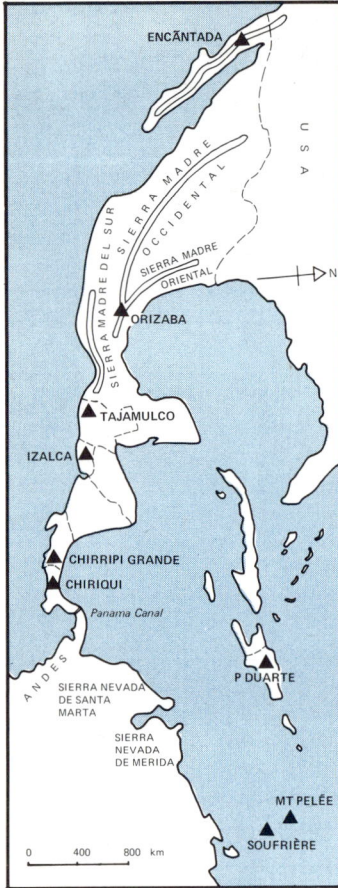

Central America

Paricutin, Mexico, in 1944 (Aerofilms)

N de Colima (V-dormant) 4339
Nauhcampatepetl (Cofre de Periote) 4282
Co Mohinora 3992
Co Ajusco (V-le 5000 BC) 3932
Co de Tancitario (V) 3859, surrounded by a crest of 250 small volcanoes.
Volcan de Colima (V-le 1965) 3850
Co Patamban 3750
Pingüicas 3191
Co Huchueto 3150
Co Nopala 3100
Co Prieto 3100
Parícutin (V-erupted 1943-52)

Jorulla (V-erupted 1759)

Climbers from Mexico City use crags at Los Monjas, Las Ventanas & Frailes. There is climbing further afield at Peñon Blanco (Durango) and Cañon Basaceachic (Chihuahua).

Sa Madre Oriental
Co Peña Nevada 3664
   There are notable caves.

San Juan de Parangaricutiro was submerged by a lava flow 10m thick, leaving only the towers of the cathedral protruding.

Only 50km away, Jorulla erupted similarly in September 1759, reaching a height of 1338m.

Barcena (300), which erupted in shallow waters off the Revillagigedo Islands in 1952, reached its full height within two weeks and had disappeared again inside six months.

There are active volcanoes in most central American states:

In Guatemala Fuego (3835) destroyed Antigua City in 1773 and Santa Maria (3768) destroyed Quezaltenango in 1902. In 1929 Santa Maria produced a plug dome which reached a diameter of 1200m and a height of 500m in two years; once it extended by 100m in a two-week period.

Earthquakes are common in El Salvador. The volcano, Izalco (2000+), began to grow from flat land in 1770 and by 1892 had reached 1885m. It is active and may still be growing.

Nicaragua was passed over as a site for the Atlantic/Pacific Canal because of the active volcanic landscape depicted on some of its postage stamps. Coseqüina (847) exploded in 1835 with the ejection of 10km³ of ash and pumice. A number of volcanoes have been active during the last decade.

In Haiti, 30km south of Cap Haitien, is a square-topped 900m mountain – the Bonnet à l'Evêque. The summit is a huge fortress built by King Henri Christophe early in the 19th century. It has been called the most awe-inspiring castle in the world.

**Sa Madre del Sur**
Teotepec 3703
Zempoaltepetl 3396
Co Yucuyácua 3376
Co Tlacotepec 3229
Sa de Oaxaca – Co Leon 3139

**Revillagigedo Islands** (370 km S of tip
  of Baja California)
Socorro 1130
Barcena (V-erupted 1952-3) 300, in
  shallow waters S of San Benedicto.

## GUATEMALA

Tajamulco (V-extinct) 4220
Tacaná (V) 4078
Alto Cuchumatanes 3993
Acatenango (V-slight activity) 3959
Fuego (V-le 1971) 3835, its 1773 erup-
  tion destroyed Antigua City.
Santa Maria (V-le 1929) 3768, the
  eruption of 1902 destroyed Quezal-
  tenango.
Agua (V-inactive) 3752
Zunil 3533
Atitlan (V-inactive) 3524
Co Quemado (V-inactive) 3179
Toliman (V-inactive) 3153
Sa de las Minas 3140
Sa de Chuacùs 2651
Pacaya (V-active) 2552

## EL SALVADOR

Santa Ana 2375, **the highest point.**
San Vicente (V) 2174
San Miguel (V-le 1970) 2153
Izalco (V-le 1966) 2000+, may still be
  growing, first appeared 1770 and
  had reached 1885 m by 1892.
San Salvador (V) 1950

## HONDURAS

Sa de Celaque 3050
Co Las Minas 2865
Sa dc Agalta 2590

## NICARAGUA

Cordillera Isabelia (Saslaya Pk) 2000
Momotombo (V-le 1905) 1258
Las Pilas (V) 1071
Telica (V) 1020
Coseqüina (V) 847, exploded 1835
  with ejection of 10 km$^3$ of ash and
  pumice.
Masaya (V-le 1970) 646
Co Negro (V-le 1971) 492, first ap-
  peared 1850.
Concepción (V), an island in Lake
  Nicaragua.

## COSTA RICA

Chirripo Grande 3820
Pico Blanco 3565
Co Buena Vista 3539
Irazú (V-le 1967) 3452
Turrialba (V) 3335
Co de las Vueltas 3087

**Fuego volcano, Guatemala (Aerofilms)**

**Mont Pelée** (around 1500) on Martinique is another notorious and lethal volcano. A disastrous eruption on 8 May 1902 killed 30 000 people in the nearby town of St Pierre in a *nuée ardente* which swept down the slopes. A spine, 150 m diameter, grew from the vent to a height of 360 m, cracking and splitting off as it cooled; at its fastest it extended 12 m in one day. After the spine was weathered away the plug dome continued to grow – 400 m in 1½ years with a diameter of 1000 m.

**Soufrière** (around 1280) on the nearby island of St Vincent, obviously connected in some way with the above, produced its own *nuée ardente* two days before Mont Pelée, killing 1600 people.

Barba (V) 2916
Poas (V-le 1972) 2739
Rincon de la Vieja (V)
Mirravalles (V)
Cordillera de Guancaste 2020

## PANAMA

Chiriqui (V) 3477
Co Echandi 3168
Co Santiago 2826
Sa de Tabascarà 1933

## CARIBBEAN ISLANDS

**Cuba** – Sa Maestra – P de Turquina
  (Turquino) 2005

**Jamaica** – Blue Mtns – Blue Mtn Pk
  2256

**Haiti** – Mt La Salle 2680, Mt La
  Fenêtre 2279

**Dominican Republic** – Cordillera Central – P Duarte (P Trujillo) 3175, P
  de Yaqui 2955

**Puerto Rico** – Co de Punta – Co Cal-
  derona Pk 1338

**St Kitts** – Mt Misery 1315, summit
  crater with lake.

**Nevis** – Nevis Pk 996

**Montserrat** – Soufrière Hills 915, the
  islands are the remains of 6 old volcanoes (some activity).

**Guadeloupe**
Soufrière (V-quiescent) 1484, destruc-
  tive eruptions in 1843.
Mt Sans Toucher (V-dormant) 1480

**Dominica**
Morne Diablotin 1447
Morne Trois Pilons 1424

**Martinique**
Mont Pelée (V-le 1929-32) around
  1500, scene of a disastrous eruption
  in 1902.

**St Lucia**
Ginie 959
Grande Piton 798
Petite Piton 750
Soufrière (V-some activity), crater 1·5
 km dia. 100 m deep with rd to edge.

**St Vincent**
Soufrière (V-le 1956) around 1280,
 scene of a disastrous eruption in
 1902, 2 days before that of Mont
 Pelée.

**Grenada** – Mt St Catherine 840

**Trinidad** – Northern Ra – Mt Aripo
 940

# South America (non-Andean)

### ECUADOR

**Islas Galápagos**
I San Cristóbal 1896
I Isabela – Wolf (V) 1707, Darwin (V)
 1280
I Fernandina (V-le 1968) 1547
I San Salvador 906
I Santa Fé 864

### CHILE

**Islas Juan Fernandez** – I Mas Afuera
 1650

**Islas de los Desventurados**
I San Félix (V-le 200 years ago)

### VENEZUELA

Tramán Tepiú 3000
Auyantepui 2953, **Angel Falls (980 m)
 nearby are the highest in the world.**
Mt Roraima 2810 (fa 1894 E. im Thurm,
 W. Perkins). Brazil, Guyana & Vene-
 zuela meet on the summit.
Mt Kukenaam 2597
Co Marahuaca 2579
Co Duida 2396 (fa 1928 G. Tate, C.
 Hitchcock)
Co Yaví 2285

Venezuela has many huge deep caves,
largely unexplored.

**SURINAM** – Juliana Top 1258 (mean)

### BRAZIL

**Sa Imeri**
P de Neblina 3014 (fa 1965), **the highest
 point in Brazil.**

**Sa Negra** – P de Itambé 2033

**Sa de Caporao** – Bandeira 2890, thought
 to be the highest point in the country
 until comparatively recently.

**Mount Roraima** (2810) in Venezuela was discovered in 1838 by R. Schomburg. It was raised to fame by Sir A. Conan Doyle who chose it as the site of his 'Lost World', where prehistoric monsters had still survived. The summit is a suitable tableland of 30 km; unfortunately, neither the first ascenders (in 1894) nor a British party, which climbed the steep prow of the plateau edge in 1973, found anything quite so exotic.

A 3000 km range of mountains runs parallel to the southern part of the 5000 km coastline of **Brazil**. Beyond the Amazon basin a further lengthy range stretches along the Venezuela/Brazil frontier. Many of the remote peaks are unclimbed on account of serious access problems.

For long it was held that the highest point of the country was **Bandeira** (2890) in the Sierra de Caporao; only recently has it been discovered that the **Pic de Neblina** (3014) on the Venezuela frontier is in fact higher.

**Pinnacle of Itabara** (nr Muqui)
Venomous spiders are said to make
this ascent possible only in the winter.

**'The Switzerland of Brazil'** (N of
 Teresópolis) – Tres Picos de Fri-
 burgo 2100

**Sa de Orgãos**
Pedro di Sino 2263
Dedo de Deus 1625, topped by a 250 m
 monolith.

**Rio de Janeiro**
Characterised by a large number of
steep hills and pinnacles (inselbergs).
Among them: Agulhas da Inhaga;
Cantagallo crag 300; Pao de Acucar
(Sugarloaf) 390, c-wy to top; Corco-
vado 704, c-ry to top, crowned by a
statue of Christ 38 m high erected 1931;
Pedra da Gávea 842; Pedra Bonita;
Agulhas da Gávea; Dois Irmães; P
da Tijucca 1021

**Sa de Mantiqueira** – Aguelas Negras
 2797

**PARAGUAY** – Co Tatug 700

**URUGUAY** – Co de las Animas 500

**TRINIDAD ISLANDS** – Il Monu-
 mento, a 250 m rock pillar.

### FALKLAND ISLANDS

**West** – Mt Adam 705

**East** – Mt Usborne 681

### ARGENTINA

**Sa de Córdoba**
Sa Grande – Co Gigante 2330
Sa de Comenchingones – Co Oveja
 2206
Sa Chica 1830
Sa de Poncho 1460
Sa de Guasapampa 1370

# The Andes

### THE ANDES OF VENEZUELA

**Sa de Norte & de Santo Domingo**
Micanón 4713 (fa 16/12/1910 A. Jahn
 & party)

The Spanish party, whose North American journey was outlined in a previous chapter (*see* p. 188), made a **South American journey** also, commencing in March 1966. During the next twelve months they climbed Orizaba in Mexico, followed by ascents, many over

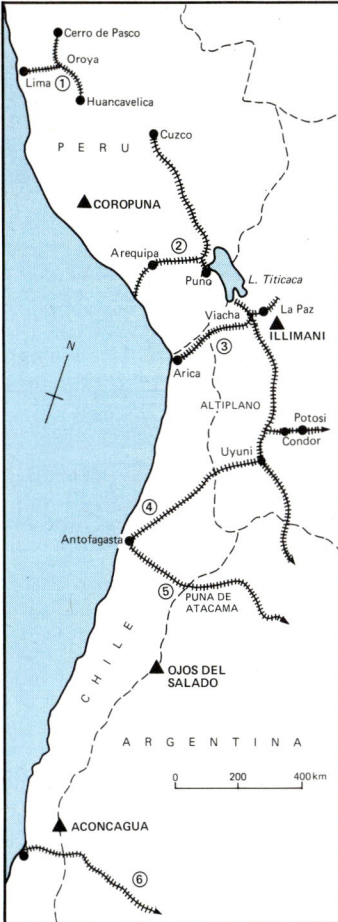

The trans-Andean railways

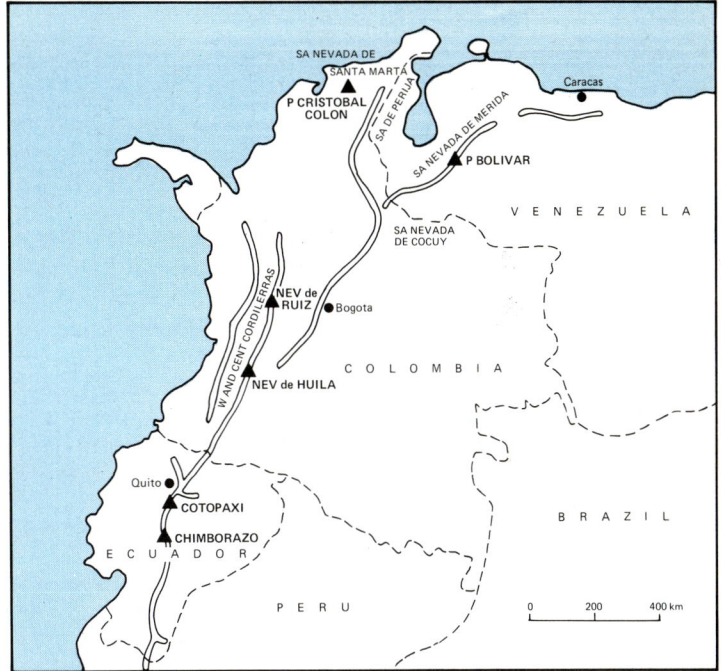

The northern Andes

Tucani 4676 (fa 17/12/1910 A. Jahn & party)
Macuñuque 4672 (fa 12/2/1922 M. Blumenthal & party)
Pan de Azúcar 4620 (fa 3/7/1885 W. Sievers)

**Sa Nevada de Mérida**
P Bolivar 4979 (fa 3/2/1936 F. Weiss & party)
El Vertigo 4950
P Humboldt 4941 (fa 4/1/1911 A. Jahn & party)
La Concha 4921 (fa 6/2/1939 F. Weiss, A. Gunther & party)
P Bompland 4883 (fa 3/2/1940 A. Gunther & party)
P Espejo 4764, c-wy to summit, **one of the highest in the world.**
Piedras Blancas 4762
P El Toro 4755 (fa 7/3/1910 A. Jahn & party)
P Leon 4740 (fa 4/1946 H. Matheus, B. Trujillo)
Páramo Cendé 3552

5000 m, of peaks in Guatemala, El Salvador, Honduras, Nicaragua, Costa Rica, Panama, Colombia, Venezuela, Ecuador, Peru, Bolivia, Chile and Argentina. Seven first ascents were made in the Nevado Aricoma of Peru; the highest peaks climbed were Chimborazo and Huayna Potosi; the hardest was thought to be Cerro Tronador. The journey of 46000 km ended on a peak of 1040 m in Tierra del Fuego.

The **trans-Andine railways** present some of **the finest examples of railway engineering in the world**. Between Lima (Peru) and Valparaiso (Chile) a series of startling mountain railways climbs up and over the Andean cordillera as follows:

1 The Central Railway of Peru climbs from sea-level to 4783 m in the Galera Tunnel (below Mount Meiggs) in 173 km and drops to Oroya (3726 m) in another 49 km. The average gradient is $4\frac{1}{2}$ per cent and there are 66 tunnels, 59 bridges and 22 zig-zags (on each of which the engine has to run round the train). Galera Station is **the highest standard gauge station in the world** and **Galera Tunnel the highest tunnel**. The line, which serves the mining area of Cerro de Pasco and Huancavelica, was engineered by an American, Henry Meiggs (1811–77) between 1867 and his death, and was completed in 1889 to his designs. The mountain above the summit is named for him. The Morococha branch reaches 4818 m, from which a spur formerly went to Volcán Mine at 4830 m.

## THE ANDES OF COLOMBIA

**Sa Nevada de Santa Marta**
P Cristóbal Colon 5775 (fa 16/3/1939
 W. Wood, A. Bakerwell, E. Praolini)
P Simon Bolivar 5775 (fa 2/2/1939 E.
 Kraus, E. Praolini, G. Pilcher), **the
 above joint highest points of Colom-
 bia.**
P Simmons 5660 (fa 24/2/1943 F. & D.
 Marmillod)
P Santander 5600
P la Reina 5535 (fa 5/3/1941 P. Petzold,
 Miss E. Knowlton, Miss E. Cowles,
 M. Eberli)
P Ojeda 5490 (fa 7/3/1941 as la Reina)
Los Vevaditos 5375
P de los Hermanos 5296 (fa 1957 H.
 Bunjé, A. & F. Cunningham, J.
 Waterlow)
El Guardian 5285 (fa, recorded, 15/3/
 1941 as la Reina; claimed for 1898)
P Juanita 5100
Tairona 5000

**Eastern Cordillera**
The higher central part, known as the
Sa Nevada de Cocuy, has been likened
to the Ruwenzori – 'lush vegetation
and dank cloudy weather'.

Alto Ritacuba 5493 (fa 1/1942 C.
 Cuenet, A. Gansser)
Subsid pks 5446, 5400

(Note: Considerable adjustments have
been made in recent years to the measured heights in this group and it now
seems likely that the above should be
5330, 5300 and 5200 respectively. In the
absence of further information the
original heights have been retained for
the present as they enable a compara-
tive assessment to be made.)

San Paulín 5389 (fa 1944)
El Chiflón 5291 (fa 1944)
No de Concavo 5268 (fa 1939)
El Puntiagudo 5266 (fa 1957)
Pt 5223 (fa 1939)
P Campanario 5197 (fa 1938 E. Kraus,
 A. Lampl)
El Concavito 5150 (fa 1924 W. Röth-
 lisberger & party)
El Pulpito 5120 (fa 1944)
P del Castillo 5090 (fa 1943)
P Blanco 5000 (fa 1944)
P Nievecitas 4880
Co El Vevado 4560
El Viejo 4200
Páramo Rico 4200
Sa de Perija 3750

**Western & Central Cordilleras** (W of
R Magdalena)
No de Huila 5350 (fa 1944 E. Kraus &
 party)
No del Ruiz (V) 5320 (fa 1936 M.
 Rapp & party), ski lodge with rd
 access at 4790m.
No del Tolima (V) 5215 (fa 1926
 German/Colombian party, but

2 The Southern Railway of Peru, which runs from Mollendo on the coast, via Arequipa, to Puno on Lake Titicaca and then on to Cuzco, reaches 4500m at Crucero Alto.
3 The Arica–La Paz Railway climbs to 4247m at General Lagos.
4 The Antofagasta–Bolivia Railway, which is metre gauge, has a summit of 3960m at Ascolán. From Ollague a metre gauge spur ran over a summit of 4826m to the copper mines of Collahuasi, whence a 13km 'aerial tram' served **the highest mine in the world on Aucanquilcha** (around 6000m). This is now disused beyond Yuma (4401m). The Condor Summit on the Potosi branch is **the highest over which passenger trains are worked** and **Condor Station is the highest in the world**.
5 The Northern Transandine Railway, which runs from Antofagasta to Salta (Argentina) across the Puna de Atacama, reaches 4475m at Chorriles, 4000m at Muñano and has four other summits of 3000 + m.
6 The Transandine Railway from Valparaiso to Mendoza (Argentina), which is metre gauge and uses the Abt System, crosses the range by the 3km La Cumbre Tunnel at 3191m.

In the **Central Cordillera of Colombia** Nevado del Tolima (5215) is the most popular peak and is often climbed. It is included in Los Nevados National Park (38000km²). A striking feature of the lower slopes is the giant groundsel, here called 'frailejones', which grows to a height of 4m.

**Chimborazo** (6267), the highest mountain in Ecuador, the 'Mountain of Snow', once called 'the Watchtower of the Universe', is an extinct volcano. It was the objective of an early Franco-Spanish expedition in 1736–44 and then for a long time was believed to be the highest mountain in the world. Since the earth is an oblate spheroid, with a slight equatorial bulge, the summit is in fact further from the centre of the earth than any other point on the surface. A. von Humboldt reached 5875m around 1800, but the actual first ascent fell to E. Whymper and his guides in 1880.

The former crater is buried beneath an ice-cap which sends large glaciers down the flanks. Since there are indications of a large crater, the mountain was presumably once much higher, but whether it was reduced by explosive eruption or by erosion is no longer clear.

Apart from Chimborazo, Ecuador has many famous and varied volcanoes within her borders:
**Cotopaxi** (5897) is **the highest active volcano in the world**. It has not been outstandingly active in recent decades, but it is known that in an early eruption a 200-tonne block was hurled 14km, ash was detected over 300km away and the noise heard at a distance of 800km. It is remarkable in erupting entirely from the crater with no secondary vents. The crater rim presents a number of points of nearly equal height and it is possible that the highest of these may not have been reached until the 1950s, even though the rim itself has been scaled on a number of occasions.

claimed for Kruger & Schmolmske
1918)
No Quindio 5151
No de Sta Isabel 5100 (fa 1943)
La Olleta (V) 4855 (fa 1868 W. Reiss)
Bravo 4800
No Cumba (V-extinct) 4764 (fa 1848
  M. Boussingault)
No Chiles (V-extinct) 4747 (fa 1869)
Pan de Azucar 4670 (fa 1929 W. Röth-
  lisberger, H. Weber)
Purace (V) 4590
Sotara (V) 4580
Sa de Coconucos 4544
Patetará 4482
P Paramillo (V-extinct) 4400
Cutanga (V) 4300
Petacas 4300
Doña Juana 4250
Co Animas (the Ghosts) 4242
Co Tamaná 4200
Co Tajumbina 4125
Galeras (V-active) 4083
Paramo Frontino 4080
Negro de Mayasquer Galeres (V)

## THE ANDES OF ECUADOR

Chimborazo 6267 (fa 4/1/1880 E.
  Whymper; J.A. & L. Carrel. A. von
  Humboldt reached 5875m around
  1880), **the highest point of Ecuador.**
Cotopaxi (V-le within the last century)
  5897 (fa 28/11/1872 W. Reiss, A. M.
  Escobar), **the highest active volcano
  in the world** and a perfect cone.
Cayambe (V) 5786 (fa 4/4/1880 E.
  Whymper & party), **the highest point
  on the Equator** is at 4875m on a
  shoulder of the mountain.
Co Antisana (V) 5705 (fa 10/3/1880 E.
  Whymper & party)
Antisanilla 5600
Sangay (V-very active) 5320 (fa 4/8/
  1929 R. & T. Moore, W. Austin, L.
  Thorne)
Altar (Capac Urcu) 5319 (fa 7/7/1963
  Italian party reached the highest
  point, Obispo), the mountain is a
  caldera and other points of nearly
  equal height on the rim had been
  reached previously.
Co Illiniza 5261 (fa 1880 J.A. & L.
  Carrel), rd to 4270m.
Carihuairazo 5028 (fa 1880 E. Whym-
  per & party)
Tungurahua (V) 5005 (fa 1873 W.
  Reiss, A. Stubel)
Co Cotocachi 4939 (fa 24/4/1880 E.
  Whymper & party)
Il Sincholagua 4901 (fa 23/2/1880 E.
  Whymper & party)
Il Quilindaña 4898 (fa 1953), the
  Ecuadoran 'Matterhorn'.
Il Corazón 4791 (fa 20/7/1738 French
  scientists La Condamine & Bouguer)
Co Pichincha (V-le 1881) 4791 (fa
  possibly 1582 Toribiode Orteguerra
  – certainly 1802 F.H.A. von Hum-
  boldt & party)
Co Rumiñahu (V-extinct) 4720
Co Ayapunga 4699

**Sangay volcano, Ecuador (Aerofilms)**

**Cayambe** (5786), a volcano of little activity, is noted for carry-
ing on its slopes at around 4875m **the highest point of the Equator.**
In the 1930s the Equator was believed to pass 24km north of
Quito and a monument was erected there. In 1949 the line was
moved 51km northwards and is indicated by a 'rock globe' –
Bola del Mundo.

**Sangay** (5320) has been called **the most active volcano in the
world**. It has not often been climbed (since the first ascent in 1929)
and any party on the upper slopes has necessarily to run the
gauntlet. In 1976 the members of a British expedition were strung
out on an ice-field just below the summit when a sudden eruption
of hot rocks and ash occurred. They were overwhelmed and
carried down for 600m. Two died, while three others were so
badly injured that getting them off the mountain became a serious
problem, solved largely with the aid of a following French party.

The highest point on the rim of **Altar** (5319) was not reached
until 1963. The first ascenders of the mountain in 1939 only
reached a lower point.

After a century of quiescence **Tungurahua** (5005), 'Black Giant',
erupted disastrously in 1886, devastating the town of Banos at the
foot. In the early part of the 19th century A. von Humboldt
reported that an eruption produced a rain of fish (from a summit
crater lake?) and by a miracle still living!

**Pichincha** (4791), 'Boiling Mountain', was known long ago,
since it appears on a list of world mountains in the late 18th
century. Looming close to Quito it is a popular ascent for
Ecuadorian mountaineers. In May 1824 it was the site of an
important battle in which the Spaniards were defeated and the
independence of Ecuador ensured.

**Reventador** is, like Sangay, continuously active.

Mountaineering in **Ecuador** really dates from 1880, from a remarkable campaign by E. Whymper, the conqueror of the Matterhorn,
and the guides J.-A. Carrel (once his rival on the same mountain)
and L. Carrel.

Il Sara Urco 4676 (fa 17/4/1880 E. Whymper & party)
Co Hermoso (V) 4638
Co Imbabura (V-extinct) 4580
Co Pasochoa (V-extinct) 4200
Reventador (V-continuously active) 3485
Sumaco (V) 3870
Llanquanate (V)
Quilotoa (V) 3981

## THE ANDES OF PERU

### Cordillera Blanca

Huascarán Sur 6768 (fa 20/7/1932 P. Borchers, W. Bernhard, H. Hoerlin, E. Hein, E. Schneider), **the highest point of Peru.**
Huascarán Norte 6655 (fa 9/1908 Miss A. Peck; R. Taugwalder, G. Zumtaugwald – there has been controversy over this claim, but it has not been completely discounted).
Huandoy Norte 6395 (fa 1932 E. Hein, E. Schneider)
Huantsan 6395 (fa 1952 T. de Booy, K. Eggeler; L. Terray)
Subsid pks 6270, 5913
Chopikalki (Chopicalqui) 6356 (fa 1932 P. Borchers, W. Bernhard, H. Hoerlin, E. Hein, E. Schneider)
Huandoy West Pk 6356 (fa 1954 L. Ortenburger)
Pallkaraju 6274 (fa 1939 W. Brecht, S. Rohrer, H. Schweizer, S. Schmid)
Subsid pk 6110
Santa Cruz (Pukaraju) 6259 (fa 20/7/1948 A. de Szepessy, F. Marmillod)
Chinchey 6222 (fa 1939 W. Brecht, H. Schweizer)
Subsid pk 5987
Copa (Pamparaju) 6188 (fa 1932 DOAV exped)
Ranrapallka 6162 (fa 1939 German exped)
Huandoy Sur 6160 (fa 1955 German exped)
Pukaranra 6147 (fa 1948 Swiss exped)
Hualcán 6125 (fa 1939 German exped)
Subsid pk 6104
Chakraraju 6113 (fa 1956 French exped)
Pukajirka 6046 (fa highest point 1955 American exped – a lower summit climbed 1936)
Kitaraju 6040 (fa 1936 Austrian exped)
Contrahierbas 6036 (fa 1939 German exped)
Toqllaraju 6032 (fa 1939 German exped)
Artesonraju 6025 (fa 1932 DOAV exped)
Carás de Parón 6025 (fa 1955)
Carás de Santa Cruz 6020 (fa 1955)
Allpamayo 5947 (fa highest point 1957 German exped – a lower summit 1951 Franco/Belgian exped)
Aguja Nevada 5886 (fa 1965 Italian exped)
Subsid pk 5840
Pyramide de Garcilaso (Paria) 5885 (fa 1957 German exped)

Previously the 1736–44 Franco-Spanish expedition had climbed Il Corazón (4791), while some 60 years later that doughty traveller, A. von Humboldt, had succeeded on Pichincha and failed narrowly on Chimborazo and Cotopaxi.

Within a period of six months in the early part of 1880, Whymper's expedition made an outstanding series of first ascents. Progressing by a number of camps, **the summit of Chimborazo** (6267) **was reached for the first time** on 4 January, the last 300m taking five hours when they had to crawl on all fours across an extensive tract of soft snow. On 18 February came the fifth ascent of Cotopaxi (5897); they spent a night in a tent pitched on the ashes close to the crater edge, which at the time was emitting steam only. Sincholagua (4901) was ascended on 23 February and Antisana (5705) on 10 March; on the latter when Whymper became snowblind the local natives suggested the application of fresh vicuna meat to his eyes as a remedy. They continued with Cayambe (5786) on 4 April, Sara Urcu (4676) on 17 April and Cotocachi (4939) on 24 April.

During May, Illiniza (5261) was climbed by the guides, while Whymper was ill in Quito. Within weeks he was fit again and in June the lower west peak of Carihuairazo (5028) was climbed by a party which included some Ecuadorians and a dog – all, including the animal, becoming temporarily snow-blind. On 3 July they climbed Chimborazo once again, this time with two Ecuadorians, finding the flag set up on their first visit still proudly standing. The same day Cotopaxi, 100km away, had begun to erupt and by the time they reached the summit the ash was falling all around them on Chimborazo in sufficient quantity to blacken

Cordillera Huayhuash, Peru (H Adams Carter)

Peru

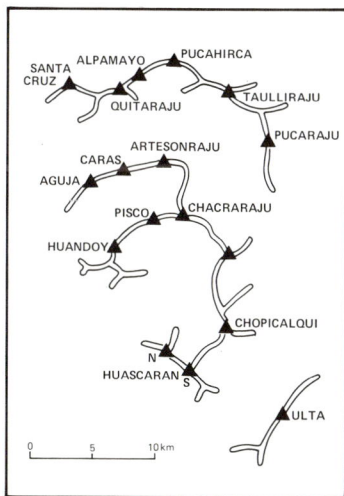

Cordillera Blanca

Oqshapallka 5881 (fa 1965)
Taulliraju 5830 (fa 1956 L. Terray)
Cajavilca 5775
Pisco 5760 (fa higher summit 1959 NZ
    exped – lower summit 1951 Franco/
    Belgian exped)
Champara 5751 (fa 1936 Austrian ex-
    ped)

the upper snows and render scientific observation, and even eating, entirely out of the question.

Opportunities for such exploratory mountaineering are rare indeed; now a century later they have almost completely vanished.

The **Andes of Peru** offer **the finest mountains and mountaineering in South America and**, indeed, **in the world outside the Himalaya/ Karakoram**. Much more readily accessible than most Asian mountains, certainly to North American mountaineers, they offer remarkably steep ice climbing of high standard on peaks up to 6768 m. Many very shapely and outstandingly steep peaks grace the Peruvian-Andean landscape. Notable features are extremely deep flutings in the steep ice, brought about by avalanching and differential melting and steep-sided and very narrow ridges sometimes carrying large and complex cornices.

The most popular and highest group is the **Cordillera Blanca** (Huascarán Sur 6768), 300 km north of Lima. The north peak of Huascarán was climbed (disputed but not disproved) by Miss A. Peck in 1908, but a majority of the summits fell during the 1930s often to German or Austrian expeditions.

In other groups further south some ascents were made in the 1930s, but most of the work has been done by post-World War II expeditions, which have come from all over the world. The Cordillera Huayhuash (Yerupaja 6634), Cordilleras Vilcabamba and Urabamba (Salcantay 6271) and Cordillera Vilcanota (Ausangate 6384) are specially noteworthy.

In the extreme **south of Peru** is another cluster of big **volcanoes**:

The highest point of **Coropuna** (6613) was reached in 1911 by H. Bingham's party. A lower summit had however been scaled a year or two previously by Miss A. Peck (of Huascarán Norte) who left there a 'Votes for Women' banner, **the highest ever Suffragette propaganda display**!

It is possible that the highest point of **Solimana** (6100) was not reached until 1973, previous claimants may well have only reached lower points on the crater rim.

**Chachani** (6084) presents a similar crater rim problem. A first ascent was claimed for 1889; an Inca grave on the summit had already been rifled when located by a local mine manager in 1901. Yet there is a first ascent claim for an Italian party in 1950. Now there is a road to the crater rim and a meteorological station.

**Ampato** (6025) also has conflicting first ascent claims.

**El Misti** (5842) received its first recorded ascent on mule back though it had certainly been climbed previously by local Indians. Now there is a road to 4000 m and a meteorological station on top, also an Inca temple; Spaniards placed an iron cross there in 1787, but no one knows what happened to it since the present cross is of railway rails and was erected by Bishop Ballon in 1900.

Between the Cordillera Occidental on one side and the Cordilleras Real and Oriental on the other, mostly in Brazil, lies the **Altiplano**, a high plateau 500 km long at an average height of 3600 m. In the

Kayesh (Cayesh) 5721 (fa 1960 NZ exped)
Pongos 5711 (fa 1951 Dutch exped)

**Cordillera Huayhuash**
Yerupaja (El Carnicero, 'the Butcher') 6634 (fa 31/7/1950 D. Harrah, J. Maxwell)
Siulá 6352 (fa 1936 A. Awerzeger, E. Schneider)
Siulá Chico 6265 (fa 1960)
Sarapo 6143 (fa 7/1954 Bachman, Lugmayer)
Jirishanca 6126 (fa 1957 T. Egger), the 'Matterhorn' of Peru.
Yerupaja Chico (El Toro) 6121 (fa 1957 T. Egger, S. Jungmeier)
Rassac 6040 (fa 1936 E. Schneider)
Carnicero 5980 (fa 1961 German exped)
Rondoy 5883 (fa 1963 British exped)
Tsacra Grande 5774 (fa 1954 German exped)
Puscanturpa 5652 (fa 1954 German exped)
Trapecio 5650 (fa 1957)
Ninashanca 5643 (fa 1954 German exped)
Espolon Tam Sur 5545 (fa 1968 NZ exped)
Jirishanca Chico 5467 (fa 1954)

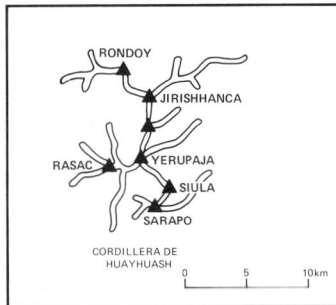

**Cordillera de Huayhuash**

**Cordilleras Huagaruncho, Huarochiri, Ruara, Millpo and Tunshu**
Cotoni 5817 (fa 1963)
Llongote 5781 (fa 1963)
Tullujuto 5752 (fa 1938 T. Dodge & party)
Huagaruncho 5730 (fa 1956 J. Kempe, J. Tucker, J. Streetly, G. Band, M. Westmacott)
Santa Rosa 5717 (fa 1957)
Tunshu 5707 (fa 1958 M. Elmslie, W. Wallace)
Yarupa 5707
Millpo Grande 5608

In this area the Sima de Millpo cavern is 407 m deep, entrance at 4000 m.

**Cordilleras Vilcabamba & Urabamba**
Salcantay 6271 (fa 5/8/1952 G. Bell, F. Ayres, D. Michael, W. G. Matthews, Mme C. Kogan, B. Pierre)
Pumasillo 6070 (fa 1957 British exped)

north is Lake Titicaca, which is the Peru frontier, and to the south Lake Poopo. The local Abominable Snow-man is called the Hualpichi, but no mountaineer has yet seen him. There are large areas of desert for him to hide in.

**Sajama** in the Cordillera Occidental **is the highest point in Bolivia. La Paz is the highest capital city in the world**, with close at hand the high peaks of the Cordillera Real – Illimani (6462), Ancohuma (6388) and Illampu (6362).

Illimani, which was first climbed in 1898 by a British party, was the scene of a plane crash in 1938. Local gossip claims that this was carrying a load of gold and for this reason a whole range of adventurers and others – guerillas, patrols, militiamen, farmers and so on – are seeking the wreck. A Spanish mountaineering party climbing the north ridge of the peak was fired on by Bolivian soldiers being mistaken for treasure seekers.

A number of **Andean summits** had been ascended, and possibly even permanently occupied, by local Indians some three centuries before the first ascent of Mont Blanc. Some of the easy volcanic peaks **around the Altiplano and the Atacama Desert**, where the snow line is at 5800 m and there are no technical climbing problems, were used as watch towers, or for religious purposes and shelters or shrines erected on the summits.

The first discovery in this field was reported by a Chilean explorer, F. J. San Román, who climbed several volcanoes in his journeys in 1883–89. In the following years isolated items – Inca relics, statuettes, pottery and metalware – were discovered in various places; then in 1954 a mummy was unearthed on Cerro Plomo; in 1964 a frozen corpse was disinterred on Cerro del Toro and a mummified body of a child on Quéhuar. Remains of buildings and altars and signs of at least temporary occupation have been noted on many sites. The discovery of the body of a guanaco, a llama-like animal, high on Aconcagua raises the speculation that the summit had been reached by locals years before the first ascent by mountaineers. The certain ascent of Llullaillaco (6723) places the highest ascent in the world firmly with the Atacama Indians, until the Schlaginweits reached 6785 m on Abi-Gamin in 1855.

The **Atacama Desert**, the Puna di Atacama, on the Chile/Argentina frontier, is a windswept upland plateau in many respects comparable with Tibet, covering tens of thousands of km² at an average height of 4000 m. It is an extensive desert surrounded by mountains which are high but present no particular mountaineering problems.

**Aconcagua** (6960) **is the highest point of Argentina, of the Andes and of South America**. It is a complex mountain with an easy side which fell in 1897 and a very steep south face, climbed Alpine-style by a French expedition in 1954 – **one of the world's first big wall climbs.** The Fitzgerald party of 1897 included the famous Swiss guide **Matthias Zurbriggen** and it was he who **first reached**

*Above:* Anticona massif and Lima-Huancayo railway (E Echevarria)

*Below:* Rio de Janeiro and the Sugarloaf (Varig Brazilian Airlines)

Central and South America—Paine Group, Patagonia (Takehide Kazami)

Sacsarayoc 5996 (fa 1963 NZ exped)
Cabeza Blanca 5940 (fa 1959 Swiss exped)
Veronica (Padre Eterno) 5822 (mean) (fa 1956 Dutch exped)
Fortaleza 5883 (fa 1968 NZ exped)
Copo de Nieve 5852 (fa 1968 NZ exped)
P Soray (Humantay) 5842 (mean) (fa 1953 Italian/Swiss exped)
Lasunayoc 5800 (fa 1956 American exped)
Saguasiray 5800 (fa 1963 Italian exped)
Quishuar 5775 (fa 1956 Jap exped)
Mitre 5750 (fa 1963 NZ exped)
Camballa 5726 (fa 1959 Swiss exped)
Chainapuerto 5700 (fa 1968)
Yanaccacca 5700 (fa 1969 Australian exped)
Panta 5670 (fa 1959 Swiss exped)
No Yucay 5650 (fa 1958 Italian exped)
Ccellacoccha 5540 (fa 1969 Australian exped)
Choquetacarpo 5520 (fa 1959 Swiss exped)
Colpachinac 5500 (fa 1956 American exped)
Helancoma (Azulcocha) 5500 (fa 1953)
Torayoc 5486 (fa 1968 NZ exped)
Pucapuca 5450 (fa 1959)
Blanco 5422
Co Jayuri 5652, faces the group across the Apurimac valley.

**Cordillera Vilcanota**
Ausangate 6384 (fa 24/7/1953 H. Harrer, J. Wellenkamp, F. März)
Subsid pks, there are 3, climbed in 1952.
Colque Cruz 6075 (mean) (fa 1953 H. Steinmetz, J. Wellenkamp)
Jatunhuma 6094 (fa 1957 German/ American party)
Jatunriti 6067 (fa 1957 C. Cronk, S. Jervis, C. Merrihue, E. Whipple)
Colquepunco 6020 (fa 22/8/1953 P. Ghiglione, F. März)
Cayangate (Callangate) 6001 (fa 1953 German exped)
Subsid pks, several.
Yamamari 6007 (fa 1957)
Zapato 5900 (fa 1957 as Jatunriti)
Mariposa 5827 (fa 1957)
Verena 5750 (fa 1952)
Cacaquiru 5750 (fa 1957)
Pachanta 5727 (fa 1957)
Ccapana 5725 (fa 1957)
Caracol 5619 (fa 1957)

**Cordillera Carabaya**
Quenamari 5850 (fa 1955 H. Katz & party)
Yanoloma 5850 (fa 1955)
Gr Chimboja 5780
No Allinccapac 5780 (fa 1960 British exped)
Alcachaya 5777 (fa 1956)
Macusani 5750
No Huaynaccapac 5678 (fa 1960 British exped)
San Vicente 5650 (fa 1955)
Tococapac 5640 (fa 1960 British exped)

**the summit**, going ahead alone after Fitzgerald had to turn back at 6700m. In the course of the same expedition a second ascent was made (S. Vines and N. Lanti) and nearby Tupungato (6550) also climbed for the first time (Vines and Zurbriggen).

Unusual ascenders have included two dogs in 1934, **possibly a canine altitude record**, and two priests in 1952 carrying a statue of Our Lady of Carmel for erection on the summit.

There was a scare twenty or thirty years ago when it was suggested that Ojos del Salado (6885), a mountain 600km to the north close to the Puna di Atacama, might possibly be higher; however, careful checks revealed that the pretender was 75m lower after all, nevertheless it remains **the highest point of Chile**. Some maps still credit Ojos del Salado with 7084m in spite of the above.

The local relief in this part of the Andres typifies that all along the western coast of South America. Aconcagua (6960) is only 150km from the shore line, while Haeckel Deep (−5667) is only 100km offshore. It is of course a typical tectonic plate boundary. On the south side of the mountain the average slope is also steep – the Pass of Uspallata (3856) is only 18km away.

There are some notable **volcanoes in southern Chile**, though generally they do not exhibit the eager activity of their counterparts further north. As early as 1848, a Pole, I. Domeyko, climbed Chillan (3180) and attempted Tinguiririca (4300). After an eruption of Quizapu (3050), a cone formed on the flanks of Cerro Azul, the ashes ejected were later detected in Africa to the east, and New Zealand to the west. All the usual volcanic phenomena are there – geysers on the slopes of Domuyo (4709), sulphur baths at the foot of Chillan and so on.

The tip of the South American Continent beyond 45°S, known as Patagonia, is divided between Chile and Argentina. The interior of much of Chilean Patagonia is occupied by two extensive ice-caps. The Hielo Patagonico del Norte, which stretches for some 200km between latitudes 46° and 48°S, was crossed by E.E. Shipton's party in 1963–4 from north-west to south-east. The highest peak here, Cerro San Valentin (4058), **was first climbed in 1952** by an Argentinian expedition. The long inlet of Canal Baker separates the two ice-caps. The Hielo Patagonico del Sur, which covers 300km between latitudes 48° and 51°S, was crossed in the south by H.W. Tilman's party in 1956 and lengthways, from north to south in 1960–1 by a team led by E.E. Shipton.

The ice-caps send down considerable glaciers on every side. Because of almost continuous bad weather any crossing trip is likely to be extremely arduous. Polar-style sledging tactics will certainly be involved.

The steep rock mountains of Patagonia offer some of the finest mixed climbing in the world.

The Parc Nacional los Glaciares incorporates a 200km stretch of the Andean chain at the heads of Lakes Viedma and Argentino. The most striking peak of this area, Cerro FitzRoy

Chichiccapac 5614 (fa 1959 Italian exped)
Uraccapac 5567 (fa 1960 British exped)
Japuma 5529 (fa 1954 British exped)

**Cordillera Apolobamba** (extending into Bolivia)
Chaupi Orco Norte (Chaupi Orco) 6100 (fa 1958)
Chaupi Orco (No de Salluyo) 6044 (fa 1/8/1957 W. Karl, H. Richter, H. Wirmer)
Cololo (ccachuca) 5915 (fa 1957)
Ananea 5842
Calijon 5837
Huellacalloc 5836 (fa 1957)
Pupuya 5816
Huanacuni 5796 (fa 1957)
Palomani Grande 5769 (fa pre 1932 Bolivian Army capt)
Ichocolo 5750 (fa 1958)
Matchu Suchi Coochi 5679
Chacnacota 5650
Katantica 5591
Ritipata 5500 (fa 1958)
Cacahuaycho 5450

**Western volcanoes**
Coropuna (V) 6613 (fa 1911 H. Bingham's party)
Solimana (V) 6100 (fa, highest point, 1973 Peruvian/Italian party – previous claimants reached lower points on the crater rim)
Subsid pks 6070, 5900, 5750
Ampato (Hualca-hualca) (V) 6025 (fa 6/4/1966 R. Culbert), at one time believed to be 6300m.
Chachani (V) 6084 (fa local Indians – claimed 1889 A. Hettner and 1950 P. Ghiglione's party), rd to 5130m.
Cordillera de Chilca 6000
Cordillera de Huanzo 6000
Tacora (V) 5988 (fa 1904 H. Hoek & party)
El Misti (V) 5842 (fa local Indians – claimed 1878 R. Falb, on muleback), rd 4000m, Harvard meteorological station on summit, also iron cross erected 1901 replacing one of 1787)
Sarasara 5959 (fa by local Indians – claimed 1889 A. Hettner)
Tutupaca (V) 5806
Sayhua 5790 (fa 1956)
Caballuni 5780 (fa 1956)
Barroso 5741 (fa 1968)
Ubinas (V) 5672 (fa 1878 R. Falb)
No de Pichu Pichu 5600
Ancasillo 5550 (fa 1956)
Huagochullo 5550 (fa 1956)
Ticsani 5391
Huaynaputina (V) 4800

## THE ANDES OF BOLIVIA AND N CHILE

**Cordillera Real**
Illimani 6462 (fa 9/9/1898 W. M. Conway & party), rd to hut at 5014m.
Subsid pks 6442 (fa 1950 H. Ertl, G. Schröder), 6400 (fa 1877 Franco/ Peruvian party)

(3375), is named for the commander of HMS *Beagle*, which in the 1830s mapped these coasts for the Royal Navy and carried Charles Darwin on his voyage of discovery. **One of the world's great mountains**, it was first climbed in 1952 by L. Terray and G. Magnone. They used three camps on the ascent, each a cross between an ice cave and an igloo. From the highest at 2800m their summit trip took 40 hours, the intermediate night being spent on a cold icy bivouac ledge. All around is a cluster of rock peaks, similarly sharp and steep – the Aiguilles de St Exupery, Raphael, Poincenot (named for a climber killed on the 1952 expedition), Mernoz and Guilland.

Four kilometres to the west are the stark outlines of the Cerro Torre group; Cerro Torre (3128), **the highest point**, was climbed from the west, the ice-cap side, by an Italian party in 1974. The claim for an ascent by the near-vertical east and north faces by C. Maestri and A. Egger in 1959 is now discounted. The latter was killed during the descent. Maestri returned in 1970 and drilled a bolt ladder up the south-east ridge; to power the drill they used a compresssor driven by an internal combustion engine, which with ancilliaries weighed 150 to 180kg. In June they reached within 400m of the top; in November they returned to complete the route. After surmounting the rock Maestri decided not to climb the topmost ice wall. Overall their efforts had involved 825m of climbing using ice techniques, 880m by traditional rock climbing methods and 365m of bolting – a truly remarkable feat of skill and endurance carried out under extreme conditions, but, because of the excessive amount of bolting, *not*, say the mountaineers, mountaineering. Similar mountaineering problems close by include Torre Egger (2900), Cerro Adela, Cerro Standhardt, Aiguille Bifida and Pier Giorgio; most of these have now been climbed.

Illampu, Bolivia (E Echevarria)

150 km further south is the Cordillera del Paine, with a similar remarkable collection of rock spires; **the highest point is Cerro Paine** (3050), first climbed in 1957. To the north-east are the three towers of Paine, mighty monoliths of rock. The first ascent of the Central Tower (1200) by a British party in January 1963 was followed a mere 20 hours later by an entirely independent Italian ascent. The east face of this tower, climbed by South Africans in 1974, is 1200m sheer, **one of the highest rock walls in the world**. Other hard rock peaks hereabouts, most of them now climbed, are Mummer, Blade, Sword, Fortress and Shield.

Attractive climbing prospects remain, but it is an area where the weather prospects are poor.

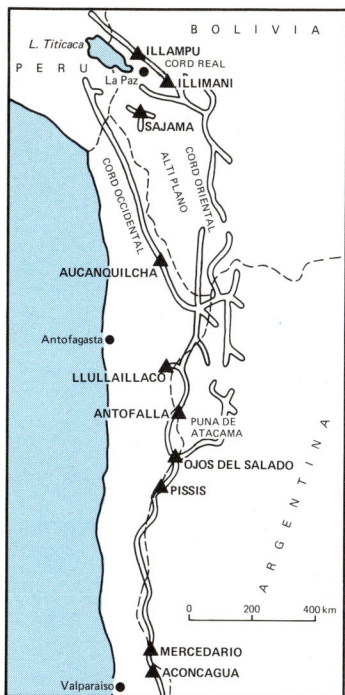

**Aconcagua, the highest mountain in South America (Aerofilms)**

**Chilean/Argentinian Andes (north)**

Ancohuma 6388 (fa 1915, & again in 1919 R. Dienst, A. Schulze)
Illampu (Sorata) 6362 (fa possibly 1919, first certain 7/6/1928 H. Pfann, H. Hortnagel, H. Horeschowsky, E. Hein)
Huakaña 6249 (fa 1919 as Ancohuma)
Chearoco 6127 (fa 1928 H. Hortnagel, H. Horeschowsky)
Cacca Aca (Huayna Potosi) 6094 (fa 1919 German exped)
Chachacomani 6074 (fa 1925 German exped)
P del Norte 6030 (fa 1928 Austrian exped)
Casiri Este 5947 (fa 1966 British exped)
Quelluani 5929
Kimsakolyo 5892 (fa 1966 British exped)
Calzada 5872 (fa 1925 German exped)
Huancopiti 5867
Casiri 5828 (fa 1928 Austrian exped)
Mururata 5775 (fa 1937)
Chamakawa 5767 (fa 1966 British exped)
Taparacu 5750
Humahallanta 5736
Condoriri 5656 (fa 1941), the 'Matterhorn' of Bolivia.
Cunatincuta 5600 (fa 1946)
Chacaltaya, carries a Cosmic Research Stn at 5180m, **the highest c-wy in the world** reaches 5029m.

**Cordillera Oriental**
Jochacunocollo 5900 (fa 1911 T. Herzog & party)

Cordillera de Lipez 5860
Huaynacunocollo 5800 (fa 1939)
Gigante 5800
Co San Juan o Altarni 5750 (fa 1939)
Atoroma 5700 (fa 1928)
Choquetanga 5700
Immaculado 5675 (fa 1911)
Co Pacuni 5666
Co Bonete 5658 (fa by local Indians)
Co Zapaperi 5655
Chorolque 5630
Co Volcan (V) 5545
Incachaca 5230 (fa 1910)
Tunari 5200 (fa 1903)
Calacruz 5100, 'Matterhornlike'.

**Cordillera Occidental** (continuing the line of the western volcanoes of Peru)
Sajama 6542 (fa 26/8/1939 J. Prem, P. Ghiglione), **the highest point in Bolivia.**
Co de Parinacota 6330
Co de Pomarata 6240 (fa 1946)
Co Aucanquilcha 6180, there are working mines at 5800m.
San Pablo (V) 6118
Huallatiri 6087 (fa 1926)
Co Ralplana 6045
Nuevo Mundo 6020 (fa 1904)
Sillajhuay 5995
Co Licancabur 5930 (fa by local Indians – claimed for 1886 J. Santdias)

Putana (V) 5890
Co Ollagüe (V) 5870 (fa 1888)
Co Cabaraya 5869
Co Maria di Piemonte 5840
Co de Tocorpuri 5833 (fa 1939)
Co Degli Alpini 5820
Co Puquintica 5746
Co Colorado 5742 (fa 1939)
Co Pacoma 5739
Co Araral 5680
Apagado (V) 5680
Miño (V) 5620
Isluga (V) 5531
No Chuquiananta 5488
Co Olca 5310
Coipasa (V) 4901, an island in Coipasa Lake

**THE ANDES OF THE ARGENTINA/CHILE FRONTIER**
(from the Argentina/Bolivia/Chile junction to the Paso de San Francisco).

**Around the Puna di Atacama**
Llullaillaco 6723 (fa by local Indians)
Co Puntas Negras 6500
Cachi 6310 (fa 1904 F. Reichart & party)
Co Pular 6225
Quehuar 6160 (fa by local Indians)
No Pastos Grandes 6157

Antofalla 6100 (V) (fa by local Indians)
Co Aracar 6080 (fa by local Indians)
Copiapó (Co Azufre) (V) (fa 1937)
Guallatiri (V-slight activity) 6060
Chani 6060 (fa by local Indians – claimed for 1901)
Co Salin (Llanin) 6060 (fa by local Indians)
Co Colorados 6049
Co Acamarachi (Pili) 6044
Co de Rio Negro 6040
Socompa (V) 6031 (fa 1905 F. Reichart solo)
Co Mojones 5990
Acay 5950 (fa by local Indians)
Dos Conos (V) 5898
Co Negro Muerto (V) 5850
Co Antuco 5791
Co Aguas Blancas (V) 5780
Co Patos 5760
Co Napa 5755
Poquis 5740
Co Peinado 5740
Co del Azufre 5680
Co Galan 5650
Lascar (V) 5641 (fa by local Indians – claimed for 16/11/1963 Chilean party)
Co Rincon 5594 (fa 1900 R. Hauthal)
Co Calalaste 5440
Co Curutú 5395
Co Lejía 5360
Sa da Aguas Calientes (Co Gordo) 5335
Sa del Hombre Muerto 5029
Sa del Aguilar 4875

**South of the Puna di Atacama (Paso de San Francisco) & north of Aconcagua River (Uspallata Pass)**

Co Aconcagua (V-extinct) 6960 (fa 14/1/1897 M. Zurbriggen solo), **the highest point of Argentina, of the Andes and of the American continent.**
Subsid pks 6955, 6750
Ojos del Salado (V-dormant or extinct) 6885 (fa 1/1937 J. Wojsznis, J. Szczepanski), **the highest point of Chile.**
Pissis 6780 (fa 1937 S. Osiecki, J. Szczepanski)
Co Mercedario 6670 (fa 18/1/1934 V. Ostrowski, A. Karpinski)
Co Sin Nombre 6637 (fa 1937 J. Wojsznis)
Tres Cruces (V) 6620 (fa 1937 S. Osiecki, W. Paryski)
Incashusi (V) 6601 (fa 1910 J. Wheelwright – possibly 1859 E. Flint)
Co del Nacimento (V) 6493 (fa 1937)
El Muerto 6476 (fa 1950)
Co Bonete 6410 (fa 1913 Penck)
Co Ramada 6410
Bonete Chico 6400
Co las Tortolas 6323 (fa by local Indians)
Co Veladero 6320
Co Alma Negra 6290
Co del Olivares 6252
Co de los Patos o Tres Quebradas 6250 (fa 1936-7)

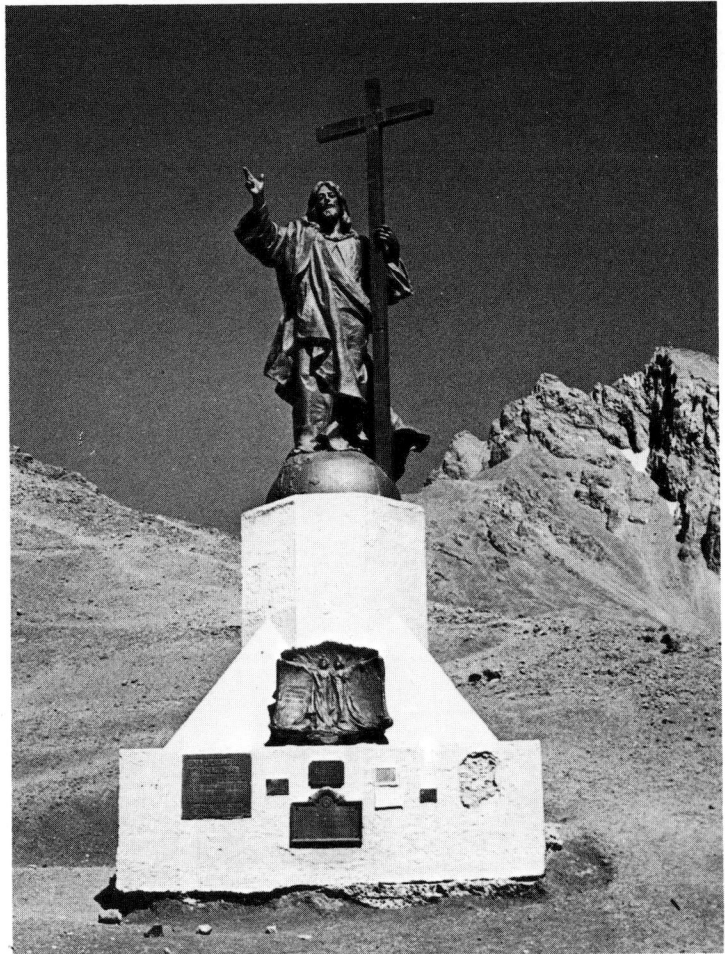

Co Solo 6190
El Ermitaño 6187
Co San Francisco 6020 (fa 1913 Penck solo)
Co del Viento (V) 6010
Co Fraile 5980
Puntiagudo 5949
Co Alto de las Leones 5930
Co de Ansilta 5885
Co Mulas Muertas 5880
Co El Toro 5850
Co de Potro 5830
Los Tambillas 5800 (fa by local Indians)
Doña Ana 5690 (fa by local Indians)
Co di Ollitas 5620
Co dos Hermanos 5540
Cordón de los Penitentes 5500
Cordón de Conconta 5315
Co Colangüil 5230
18 km S of Aconcagua is the Pass of Uspallata 3856, carrying rd and ry crossing of the range from Mendoza (Argentina) to Valparaiso (Chile). It is crowned by the lofty figure of 'Christ of the Andes' dedicated 1904. The ry utilises a 3 km tunnel at 3170 m.

**Parallel ranges to the East**

*No de Famatina* – Cumbre de la Mejicana 6250 (fa 1895 F. Hauthal)

*No del Aconquya* – 5485

*Sa de Quilmes* – 4875

*Sa de Velasco* – 3950

*Sa de Ambato* – 3350

*Sa Changoreal* – 4575

*Sa Léon Muerto* – 5335

The Chile-Argentina frontier (Argentina Embassy)

## Aconcagua River to Patagonia (Pto Aisen)

Tupungato (V-dormant) 6550 (fa 12/4/1897 S. Vines; M. Zurbriggen)
Co Pabellón 6152
Juncal 6110 (fa 1910 F. Reichart, R. Helbling: D. Beiza)
Marmolejo 6100 (fa 1928 Germans living in Chile)
No del Plomo 6050 (fa 1909 F. Reichart, R. Helbling)
Los Piuquenes 6012 (fa 1933)
Co Polleras 5947 (fa 1908)
Sin Nombre 5913 (fa 1944)
San José (V) 5880 (fa 1931)
Co El Plata 5850 (fa 1925)
Co los Vallecitos 5756 (fa 1937)
Bravard 5750
Tupungatito (V) 5640 (fa 1897)
Co Trono 5600 (fa 1960)
Co Pirámide 5600 (fa 1937)
Co Tronco 5600 (fa 1960)
Gemelo Grande 5486 (fa 1968)
Co del Castrillo 5485 (fa 1953)
Maipó (V) 5290 (fa 1883 P. Güssfeldt)
Co Sosneado 5189
P de las Damas 4860

### Chilean/Argentinian Andes (south)

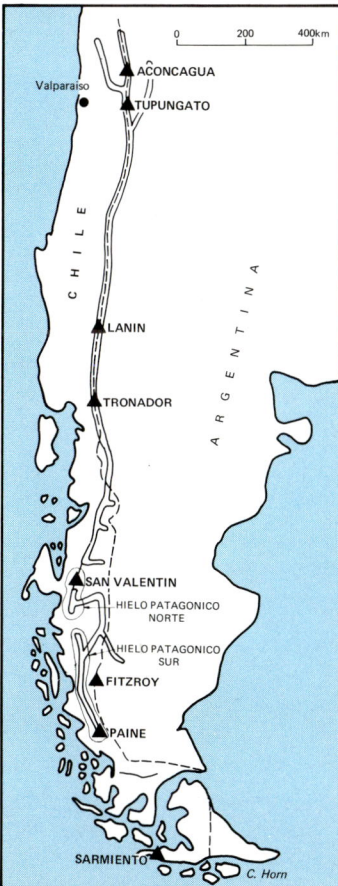

Condors hover above Lanin volcano, Chile (Barnaby's)

Llaima volcano, Chile (Barnaby's)

Co del Palomo 4855
Overo (V) 4765 (fa 1924)
Domuyo (V) 4709 (fa 1949), geysers on the slopes.
Torre Pangol 4520 (fa 1963 Chilean party), a 'Matterhorn'.
Volcan Tinguiririca 4300 (fa 1930)
Tromen (V) 4115
Peteroa (V) 4090 (fa 1900)
Co de Campanario 4020
Descabezada Grande (V) 3830
Co Azul (V-le 1932, see below) 3810
Lanin (V) 3774 (fa 1897 R. Hauthal)
Sa Velluda 3560 (fa 1940)
Tronador 3470 (fa 30/1/1934 G. Clausen), the 'Thunderer'.

Chillan (V) 3180 (fa 16/2/1848 I. Domeyko)
Llaima (V) 3124
Callaqui (V) 3089
Quizapu (V-le 1932) 3050, formed on the flanks of Co Azul.
Antuco (V) 2990
Los Copahues (V) 2987
Lonquimay (V) 2890
Villarica (V-active) 2842
Osorno (V) 2660 (fa 1848), a symmetrical cone.
Minchinmavida (V) 2470 (fa 1953)
Corcovado (V) 2300
Puyehue (V) 2240
Calbuco (V) 2015

## The Patagonian Ice-caps

There are two. The Hielo Patagonico del Norte, stretching for some 200 km between latitudes 46° & 48°S, is separated by the long inlet of Canal Baker from Hielo Patagonico del Sur, stretching some 300 km between latitudes 48° & 51°S. Crossing these ice-caps from N to S, or even from W to E, involves a major expedition. The principal peaks are:

Co San Valentin 4058 (fa 18/12/1952 O. Meiling & party)

Co San Lorenzo 3700 (fa 17/12/1944 Pere Agostini, H. Schmoll; A. Hemmi)

Co Francisco Moreno 3536

Co Titlis 3500

Co Arenales 3437 (fa 1958 Jap/Chilean party)

Co Lautero (V-slightly active) 3380 (fa 29/1/1964 P. Skvarca, L. Pera)

Co FitzRoy 3375 (fa 2/2/1952 G. Magnone, L. Terray), one of the great rock peaks of the world, a number of satellite pks have also been ascended.

Co Pio XI 3300

Co Mellizo Sur 3292

Co Bertrand 3270

Co Torre 3128 (fa 1974 Italian party from the back. The claim for an ascent of the steep front in 1959 by C. Maestri & A. Egger has been discounted.

Co Murallon 3060

Co Paine 3050 (fa 1957 Italian party), nearby is a series of splendid climbing pinnacles – the Towers of Paine. The Central Tower (fa 1/1963) has a sheer E face of 1200 m, **one of the highest rock walls in the world** (fa 1974 S African party)

Co Arco 3010 (fa 1963-4 E. E. Shipton & party)

Co Agassiz 2940

Co O'Higgins 2910 (fa 16/1/1960 Chilean party)

Torre Egger 2900 (fa 1976 Anglo-American party)

## Hielo Patagonico del Sur to Cape Horn

Co Yogan (Mte Luigi di Savoia) 2469 (fa 1962 British/Argentino party)

Mt Darwin 2438 (fa 1962 E. E. Shipton & party)

Sarmiento 2404 (fa 7/3/1956 C. Mauri, C. Maffei)

Mt Buckland 1800 (fa 1966 Italian party)

Mt Burney (V) 1750 (fa 1962 E. E. Shipton & party)

*Top right:* Cerro Torre (*left*) and FitzRoy (*right*), Patagonia (J Cleare)

*Right:* The Paine Group, Patagonia (AC Collection)

# Polar Regions

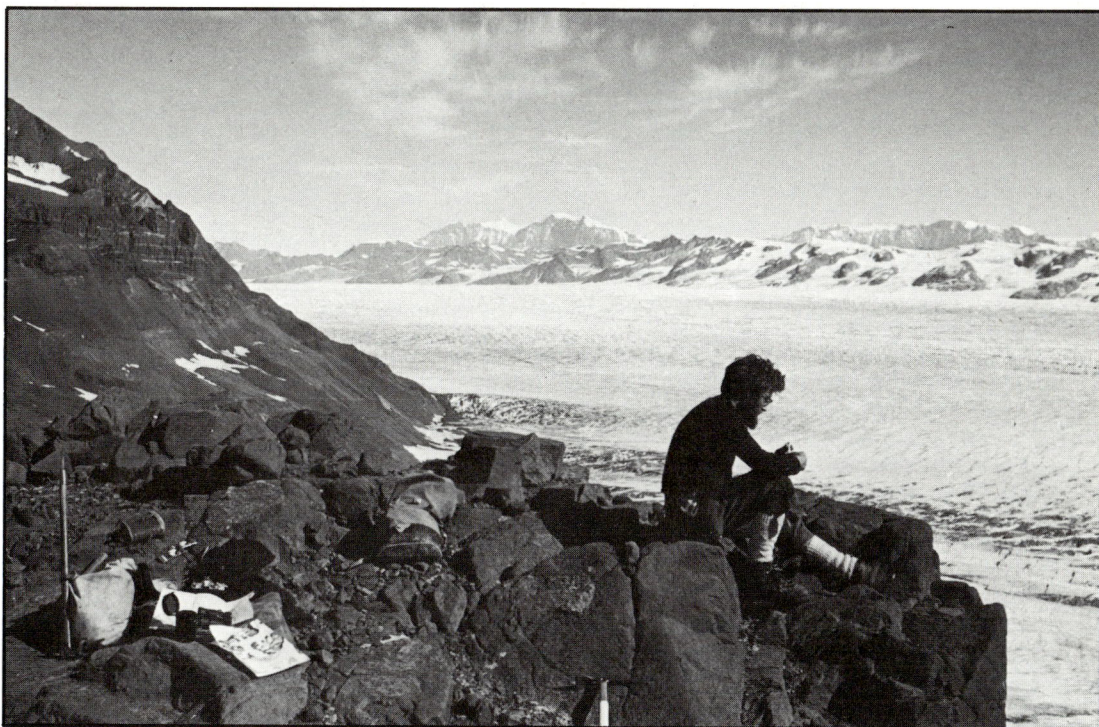

Gunnbjornsfjeld, the highest point of Greenland, seen across the glacier (D W Matthews)

## *Antarctic Continent*

**FORMER US SECTOR** (80°W to 150°W) (Marie Byrd Land, Ellsworth Land)

**Ohio Ra** – Mt Glossopteris 2867

**Executive Committee Ra**
Mt Sidley 3658
Mt Siple 3100, on an offshore island.

**Kohler Ra** – Mt Murphy 2446

**Thiel Mtns** – Mt Walcott 2154

**Inland** – Mt Seelig 3022

**Ellsworth Mtns**
Vinson Massif 5140 (fa 18/12/1966 W. Long, P. Schoening, B. Corbet, J. Evans), **the highest point on the Antarctic Continent.**
Mt Tyree 4965 (fa 6/1/1967 B. Corbet, J. Evans)
Mt Shinn (V) 4800 (fa 1966)
Mt Gardner 4690 (fa 1966)
Mt Epperly 4602
Mt Ostenso 4180 (fa 1967)

Antarctica

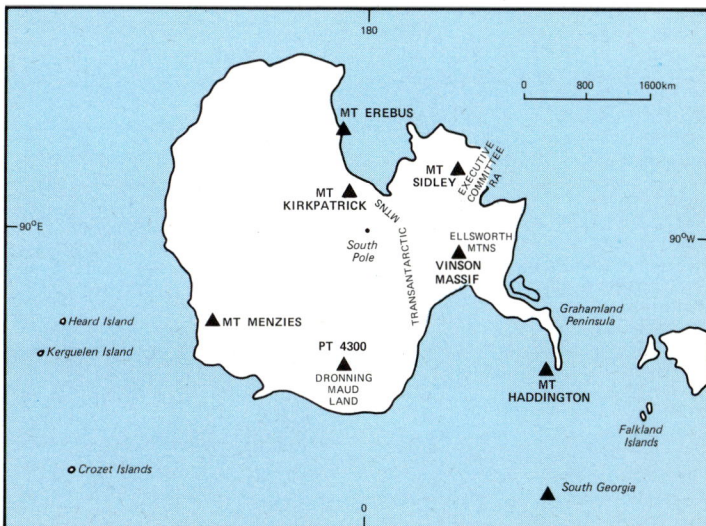

Mt Long Gables 4150 (fa 1967)
Mt Giovinetto 4087

**FORMER BRITISH SECTOR** (20°W
to 80°W) (Graham Land, Palmer
Land, Coats Land, around the Weddell Sea)

**Graham Land Peninsula** – Mt Francais
2762

**James Ross Island** – Mt Haddington
1620

**Adelaide Island** – Mt Gaudry 2453 (fa
1963), Mt Liotard 2440 (fa 1962)

**Palmer Land** – Mt Andrew Jackson
3950 (mean) (fa J. Cunningham)

**Alexander Island** – Mt Paris 2900, Mt
Charity 2740 (fa R. Collister, M.
McArthur)

**Pensacola Mtns** – Mt Hawkes 3660, Mt
Coman 3640

**DRONNING MAUD LAND** (45°E to
20°W)

**Inland** – Pt 4300, **the highest on the
ice-cap.**

**Thorshavnfjella**
Sor Rondane Mtns 3180
Mt Victor 2588
Dronning Fabiolafjella 2470
Birger Bergersenfjellet, has 800 m preci-
pices on its NE face.

**Tottanfjella** – Mt Krone 2300 (fa 1961,
within 15 m of top)

**FORMER AUSTRALIAN (and
FRENCH) SECTOR**
(Enderby Land, Kemp Land, Mac-
Robertson Land, Princess Elizabeth
Land, Queen Mary Land, Wilkes
Land, Terre Adélie, King George V
Land), scene of epic polar journeys
by Douglas Mawson in 1911-12.
Mt Menzies 3355 (fa 20/12/1961 Aus-
tralian party)
Summers Pk 2227

**FORMER NEW ZEALAND
SECTOR** (150°W to 160°E) (Vic-
toria Land, King Edward VII Land
around the Ross Sea)
Mt Kirkpatrick 4511
Mt Elizabeth 4480
Mt Markham 4350
Mt MacKellar 4290
Mt Kaplan 4250
Mt Minto 4160
Mt Miller 4145
Mt Falla 4115
Mt Lister 4070 (fa 1963 New Zealand
party)
Mt Fisher 4066
Mt Wade 4063

**Motor sledge hauling on the Beardmore glacier, Antarctica (C Swithinbank)**

The **Antarctic Continent** carries the largest ice mass in the world having a greatest thickness of 4275 m, an average thickness of 2000 m and a total volume of 28 million km³. Seventy-four per cent of the world's fresh water is here. There are really two ice sheets, Greater and Lesser Antarctica, separated by the trans-Antarctic mountain chain, some summits of which rise above the ice though most are beneath it. At the centre of the land mass of Greater Antarctica, equidistant from the surrounding seas is the (so-called) Pole of Inaccessibility or Pole of Cold (3719) where the average temperature is −60°C and winter values of −85°C have been recorded. In places the land surface is below sea-level; a point in Marie Byrd Land (the deepest depression on the land surface of the globe) reaches down to −2468 m. Some of the features below the ice are Alpine-type peaks, some exhibit smooth contouring. Superimposed on a map of Europe the ice-sheet would stretch from the North Cape to the Mediterranean Sea and from Portugal to the Caspian Sea. If it melted the world sea-level would rise by 60 m.

There are tremendous glaciers; the Lambert, for example, flows in a trough 3500 m deep and 60 km wide; the Beardmore is 128 km long. Researchers are still seeking the line of the tectonic plate boundary which traverses the Antarctic Continent beneath the ice.

**Super Constellation landed below Mount Erebus (C Swithinbank)**

Mt Fridtjof Nansen 4010 (fa 1961)
Mt Don Pedro Christopherson 3925
Mt Huggins 3920 (fa 1958 F. R. Brooke, B. Gunn)
Mt Anne 3875
Mt Sabine 3850
Mt Erebus (V-le 1947) 3794 (fa 1908 Shackleton expedition)
Mt McClintock 3510
Mt Herschel 3498 (fa 1967 New Zealand party)
Mt Terror 3093 (fa 1959 New Zealand party)
Mt Harmsworth 2771 (fa 1957 B. Gunn, A. Heine, G. Warren)
Mt Discovery 2770 (fa 1959)
Mt Melbourne (V) 2700

## ANTARCTIC ISLANDS

### S Shetland Islands
Clarence Island 2300
Smith Island – Mt Foster 2100
Deception Island (V-le 1970) 576

### S Orkney Islands
Coronation Island 2195
Mt Niven 1274 (fa 1957 British party)

### S Georgia (discovered by Capt Cook, 1775)

*Isolated* – Mt Paget 2915 (fa 1964 M. K. Burley party)

**Big Ben, Heard Island (Australian News & Information Bureau)**

The South Pole (2866) was first reached by R. Amundsen's Norwegian expedition in December 1911. Captain Scott's British party reached it a few weeks later and perished on the march out. A scientific and meteorological station is now installed there permanently.

A great embayment in the coastline of the Antarctic Continent is the Weddell Sea, lying between 20°W and 60°W. It is bounded on the west side by the mountainous peninsula which is Graham Land and Palmer Land, **the highest point of which is Mount Andrew Jackson** (3950). Further south at the south-west corner of the embayment are the Ellsworth Mountains, discovered from the air as late as 1935, where **Vinson Massif** (5140) **is the highest point of the Continent**. Nearby **Mount Shinn** (4800) **is the highest Antarctic volcano**. Several of the mountains in this area were climbed during the course of one expedition by the Americans in 1966–7.

**Sir Vivian Fuchs's party**, which **crossed the Antarctic Continent for the first time in 1957-8**, started from Vahsel Bay on the shores of the Weddell Sea and reached the South Pole on 19 January 1958. There they met a group led by Sir Edmund Hillary (of Mount Everest), which had started from McMurdo Sound on the Ross Sea and had arrived on 4 January 1958. Once united the whole party retraced the Hillary route back to the Ross Sea.

The Graham Land peninsula is joined to the tip of South America by the **Scotia Ridge**, which breaks the sea surface as the islands of **South Georgia**, **South Sandwich**, **South Orkneys** and **South Shet-**

*Allardyce Ra* – Nordendskjöld 2355,
Mt Sugartop 2324

*Salvesen Ra* – unnamed 2151, Mt
Gregor 1881

**Bouvet Island** (V) 935

**Prince Edward Islands**
Jan Smuts Pk (Marion Island) 1186

**Crozet Island**
Î de l'Est 1980
Î de la Possession 960
Î aux Cochons 600

**Kerguelen**
Mt Ross 1850
Le Dome ice-cap 1049

**Heard Island**
Big Ben (V) 2745 (fa 25/1/1965 W.
Deacock's party)
Mawson Pk 2415

**Amsterdam Island**
Mt de la Dives 881

**St Paul Island,** remains of a large
caldera.

**Scott Island** (Ross Sea), the summit of
a mountain rising 3000m above the
ocean floor.

# The Arctic

**ICELAND**

**Icefields**
Öraefajökull – Hvannadalshnukur (V)
2119 (fa uncertain: modern recorded
ascent 1963 A. Bignami, P. Bernas-
coni), **the highest point of icefield and
of Iceland.**
Vatnajökull – Baroarbunga 2000, high-
est point Grimsvötn (V) 1400 is be-
neath the ice.
First crossing of the icefield by W. L.
Watts & 5 Icelanders in 1875. Crossed
solo in 1959 by O. Woitsch. Area of
ice 21 000 km².
Hofsjökull 1765
Eiriksjökull 1675
Eyjafjellajökull (V) 1666
Tungnafellsjökull 1540
Myrdalsjökull 1480 – Katla (V) is be-
neath the ice.
Snaefellsjökull 1446
Langjökull 1400
Tungnahryggs Jökull 1384
Drangajökull 925

**Peaks**
Snaefell 1833
Herdhubreidh 1682
Askja (V-le 1961, violent activity 1875)
1510

**lands**. Close to the South Sandwich Islands the ridge bends so
sharply as to reverse direction; round the outside of the bend is the
South Sandwich Trench (Meteor Deep, −8264). It is no surprise
therefore to find volcanoes on some of the islands.

The **Ross Sea**, named for Sir J. Clark Ross who discovered it in
1841, is a big embayment in the Antarctic coastline lying between
160°E and 150°W. This same expedition **first saw the active volcano
of Mount Erebus** (3794) on Ross Island and named it for their
leading ship. It was **first climbed by Shackleton's expedition** in
1908. The crater is 275m deep and 1000m across.

The trans-Antarctic mountain chain runs along the west coast
of the Ross Sea. The most northerly part, called the Admiralty
Range, was first sighted in 1841. On southwards there are many
other named ranges and to quote them all would only lead to
confusion. **The highest point here is Mount Kirkpatrick** (4511).

The southerly part of the embayment is occupied by the Ross
Ice Shelf. This is the nearest point to the South Pole and the expeditions of both Amundsen and Scott started from here their
attempts to reach it, the former by the Axel Heiberg glacier, the
latter by the Beardmore glacier.

**Scott Island** is the summit of a mountain rising 3000m above
the ocean floor. It lies at the junction of a ridge from the southern
tip of New Zealand, which carries Macquarie and Balleny
Islands, and the Pacific–Antarctic Ridge which crosses the ocean
to Easter Island and the Galapagos Islands.

**Iceland** rises directly on the mid-Atlantic Ridge and is thus being
split apart as the ocean floor spreads. One of the active axes runs
north-east, then north, and carries the active calderas of Askja
and Krafla. There is a long record of vigorous volcanic activity.
Parts of the land surface are covered by small ice sheets, jökulls,
sometimes with fissures and vents buried deep beneath the ice.
Under these conditions it is possible that eruptions under the ice
may melt large amounts of it, producing disastrous floods, known
as jökulhlaups, which do widespread damage.

**Hvannadalshnukur** (2119), on Öraefajökull, **is the highest point
of Iceland**. In the eruption of 1562 it is estimated that 10km³ of
material was ejected.

Grimsvötn (1400) lies beneath Vatnajökull. The jökulhlaups of
the 1934 and 1938 eruptions produced melt water flow rates of
50 000 m³/second (compare the flow rate of the Amazon – 10 000
m³/second).

Askja (1510), with a caldera 7·5 km diameter having walls 400 m
high, is of the type where subsidence into a magma chamber has
occurred. There is a crater lake.

Hekla (1447), perhaps the best known Icelandic volcano, is
called locally 'the gateway to hell'. The 1300 eruption is said to
have 'split the mountain open'; that of 1766 led to great loss of

Hekla (V-le 1970, violent history of eruption) 1447, highest point of a ridge 27 km long, 2 to 5 km wide.
Kerlingarfjöll 1478
Trolladyngja 1460
Krafla (V) 818
Laki, the site of the biggest basalt flood of modern times (1783).

**Islands**
Surtsey (V) 174, born 1963
Heimaey (V)

## ARCTIC ISLANDS

**Faeroe Islands** – 882, volcanic

**Jan Mayen Island**
Beerenberg (Haakon Pk) (V) 2277 (fa 1921 J. M. Wordie & party), 50 km round the base with 15 glaciers.

**Spitzbergen (Svalbard)**
Newtontoppen 1712 (fa 1900 Russian exped)
Perriertoppen 1712 (fa 1950 C. Maillard, F. Gendron)
Chadwickryggen 1636
Mt Irvine 1591
Eidsvollfjellet 1454
Hornsundtind 1431 (fa 1897 Garwood of Conway's exped)
Tre Kroner 1225 (fa 1897 Conway's exped)
Mt Terrier 1097 (fa 1921)

**Franz Josef Land (Zemlya Frantsa Iosifa)** – 606

**Novaya Zemlya**
Ice-cap – G. Blednaya 1052
Peak – Pt 1342

**Severnaya Zemlya** – 800

**Wrangel Island (O Vrangelya)**
G Sovetskaya 1097

**Axel Heiberg Island** – 2130
Some geographers hold that from here through Ellesmere Island to the Torngats of Labrador is one mtn ra – the Inuitian Alps.

**Devon Island** – 1890

**Baffin Island**

*Cumberland Peninsula*
Tête Blanche 2156 (fa 1953 Swiss exped)
Mt Queen Elizabeth 2138 (fa 1953 Swiss party with P. D. Baird)
Mt Odin 2130
Mt Friga 2030
Mt Asgard 2011 (fa 1953 Swiss exped)
Adluk Pk 1980
Ungardaluk 1950
Mt Pingo 1920
Freya 1920
Tête des Cirques 1885
Turnweather Pk 1845

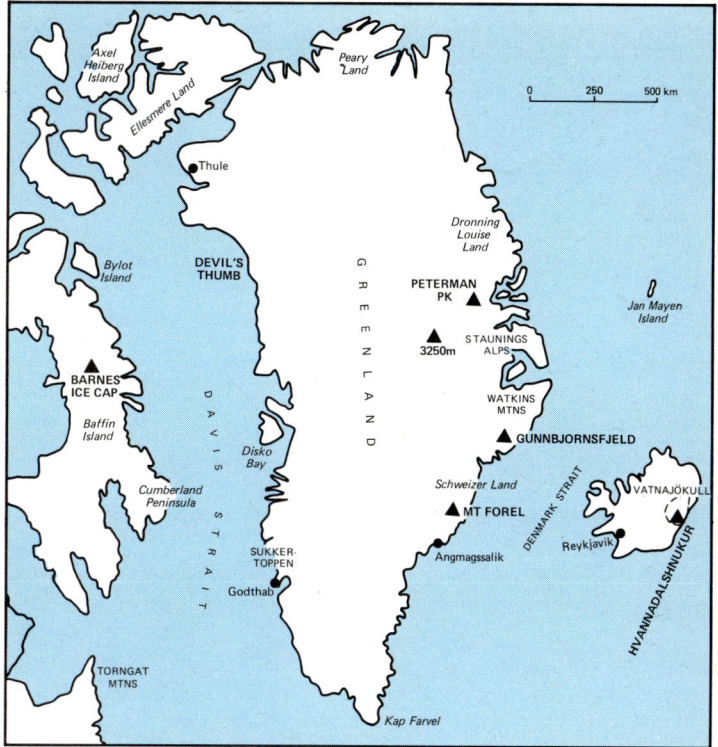
Greenland and adjacent islands

life; that of 1947 ejected one km³ of material. The word 'geyser' comes from the Icelandic word 'geysir', the actual name of a famous volcanic water/steam spout in a broad valley north-west of Hekla.

In 1783 **Laki was the site of the biggest basalt flow of modern times**. A 30 km fissure produced 12 km³ of lava covering 600 km² (as a result one-fifth of the humans, three-quarters of the sheep and horses and one-half of the cattle of the island died in the ensuing famine).

**Eldgjá fissure**, only a short distance south-west, produced 9 km³ of lava in 950.

The last two decades have seen continued activity. **The Island of Surtsey** has risen from the sea to 174 m and volcanic emissions on **Heimaey** have buried the local settlement.

The Cumberland Peninsula of **Baffin Island** was visited by John Davis in 1585. Of recent years there have been striking developments in mountaineering on peaks up to 2156 m with steep granite faces – **some of the most impressive rock walls in the world**. The mountains near Pangnirtung are now in the Auyuittuq National Park.

**Bylot Island** was **first traversed from north to south in 1963** by H. W. Tilman. In 1977 a Canadian expedition accomplished the traverse

Mt Raleigh 1600 (fa 1962 H. W. Tilman's party), named by John Davis in 1585.

Several unnamed pks round Stewart Valley were climbed by a Canadian exped in 1977.

*North-west*
Ice-cap 2600
Barnes Ice-cap
At Buchans Gulf there are sea-cliffs of 1500 m.

**Bylot Island**
Mt Mitima 2530
Angilaak Mtn 2072
Mt Mallik 2012
Mt Aktinaq 1980

**Ellesmere Island**
Mt Whistler 2590 (fa 1961 British exped), first sighted and named by Greely in 1882.
Commonwealth Mtn 2290
Mt Oxford 2210 (fa 1935)
Mt Arrowhead 2140 (fa 1958)
Mt Grant 1830 (fa 1953)
Mt Nukap 1780 (fa 1958)

**GREENLAND**

**Melville Bay area**
Devil's Thumb (Djavlens Tommelfinger) (fa 1934)

**Disko Bay area** (including Upernivik Island)
Perserajoq Ghiglione (Punta Italia) 2310 (fa 1960 Italian exped)
Snepyramiden 2236 (fa 1961 Danish/Italian party)
Palup qaqâ (Paulus Pk) 2101 (fa 1939 Scottish exped)
Kakartoq Napasudlerâq 2089 (fa 1950 Scottish exped)
Qilertinguit tunuliat 2070
Merendi 2057 (fa 1965 Italian party)
Horns of Upernivik 1905 (fa 1967 Scottish exped)
Qioqe 1803 (fa 1967 Italian exped)
Umanakfels 1200 (fa 1929 Sorgi, Georgi), the Greenland 'Matterhorn'.

**Sukkertoppen Ice-cap area**
Mt Atter 2799 (fa 1956 British exped)
Agssaussat 2048
Qingarssak 1908

**S Greenland – Cape Farewell area**
Mt Patuersoq 2740 (fa 1966)
Apostelens Tommelfinger 2300
Igdlorsuit qáqâ 2242
Kingitoarsuk 2150 (fa 1909)
Cathedral 2130
Pingasut 2100
Nalumasortoq 2088
Akerne 2048
Minster 2010

**Schweizer Land area**
Mt Forel 3360 (fa 1938 K. Baumann, G. Pidermann, A. Roch)

of the length of the island and also succeeded in making **19 first ascents of mountains**, including that of Angilaak Mountain (2072).

**The highest point of Ellesmere Island, Mount Whistler** (2590), was **first sighted by Greely in 1882** and **first climbed by a British expedition in 1961**. This forms part of the Innuitian Alps, which some geographers see as a very long continuous mountain chain running from Axel Heiberg Island to the Torngats of Labrador ($2\frac{1}{2}$ times longer than the European Alps – 3000 km long, 300 km wide and covering 800000 km$^2$). In 1909 R. Peary set out from Ellesmere Island for the **North Pole** ( – 4087 m, but sea permanently ice covered), which **he claimed to have reached on 6 April**.

After Antarctica, **Greenland** carries the second great ice-cap of the world having a volume of 3·7 million km$^3$ and covering 1·25 million km$^2$. It reaches 3000 + m in the interior with a greatest measured ice thickness of 3350 m. The **first crossing was made by F. Nansen in 1888** from Umivik to Godthaab at the south end of the country. This was one of the earliest journeys to demonstrate the

*Left:* Umanak and its mountain, Greenland (M Heller)

*Right:* Upernavik Island, Greenland (M Heller)

Stephenson Bjorg 3240
Dabbel Haernet 3240
Fruebjerg 3100 (fa 1938 Swiss exped)
Bellepheron 3090
Perfekt 3000 (fa 1938 Swiss exped)
Pharoah 3000
Pte de Harpon 2734
Quervains Bjoerg 2697
Laupersbjoerg 2580 (fa 1938)
Ingolfsfjeld 2560 (fa 1971 Yugoslav exped)
Stockenbjoerg 2520 (fa 1966 British exped)
Pusugssivit 2072
Sølverbjoerg 2050 (fa 1938 Swiss exped)
Rytterknaegten 2010

**Watkins Mtns area**
Gunnbjorns Fjeld 3700 (fa 1935 A. Courtauld, L. Wager, J.L. Longland, Fontaine, E. Munke), **the highest point in Greenland.**
Borgetinde 3348 (fa 1971 British exped)
Ejnar Mikkelsens Fjeld 3261 (fa 1970 British exped)

**Scoresby Land & Staunings Alps**
Danmarkstinde 2930 (fa 1954)
Hjornespitze 2880 (fa 1960)
Stortoppen 2870 (fa 1954)
Norsktinde 2789
Pyramidefjeld 2770
Prometeus 2770
Frihedstinde 2620 (fa 1950-1)
Elisabethtinde 2260 (fa 1950)
Payer Spitze 2133 (fa 1870 J. Payer)

**King Christian X Land**
Petermann Pk 2975 (fa 1929 J.M. Wordie, A. Courtauld, V. Fuchs)
Mt Shackleton 2900 (fa 1953)
Gog 2667 (fa 1933)
Magog 2580 (fa 1948)
Nathorsts Tinde 2390 (fa 1933)
Mt Gore (Strawberry Pk) 2256 (fa 1933)
Mt Mona 2134 (fa 1933)
Teufelsschloss 1311 (fa 1933)

**Dronning Louise Land**
Gefionstinde 2798 (fa 1952 F.R. Brooke party)

**N Greenland** (Peary Land, Daugaard Jensens Land)
Swiss Pk 1920

potential of skis for long-distance mountain travel. Rasmussen crossed in the far north in 1912 and J.P. Koch at the widest part in 1913.

Around the periphery there are many ranges of mountain peaks reaching to 3700m at **Gunnbjorns Fjeld** in the Watkins Mountains, **the highest point in Greenland**. There are almost unlimited mountaineering prospects.

The ice sheet sends down glaciers between the peaks, some of them of vast proportions. For example, the Daugaard–Jensens glacier in the far north-west is said to discharge enough ice every year to meet all the freshwater requirements of the USA. The icebergs which menace the North Atlantic shipping lanes mostly originate in the glaciers of Greenland.

Another striking topographical feature is Nordvestfjord in Scoresby Land. This has been carved by ice in a 2000m high plateau down to a depth of −1450m below sea-level (ie an estimated 3500 km³ of rock has been removed).

The **Watkins Mountains** were located during the 1930–1 expedition led by Gino Watkins, the primary objective of which was to study conditions in Greenland relevant to the expansion of world air travel. After Watkins's death the range was named after him. The highest peak was located by M. Lindsay when he sledged across the ice-cap in 1934 and it was **climbed for the first time in 1935** by members of L.R. Wager's expedition. Known as **Gunnbjorns Fjeld** (3700), **it is the highest mountain in the Arctic.**

The third highest peak of the range, Ejnar Mikkelsens Fjeld (3261) was not climbed until 1970. The explorers found that the actual summit was an obelisk of rock which they could not climb.

# Appendix I *A Mountain Calendar*

## JANUARY

**1** 1882, St Gotthard Tunnel opened for goods; 1914, Fletschhorn fa(winter); 1932, Mt Evans (New Zealand) fa & traverse.

**2** 1875, Gross Glockner fa(winter); 1880, Königspitze fa(winter).

**3** 1896, Finsteraarhorn fa(winter).

**4** 1880, Chimborazo fa; 1888 Mt Blanc first winter traverse Courmayeur to Chamonix.

**5** 1888, Gr Lauteraarhorn fa(winter); 1968, The Fortress (Patagonia) fa.

**6** 1899, Breithorn fa(winter); 1929, Mt Kenya Nelion pk fa, Batian pk 2nd a.

**7** 1890, Eiger fa(winter).

**8** 1880, Mt Cevedale fa(winter).

**9** 1910, Aig d'Argentière fa(winter).

**10** 1902, Weisshorn fa(winter).

**11** 1888, Gr Fiescherhorn fa(winter).

**12** 1832, Strahlegg first traverse.

**13** 1894, Dom fa(winter); 1911, Dent Blanche fa (winter); 1974, Torre Egger (Patagonia) fa.

**14** 1847, Dachstein fa(winter); 1891, Grandes Jorasses fa(winter); 1897, Aconcagua fa.

**15** 1874, Wetterhorn fa(winter).

**16** 1858, Jungfrau first traverse & descent to Wengernalp; 1963, Central Tower of Paine fa.

**17** 1893, Rimpfischhorn fa(winter); 1911, Cerro Juncal fa; 1958, North Tower of Paine fa.

**18** 1925, Mt Pelvoux fa(winter); 1934 Mercedario fa.

**19** 1897, Oberaarjoch reached in first ski traverse of the Bernese Oberland.

**20** 1897, Grünhornlücke crossed in first ski traverse of the Bernese Oberland.

**21** 1889, 2nd winter ascent of Gran Paradiso by 16 climbers and 11 guides; 1974 east face of Central Tower of Paine fa.

**22** 1874 Jungfrau fa(winter); 1878 Mte Viso fa (winter).

**23** 1877, Ciamarella fa(winter).

**24** 1556, the most destructive earthquake ever in Shensi Prov, China; 1895, Sealy (New Zealand) fa.

**25** 1965, Big Ben (Heard Island) fa.

**26** 1884, Dufourspitze of Mte Rosa fa(winter).

**27** 1879, Gr Schreckhorn fa(winter).

**28** 1907, Douglas (New Zealand) fa.

**29** 1705, Start of a big eruption of Vesuvius; 1873, The Duke of the Abruzzi born.

**30** 1882, Mt Blanc 2nd winter ascent; 1934, Tronador fa.

**31** 1876, Mt Blanc fa(winter); 1894 Footstool (New Zealand) fa.

## FEBRUARY

**1** 1906, La Pérouse (New Zealand) fa; 1935, Ilaman (Algeria) fa; 1973, El Capitan (Yosemite) descended by skis and parachute.

**2** 1939, P Simon Bolivar fa; 1952, Co Fitzroy fa.

**3** 1920, Ober Gabelhorn fa(winter); 1962, Matterhorn north face fa(winter).

**4** 1797, 40 000 killed in earthquake associated with Volcan Tunguragua (Ecuador); 1880, Piz Bernina fa(winter); 1907 Torres fa.

**5** 63, Beginning of eruption of Vesuvius which destroyed Pompeii and Herculaneum.

**6** 1895, Mt Tasman fa; 1896, Snowdon Railway opened.

**7** 1914, Zinal Rothorn fa(winter); 1920, Täschhorn fa(winter).

**8** 1895, Haidinger fa.

**9** 1925, Gran Paradiso first ski traverse; 1934, D. W. Freshfield died.

**10** 1929, Mt Teichelmann fa; 1940, Swiss Foundation for Alpine Research founded.

**11** 1869, Vignemale fa(winter).

**12** 1894, Mt de la Bèche (New Zealand) fa.

**13** 1962, Ngga Pulu (Carstensz Pyramide) fa; 1973, Noshaq fa(winter).

**14** 1895, Sefton fa; 1965, Ball's Pyramid fa.

**15** 1882, Aig du Moine & Tinzenhorn both fa (winter).

**16** 1848, Nev Chillan fa; 1896, Mte Disgrazia fa (winter).

**17** 1890, Mt Elgon fa; 1928, Himalayan Club founded.

**18** 1880, Cotopaxi fa.

**19** 1889, Dufourspitze of Mte Rosa first traverse Gressoney to Zermatt.

**20** 1912, Co La Paloma (Chile) fa; 1943, Volcano Parícutin born.

**21** 1926, Les Ecrins fa(winter).

**22** 1882, Mte Cristallo fa(winter).

**23** 1880, Sinchologua (Ecuador) fa.

**24** 1943, P Simmons (Venezuela) fa.

**25** 1904, Mt Blanc fa(skis); 1928, Aigs des Drus fa (winter); 1934 Aig de Triolet fa(winter).

**26** 1907, Haast & Lendenfeld fas; 1937, Ojos del Salado fa.

**27** 1878, notable eruption of Hecla (Iceland); 1890, P Palu fa(winter).

**28** 1880, the headings met in the St Gotthard Tunnel.

**29** 1918, Dent d'Hérens fa(skis); 1960, Agadir (Morocco) earthquake.

## MARCH

**1** 1960, Co O'Higgins (Patagonia) fa.
**2** 1882, Mt Cook attempt by W.S. Green *et al*;
1885, Grand Paradis fa(winter).
**3** 1965, Huamanrazo (Cordillera Chonta) fa.
**4** 1924, Mt Tutoko (New Zealand) fa.
**5** 1913, Castor fa(winter).
**6** 1931, San José fa; 1958, Co Arenales (Patagonia) fa; 1967, Gust of 232 km/hr recorded in the Cairngorm Mountains of Scotland.
**7** 1894, Malte Brun fa; 1908, Alpine Ski Club founded; 1913, Pollux fa(winter); 1956, Sarmiento fa.
**8** 1887, Finsteraarhorn fa(winter).
**9** 1895, Hochschwab fa(skis).
**10** 1880, Antisana fa; 1908, Mt Erebus fa.
**11** 1929, Meteor Pk (New Zealand) fa.
**12** 1910, D'Archiac (New Zealand) fa.
**13** 1892, Ruapehu fa.
**14** 1939, Co Pili o Acaramachi fa.
**15** 1903, Aig Verte fa(winter).
**16** 1926, La Meije fa(winter); 1939, P Cristóbal Colon fa.
**17** 1882, Matterhorn fa(winter).
**18** 1876, Old Weissthor first winter traverse.
**19** 1914, McKerrow (New Zealand) fa.
**20** 1953, Matterhorn Furggen Ridge fa(winter).
**21** 1939, Co Degli Alpini (Bolivia) fa.
**22** 1885, Lyskamm fa(winter); 1894, Mt Darwin fa; 1918, Laquinhorn fa(winter).
**23** 1880, Pinchincha erupted.
**24** 1893, Rincón (Andes) fa; 1937, Mte Italia (Tierra del Fuego) fa.
**25** 1894, Lenzspitze & Nadelhorn fa(winter); 1951, Thabana Ntlenyana fa.
**26** 1907, Aig du Chardonnet fa(skis).
**27** 1914, Ruareka & Dilemma (New Zealand) fas.
**28** 1964, World's record tsunami (67 m) off Valdez, south-west Alaska.
**29** 1910, Alphubel fa(winter).
**30** 1916, Mt Drummond (New Zealand) fa; 1917, The Acolyte (New Zealand) fa.
**31** 1907, Grand Combin fa(winter); 1912, Mt Dampier fa.

## APRIL

**1** 1907, Allalinhorn fa(winter).
**2** 1874, Foundation of Club Alpin Français.
**3** 1966, Mismi (Cordillera Occidental, Peru) fa.
**4** 1880, Cayambe fa.
**5** 1931, Nev Mururata (Bolivia) fa.
**6** 1966, Hualca-hualca fa.
**7** 1915, Mt Edgar Thompson (New Zealand) fa.
**8** 1906, Cone blown off Vesuvius and Thomas Cook's funicular destroyed.
**9** 1935, Mt Warner (New Zealand) fa; 1959, Yocum Ridge of Mt Hood fa.

**10** 1964, Gyachung Kang fa.
**11** 1850, Volcano born on the Plain of Leon (Nicaragua).
**12** 1897, Tupungato fa; 1934, Gust of 371 km/hr recorded on Mt Washington.
**13** 1936, Mts Glenisla & Onslow (New Zealand) fas.
**14** 1921, World record one day snow fall (187 cm) at Silver Lake, Colorado.
**15** 1953, Camp 2 established on Mt Everest by John Hunt's expedition.
**16** 1960, Co Catedral Sur (Central Andes) fa.
**17** 1880, Mt Sara Urco (Ecuador) fa.
**18** 1906, San Francisco earthquake.
**19** 1903, Tolosa (Argentina) fa.
**20** 1924, Mt Blanc first ski traverse, Courmayeur to Chamonix.
**21** 1962, Marisember (Chile) fa.
**22** 1934, Apennine Tunnel opened.
**23** 1828, Birth of F.J.A. Hort, one of the principal founders of the Alpine Club.
**24** 1880, Cotacachi fa; 1965, Ngojumba Ri II fa.
**25** 1974, Lamjung Himal fa.
**26** 1336, Mt Ventoux fa.
**27** 1840, Edward Whymper born; 1916, Brass Pk (New Zealand) fa; 1962, Jannu fa.
**28** 1893, Death of F.W. Jacomb; 1969, Dhaulagiri disaster – 5 Americans and 2 Sherpas killed.
**29** 1970, Churen Himal fa.
**30** 1957, Nev Quichas (Peru) fa.

## MAY

**1** 1975, Dhaulagiri V fa.
**2** 1964, Gosainthan fa.
**3** 1962, Lönpo Gang fa.
**4** 1880, Illiniza (Ecuador) fa.
**5** 1965, Ngojumba Ri fa; 1972, Nampa fa.
**6** 1902, Disastrous eruption of Soufrière (St Vincent); 1961, Annapurna III fa; 1965, Gangapurna fa.
**7** 1959, North-west face of Great White Throne (Zion NP) fa completed.
**8** 1902, Disastrous eruption of Mt Pelée (Martinique); 1926, W.A.B. Coolidge died; 1935, Iharen fa; 1978, Mt Everest fa(without oxygen).
**9** 1956, Manaslu fa; 1975, Dhaulagiri IV fa.
**10** 1960, Api fa(certain).
**11** 1942, Monk's Cowl (Drakensberg) fa.
**12** 1970, Lhotse Shar fa.
**13** 1960, Dhaulagiri I fa.
**14** 1973, Yalung Kang fa.
**15** 1955, Makalu fa.
**16** 1961, Nuptse fa; 1975, Mt Everest fa(by a woman).
**17** 1960, Annapurna II fa; 1962, Pumori fa.
**18** 1956, Lhotse fa; 1971, Dhaulagiri II fa.
**19** 1912, Mt Blackburn fa; 1964, Talung Pk fa.

**20** 1962, Nupchu fa.
**21** 1937, Chomo Lhari fa.
**22** 1954, Nev Quehuar (Argentina) fa.
**23** 1968, Chanconcurane (Peru) fa.
**24** 1949, Agathla Pk fa; 1960, Himal Chuli fa.
**25** 1897, Lanin fa; 1955, Kangchenjunga fa; 1964, Mt Huntington fa.
**26** 1802, Pichincha (Ecuador) fa; 1953, Mt Everest South summit fa; 1973, Panch Chuli fa.
**27** 1939, Nepal Pk – North-east summit fa.
**28** 1953, Mt Everest – Hillary and Tenzing installed in Camp 9 at 8500m.
**29** 1939, Tent Pk fa; 1953, Mt Everest fa.
**30** 1954, Baruntse fa; 1955, Annapurna IV fa.
**31** 1962, Chamlang fa; 1970, disastrous earthquake in Peru.

## JUNE

**1** 1882, St Gotthard Tunnel opened for passenger traffic; 1906, Simplon Tunnel opened.
**2** 1960, Amne Machin fa.
**3** 1870, Cimone della Pala fa; 1930 Jongsong Pk fa; 1950, Annapurna I fa; 1974, Shivling fa.
**4** 1893, Ago di Sciora fa; 1961, Mt Ghent fa; 1974, Changabang fa.
**5** 1912, Great eruption of Mt Katmai (Alaska); 1973, Saser Kangri fa.
**6** 1964, Pongos Norte (Peru) fa.
**7** 1925, Illampu (Sorata) fa(certain); 1967, Miangul Sar fa.
**8** 1783, Laki (Iceland) fissure eruption began; 1924, Mallory & Irvine lost on Mt Everest; 1931, Mt Fairweather fa; 1955, Istor-o-Nal fa.
**9** 1957, Broad Pk fa; 1960, Disteghil Sar fa.
**10** 1962, Nev Mitra (Peru) fa.
**11** 1924, Jungfrau first ski traverse.
**12** 1907, Trisul fa.
**13** 1952, Chaukamba fa; 1961, Qungur I climbed by a party including 2 women.
**14** 1954, Api fa claimed, party lost.
**15** 1794, large scale eruption of Vesuvius; 1868, Fell Railway opened across Mt Cenis Pass.
**16** 1865, Grand Cornier fa; 1911, Pauhunri fa; 1932, Rakhiot Pk fa.
**17** 1936, Kitaraju fa.
**18** 1859, Aletschhorn fa; 1906, Mt Stanley (Ruwenzori) fa.
**19** 1957, Skil Brum fa.
**20** 1873, Roche Faurio fa.
**21** 1931, Kamet fa.
**22** 1741, Windham expedition reached Montanvert and the Mer de Glace; 1866, P Scalino fa.
**23** 1870, Aigs Rouges d'Arolla fa; 1906, Mt Speke (Ruwenzori) fa; 1925, Mt Logan fa.
**24** 1865, Grandes Jorasses (Pte Whymper) fa.
**25** 1864, Barre des Ecrins fa; 1957, Rakaposhi fa.

**26** 1492, Mt Aiguille (Vercors) fa; 1865, Col du M.T. Dolent first traverse.
**27** 1892, Punta Rasica fa.
**28** 1787, Col du Géant first traverse; 1865, P Roseg fa; 1912 Kolahoi fa.
**29** 1865, Aig Verte fa; 1935, Grandes Jorasses – North face fa completed; 1964, Momhil Sar fa.
**30** 1868, Grandes Jorasses fa.

## JULY

**1** 1868, Kasbek fa.
**2** 1939, Nanda Devi East fa.
**3** 1953, Nanga Parbat fa.
**4** 1874, Roccia Viva fa.
**5** 1865, Trifthorn fa; 1939, Dunagiri fa; 1957, Gasherbrum I fa.
**6** 1865, Obergabelhorn fa; 1878, P Gaspard fa; 1960, Masherbrum fa.
**7** 1870, Ailefroide fa; 1956, Muztagh Tower & Gasherbrum II fas.
**8** 1964, Koh-i-Langar fa.
**9** 1864, Mt Dolent fa; 1937 Mt Lucania fa.
**10** 1869, Gspaltenhorn fa.
**11** 1886, Fleischbank fa; 1951, Mukut Parbat fa.
**12** 1864, Aig de Trelatête fa.
**13** 1866, Mt Pleurer fa; 1966, Shingeik Zom fa; 1977, Ogre fa.
**14** 1865, Matterhorn fa.
**15** 1864, Aig d'Argentière fa.
**16** 1866, Serpentine fa; 1876, Les Droites – North summit fa.
**17** 1865, Matterhorn fa(by Italian ridge).
**18** 1862, Dent Blanche fa; 1935, Khartaphu fa.
**19** 1959, Kanjut Sar fa.
**20** 1873, Ayer's Rock (Australia) fa; 1932, Huascaran Sur fa.
**21** 1876, Store Skagastølstind fa; Mt Alberta fa; 1936, Mt Waddington fa; 1950, Tirich Mir fa.
**22** 1878, Aigs d'Arves fa.
**23** 1862, Gr Fiescherhorn fa.
**24** 1888, Dykh Tau fa; 1953 Ausangate fa; 1962, Saltoro Kangri fa.
**25** 1964, Tirich Mir East fa.
**26** 1903, Ushba – South Pk fa.
**27** 1866, Gr Litzner fa.
**28** 1800, Gr Glockner fa; 1874, El'Brus – West Pk fa.
**29** 1859, Muttler fa.
**30** 1828, Mt Pelvoux fa; 1859, Grand Combin fa; 1862, Täschhorn fa.
**31** 1843, Demavend fa; El'Brus – East Pk fa; 1897, Mt St Elias fa; 1954, K 2 fa; 1956, Muztagh Ata fa.

## AUGUST

**1** 1855, Mte Rosa fa; 1931, Matterhorn – North face fa completed.

**2** 1959, Malubiting East fa.

**3** 1811, Jungfrau fa; 1913, Kun fa.

**4** 1925, Civetta – North-west face fa; 1929, Sangay fa; 1957, Chogolisa & Haramosh fas.

**5** 1877, Aig Noire de Peuterey fa; 1881, Aig de Grépon fa; 1952, Salcantay fa.

**6** 1957, Gasherbrum IV fa.

**7** 1876, Les Droites fa.

**8** 1786, Mt Blanc fa; 1860, Alphubel fa.

**9** 1889, Koshtantau fa.

**10** 1848, Ulrichshorn fa.

**11** 1858, Eiger fa; 1898, Grand Teton fa(undisputed).

**12** 1863, Dent d'Hérens fa; 1937 Mana fa; 1938, Mt Forel fa.

**13** 1813, Breithorn fa; 1859, Bietschhorn fa.

**14** 1861, Schreckhorn fa.

**15** 1842, Mt Woodrow Wilson (Wind River Ra, Wyoming) fa(probably); 1857, Mönch fa.

**16** 1812, Finsteraarhorn fa; 1877, Meije fa; 1956, Qungur I fa.

**17** 1870, Mt Rainier fa; 1960, Trivor & Noshaq fas; 1964, Shachaur fa.

**18** 1879, Argentera fa.

**19** 1861, Weisshorn & Lyskamm fas; 1964, Udren Zom fa.

**20** 1882, Aig du Géant fa.

**21** 1963, Noshaq West & Noshaq East fas.

**22** 1864, Zinal Rothorn fa; 1873, Herbetet fa.

**23** 1859, Grivola fa.

**24** 1853, Mt St Helens fa; 1854, Gran Zebru fa; 1862, Mte Disgrazia fa; 1959, Saraghrar fa.

**25** 1892, Pioneer Pk (Karakoram) fa.

**26** 1856, Lagginhorn fa; 1883, catastrophic eruption of Krakàtau (Indonesia).

**27** 1820, Zugspitze fa; 1868, Ebnefluh fa.

**28** 1854, Fletschhorn fa; 1856, Allalinhorn fa; 1953, Nun fa.

**29** 1936, Nanda Devi fa.

**30** 1861, Mte Viso fa; 1956, P Pobeda fa.

**31** 1844, Wetterhorn fa.

## SEPTEMBER

**1** 1358, Roche Melon fa.

**2** 1866, Albaron fa; 1875, Grande Dent de Veisivi fa.

**3** 1841, Gr Venediger fa; 1933, P Kommunisma fa.

**4** 1860, Gran Paradiso fa; 1904, Oldoinyo Lengai fa; 1963, Urgend fa.

**5** 1866, Blindenhorn fa; Mt Mallet fa.

**6** 1871, Dent des Bouquetins fa.

**7** 1888, Shkhara fa.

**8** 1860, Grande Casse fa; 1868, Grosshorn fa.

**9** 1859, Rimpfischhorn fa; 1898, Illimani fa.

**10** 1843, Wildhorn fa; 1921, Eiger – Mittellegi Ridge fa.

**11** 1858, Dom fa; 1930, catastrophic eruption of Stromboli.

**12** 1878, Grand Dru fa; 1912, Cathkin Pk fa & Hellenic Alpine Club founded; 1965, Darban Zom fa.

**13** 1850, P Bernina fa; 1877, P Scerscen fa.

**14** 1865, Mte Cristallo fa.

**15** 1864, Adamello fa.

**16** 1858, Nadelhorn fa; 1911, Edward Whymper died.

**17** 1854, Historic ascent of the Wetterhorn by Alfred Wills; 1864, Presanella fa.

**18** 1863, Antelao fa; 1865, Nesthorn fa.

**19** 1857, Mte Pelmo fa.

**20** 1865, Aig de Chardonnet fa.

**21** 1853, opening of a hut on Grands Mulets; 1876, Fusshorn fa.

**22** 1879, Mitre de l'Évêque fa.

**23** 1876, Petites Jorasses fa; 1936, Siniolchu fa.

**24** 1865, P de Tenneverge fa; 1975, Mt Everest – South-west face fa.

**25** 1928, P Lenin fa.

**26** 1864, Berglistock fa.

**27** 1804, Ortler fa; 1829, Mt Ararat fa.

**28** 1538, Mte Nuovo born; 1864, Marmalado fa.

**29** 1759, Jarullo volcano born.

**30** 1877, Mare Perci (Gran Paradiso) fa; 1895, Pte des Lanchettes (Mt Blanc) fa.

## OCTOBER

**1** 1893, Sudlicher Manndkogel (Dachstein) fa.

**2** 1973, Vasuki Parbat fa.

**3** 1870, Presolana (Bergamesque Alps) fa; 1898, early traverse of the Alps in a balloon.

**4** 1861, Mt Pourri fa.

**5** 1977, Bomba Dhura fa.

**6** 1889, Kilimanjaro fa.

**7** 1967, Nev Capurata (Chile) fa.

**8** 1893, Gr Oedstein (Ennsthaler Alps) fa.

**9** 1875, Bric de Rubren fa; 1966, Tirsuli fa.

**10** 1977, Nampa South fa.

**11** 1960, Punta Canaletas (Andes) fa.

**12** 1973, Ayapungo fa.

**13** 1866, The Arkwright accident on Mt Blanc.

**14** 1965, Nev Acotango (Chile) fa.

**15** 1911, Coropuna fa; 1964, Modi Pk fa.

**16** 1964, Glacier Dome fa; 1968, Mt Blanc du Tacul – first descent of Gervasutti Couloir on skis.

**17** 1972, Panchchuli I fa; 1977, Nar Parbat fa.

**18** 1905, Akademischer Alpenclub, Bern founded.

**19** 1954, Cho Uyo fa; 1970, Pk 29 fa.

**20** 1973, Dhaulagiri III fa.

**21** 1963, Saipal fa.

**22** 1954, Makalu II fa.

**23** 1863, Italian Alpine Club founded.

24  1902, big eruption of Santa Maria, Guatemala; 1955, Ganesh Himal fa.
25  1964, Manaslu North fa; 1974, Swargarohini fa.
26  1969, Pt 5055 (Cordillera Vilcanota) fa.
27  1967, Mt Herschel fa.
28  1932, Minya Konka fa.
29  1942, Matterhorn – Furggen Ridge fa(direct).
30  1954, Chomo Lonzo fa.
31  1966, Co Quilpué (Chile) fa.

## NOVEMBER

1  1755, great Lisbon earthquake; 1882, Aig du Tour fa(winter); 1969, Gurja Himal fa.
2  1866, Dreiherrenspitze fa; 1884, Dent Parrachée fa(winter).
3  1889, Ixtacclhuatl fa; 1955, Rimutaka Tunnel opened.
4  1964, Thamserku fa.
5
6  472, start of an eruption of Vesuvius which threw ash as far as Istanbul; 1919, Ancohuma fa.
7  1866, P Platta fa; 1970 Mt Cook – Caroline face fa.
8  1892, Akademischer Alpenverein, Munich founded.
9  1970, Mt Cook – Caroline face 2nd ascent.
10  1965, Cielo Amarillo (Andes) fa.
11  1954, Putha Hiunchuli fa.
12  1881, P d'Aela fa.
13  1967, Pariahuachuco (Cordillera Huaytapallana) fa.
14  1892, Cima Tosa fa(winter).
15  1697, an eruption of Vesuvius; 1802, an eruption of Etna; 1867, more Vesuvius.
16  1884, Kesselkogel (Rosengarten) fa(winter); 1963, Lascar fa.
17  1935, Sosneado fa; 1939, Limitajo (Andes) fa.
18  1935, Kabru fa.
19  1907, Columtucsa (Chile) fa.
20  1939, Crucesnioj volcano (Andes) fa.
21  1944, Macá (Andes) fa.
22  1896, Mt Morrison (Taiwan) fa.
23  1910, Mt Aspiring fa.
24
25  1939, Co Cuevas (Argentina) fa.
26  1881, Sorapis fa(winter).
27  1883, Marmolada fa(winter).
28  1832, Leslie Stephen born; 1872, Cotopaxi fa.

29  1939, Co Panizos (Argentina) fa.
30  1937, Mt Underwood (New Zealand) fa.

## DECEMBER

1  1909, Mt Hamilton (New Zealand) fa.
2  1759, Jorullo volcano, Mexico; born.
3  1970, Co Torre 2nd (disputed) ascent.
4  1895, Cimone della Palla fa(winter).
5  1936, Ngga Pulu fa.
6  1964, Yuracmayo (Cordillera Huaytapallana) fa.
7  1968, Co Planchón (Patagonia) fa.
8  1914, Melchior Anderegg died.
9  1962, Becker (Chile) fa.
10  1891, Croda da Lago fa(winter).
11  1961, Punta Equivocados fa.
12  1897, Mt Clapier fa(winter).
13  1916, 100 avalanches in one valley in the Dolomites; 1952, Co San Valentin fa.
14  1867, Breche de la Meije first winter traverse.
15  1949, Matterhorn fa(winter-nocturnal).
16  1631, disastrous eruption of Vesuvius (4000 deaths).
17  1634, Mt Etna started an eruption which lasted a year and a half; 1944, Co San Lorenzo fa.
18  1966, Vinson Massif (Antarctica) fa.
19  1915, the headings met in the Connaught Tunnel below Rogers Pass; 1971, The Sword (Patagonia) fa.
20  1882, Col du Tacul first traverse.
21  1945, Twilight Pk (New Zealand) fa.
22  1839, Mt Egmont fa; 1859, first moves towards the foundation of the Alpine Club.
23  1900, Mt Jalovic (Julian Alps) fa(winter).
24  1866, Finsteraarjoch & Strahlegg first winter traverses.
25  1870, the headings met in the Mont Cenis Tunnel; 1894, Mt Cook fa.
26  1898, Levanna Occidentale fa(winter).
27  1836, an avalanche killed 8 people at Lewes, England; 1957, Paine Grande fa.
28  1892, Gr Zinne fa(winter); 1893, P Roseg fa (winter); 1959, Gannett Pk fa(winter).
29  1914, Mt Jean (New Zealand) fa; 1934, Mts Loughnan & Gorrie (New Zealand) fas; 1945, Mono Blanco (Andes) fa.
30  1964, Mt Paget fa.
31  1901, Strahlhorn fa(winter).

# Appendix II — *Mountain Data by Countries*

| Country | Highest peak or point | National Tourist Office | Principal mountaineering clubs and organizations | Principal mountaineering periodicals and their frequency |
|---|---|---|---|---|
| **Algeria** | Mt Tahat 2918 | Office National Algérien de Tourisme, 25 rue Khélifa Boulchalfa, Algiers | | |
| **Argentina** | Aconcagua 6960 | | Federacion Argentina de Montanismo y Afines, José P Varela 3948, Buenos Aires | *La Montana* (FA de M) (1) |
| **Australia** | Kosciusko 2230 | | Melbourne University Mountaineering Club | *Mountaineering* (Melbourne UMC); *Thruch* (privately) |
| **Austria** | Gross Glockner 3798 | Österreichische Fremdenverkehrswerbung, Höhenstaufengasse 3–5, Wien 1 | Verband Alpiner Vereine Österreichs, Bäckerstrasse 16/11, 1010 Wien 1 Österreichischer Alpenverein, Wilhelm Greil Strasse 15, 6010 Innsbruck 1 Österreichischer Alpen Klub, 6 Getreidemarkt 3, Wien | *Österreichische Alpen Zeitung (OAK)* (6); *Der Bergsteiger (OAV)* (12); *Gebirgsfreund (OAV)* (6); *Mitteilungen des Österreichische Alpenvereins (OAV)* (6); *Österreichische Bergsteiger Zeitung* (12); *Alpenvereins Jahrbuch (OAV)* (1) |
| **Belgium** | Botrange 694 | Belgian National Tourist Office, rue Marché aux herbes 61–63, Brussels 1000 | Club Alpin Belge, 19 rue de l'Aurore, 1050 Bruxelles | *Revue d'Alpinisme* |
| **Bolivia** | Sajama 6542 | Dirección Nacional de Turismo, Central Island of Prado, La Paz | Club Andino Bolivano, Avenida 16 de Julio 1473, Casilla 1346, La Paz | |
| **Brazil** | Pde Neblina 3014 | Brazilian Tourist Information Centre, Barata Ribeiro 272, Copacabana | Federacion de Montanhismo do Estado do Rio de Janeiro, Av Almirante Barroso 2/8, 2000 Rio de Janeiro. | |

| Country | Highest peak or point | National Tourist Office | Principal mountaineering clubs and organizations | Principal mountaineering periodicals and their frequency |
|---|---|---|---|---|
| **Bulgaria** | Musala 2925 | Balkantourist, Pl Lenin 1, Sofia | Federation d'Alpinisme Bulgare, Bd Tolboukhine 18, Sofia 1 | |
| **Canada** | Mt Logan 6050 | Canadian Govt Office of Tourism, 240 Sparks St, Ottowa | Alpine Club of Canada, POB 1026, Banff, Alberta<br><br>Federation Quebecoise de la Montagne, 1415 East Jarry St, Montreal, Quebec H2E 2Z7 | *Canadian Alpine Journal* (AC of C) (1) |
| **Chile** | Ojos del Salado 6870 | Servicio Nacional de Turismo, Calle Catedral 1165, 3°Santiago | Federacion de Andinismo de Chile, Vicuna Mackenna 44, Casilla 2239, Santiago | *Annuario de Montaña* (F de A de C); *Rivista Andina* (AC of C) (1) |
| **China, People's Republic of** | Mt Everest 8848 | China Internat Travel Service (Luxingshe), Hsitan Bldg, Pekin | | |
| **Colombia** | P Cristobal Colon 5775 | Corporacion Nacional de Turismo, Calle 28, No 13-A-15, 16 Apdo Aéreo 8400, Bogotá | Club Los Yetis, Ibagué | *Campo Abierto* |
| **Czechoslovakia** | Gerlachovsky Stit 2655 | Čedok, Na Příkopě 18, 11135 Praha 1 | Ceskoslovenski Horolezecky Svaz, Na Porici 12, 11530 Praha 1<br><br>Section Tchécoslovaque d'Alpinisme (address as above) | |
| **Denmark** | Yding Skovhøj 173 | Danish Tourist Board, Vesterbrogade 6D, 1620 Copenhagen V | Dansk Bjergklub, c/o Jens Andersen, Ved Ermelunden 12, 2820 Gentofte | |
| **Ecuador** | Chimborazo 6267 | Dirección Nacional de Turismo, Reina Victoria 514 y Roca, Apdo 2454, Quito | Associacion de Excursionismo y Andinismo de Pichincha, Concentracion Deportiva, Apdo A 108, Quito | |
| **Egypt** | J Katarina 2609 | General Organization for Tourism and Hotels, 4 Latin America St, Garden City, Cairo | | |
| **France** | Mt Blanc 4807 | French Govt Tourist Office, 127 ave des Champs Elysées, 75008 Paris | Federation Française de la Montagne, 20 bis rue la Boétie, 75008 Paris<br><br>Club Alpin Français, 9 rue la Boétie, 75008 Paris | *La Montagne et Alpinisme* (CAF) (4); *Montagne et Sports* (L'ecole Nationale de Ski et d'Alpinisme, Chamonix) (4); *Alpinisme et Randonnée* (Commercial) (12); several sections of the CAF produce regular printed bulletins |

| | | | |
|---|---|---|---|
| **Germany (East)** | Fichtelberg 1214 | Reisebüro der Deutschen Demokratischen Republik, 102 Berlin, Alexanderplatz 5 | |
| **Germany (West)** | Zugspitze 2963 | Duetscher Alpenverein, Praterinsel 5, 8 München 22 | *Mitteilungen des Deutschen Alpen Vereins (DAV)* (6); *Alpinismus* (Commercial) (12); *Der Bergsteiger* (Commercial) (12) |
| | | Deutsche Zentrale für Tourismus e V., 6000 Frankfurt a M1, Beethovenstrasse 69 | |
| **Great Britain** | Ben Nevis 1343 | British Mountaineering Council, Crawford Hse, Precinct Centre, University, Booth St E, Manchester 13 | *The Alpine Journal* (AC) (1); *Alpine Climbing* (AC) (1); *Scottish Mountaineering Club Journal* (1); *Mountain* (Commercial) (6); *Climber and Rambler* (Commercial) (6); *Crags* (Commercial) (6) |
| | | British Tourist Authority, Queens House, 64 St James's St, London SW1 | |
| | | The Alpine Club, 74 South Audley St, London W1Y 5FF | |
| **Greece** | Olympus 2911 | Ellinikos Organismos Tourismou, Odos Amerikis 2, Athens | Federation Hellenique de Montagne et de Ski, Karageorgi Servias 7, Athens 126 |
| | | | Club Alpin Hellenique (address as above) |
| **Guatemala** | V Tajamulco 4220 | Instituto Guatemalteco de Turismo, 7a Avenida 1–13 Centro Cívico, Guatemala City | Federacion de Andinismo de Guatemala, Palacio de los Deportes, 2 nivel zona 4, Guatemala City |
| **Hungary** | Kékes 1015 | IBUSZ, H1053 Budapest, Apaczai Csere Janos, ull Hungarian National Council for Tourism, H1371 Budapest, Deák Ferenc u23, POB 422 | Magyar Hegymaszo Klub, Baçjcsy Zsilinszky ut 31.11, 1065 Budapest 6 |
| **Iceland** | Hvannadalshnukur 2119 | Icelandic Tourist Bureau, Reykjanesbraut 6, Reykjavik | Iceland Alpine Club, Post Box 4186, Reykjavik |
| **India** | Nanda Devi 7817 | Dept of Tourism of the Govt of India, Min of Tourism & Civil Aviation, 1 Parliament St, Transport Bhawan, New Delhi | Himalayan Club, PO Box 9049, Calcutta Himalayan Mountaineering Institute | *Himalayan Journal* (HC) (1) |
| **Indonesia** | Ngga Pulu 5030 | Dewan Pariwisata Indonesia, Jalan Diponegoro 25, Jakarta | Wisata Ria Remaja, Jalan Alhambra 63, Jakarta-Barat |
| **Iran** | Demavend 5604 | Min of Tourism & Information, 174 Elizabeth Blvd, Teheran | Iranian Mountaineering Federation, PO Box 11 1642 Teheran |

| Country | Highest peak or point | National Tourist Office | Principal mountaineering clubs and organizations | Principal mountaineering periodicals and their frequency |
|---|---|---|---|---|
| Ireland | Carrantouhill 1041 | Irish Tourist Board, Baggot St Bridge, Dublin 2 | Federation of Mountaineering Clubs of Ireland, Sorbonne 7, Ardilea Estate, Dublin 14 | *Irish Mountaineering* |
| Israel | Mt Atzmon 1208 | Min of Tourism, PO Box 1018, Jerusalem | Club Alpin Israelien, Blvd Rothschild 60, Tel-Aviv | |
| Italy | Mt Blanc de Courmayeur 4748 | ENIT, Via Marghera 2, Rome | Federazione Italiana Sport Invernali, Via Cerva 30, 20122 Milano<br><br>Club Alpino Italiano, Via Ugo Foscolo 3, 20121 Milano<br><br>Alpenverein Sudtirol, Sernesiplatz 34/1, 39100 Bolzano | *Revista Mensile del CAI* (CAI) (6); *Rasegna Alpina* (10); *Revista della Montagna* (Commercial) (4); several sections of the CAI produce regular printed bulletins |
| Japan | Fuji 3776 | Japan National Tourist Organization, Tokyo Kotsu Kaikan Bldg 2-10-1 Yuraku-cho, Tokyo | Japanese Alpine Club, 5-4 Yonbancho, Chiyoda-ku, Tokyo 102 | *Sangaku* (JAC) (1); *Yama to Keikoku* (12); *Iwa to Yuki*; *Gakujin* |
| Kenya | Mt Kenya 5199 | Min of Tourism & Wildlife, PO Box 30027, Nairobi | Mountain Club of Kenya, PO Box 5741, Nairobi | *Kenya Mountain Club Bulletin* (1) |
| Korea, North (Democratic Peoples' Republic) | Paektu San 2744 | Korean International Tourist Bureau, Ryohaengsa, Pyongkang | | |
| Korea, South (Republic) | Halla-san 1950 | Korean National Tourism Corporation, 198-1 Kwanhoondong, Chongno-ku, Seoul | Korean Alpine Federation, CPO Box 6528, Seoul or 29-1 4-KA Myung-Yun-Dong, Chong-Ro-Ku, Seoul<br><br>Korean Alpine Club, New Pagoda Bldg Rm 506, 39 2-Ga Jongro-Gu, Seoul | |
| Liechtenstein | Grauspitze 2599 | Liechtenstein National Tourist Office, Postfach 139, 9490 Vaduz | Liechtenstein Alpenverein, FL-9490, Vaduz | |
| Luxembourg | Bourgplatz 559 | Office National du Tourisme, 51 ave de la Gare, Luxembourg | Groupe Alpin Luxembourgeois, Place d'Armes 18, Boite Postale 363, Luxembourg | |
| Malaysia | Kinabalu 4101 | Tourist Development Corp of Malaysia, Min of Trade & Industry, PO Box 328, Kuala Lumpur | | |

| | | | | |
|---|---|---|---|---|
| **Mexico** | Orizaba (Citlaltépetl) 5750 | Consejo Nacional de Turismo, Mariano Escobeda 726, México 5 | Federacion Mexicana de Excursionismo, c/o Confederacion Deportiva Mexicana, Puerta 9 Ciudad Deportiva, Magdalena Mixhuca, Mexico 8 DF | |
| **Morocco** | Toubkal 4165 | Office National Morocain de Tourisme, BP 19, 22 ave d'Alger, Rabat | Federation Royale Marocaine d'Alpinisme et de Montagne, Maison des Sports, Parc Lyautey, Casablanca | |
| **Nepal** | Mt Everest 8848 | Dept of Tourism, HM Govt of Nepal, Ram Shah Path, Kathmandu<br><br>Dept of Information, Min of Communications, Ghantaghar, Kathmandu | Nepal Mountaineering Association, Sports Council Bldg, PO Box 1435, Kathmandu | |
| **The Netherlands** | Valserberg 321 | Stichting Nederlands National Bureau voor Toerisme, The Hague, Bezuidenhoutseweg 2, PO Box 90415 | Koninklijke Nederlandse Alpenvereinigung, Lange Voorhout 16, 's Gravenhage | *De Berggids* (KNAV) (6) |
| **New Zealand** | Mt Cook 3764 | New Zealand Tourist & Publicity Dept, PO Box 95, Wellington | New Zealand Alpine Club, PO Box 41–038, Eastbourne | *New Zealand Alpine Journal* (NZAC) (1); *Canterbury Mountaineer* (1); *Tararua* |
| **Norway** | Galdhoppigen 2469 | Norway Travel Association, H Heyerdahlsgate 1, Oslo 1<br><br>Norwegian Touring Federation (DNT) Den Norske Turistforening, Stortingsgaten 28, Oslo 1 | Norsk Tindeklub, Post Boks 1727, Vika, Oslo | |
| **Pakistan** | K2 8611 | Pakistan Tourism Development Corp, Hotel Metropole, Karachi 4 | Alpine Club of Pakistan, 228 Peshawar Rd, Rawalpindi | |
| **Peru** | Huascaran 6768 | Empresa Nacional de Enturperu-Turismo, Conde de Superunda 298, Lima | Club Andino Peruano, Las Begonias 630–11, San Isidro, Lima<br><br>Grupo Andino de Cordillera Blanca, Huarez Club Socorro Andino, c/o Concejo de la Victoria, Plaza Manco Capac, La Victoria, Lima<br><br>Club Andinista Cordillera Central, Calle Teofillo, Castillo 257, Lima | *Revista Peruana de Andinismo* |
| **The Philippines** | Mt Apo 2953 | Dept of Tourism, Agrifina Circle, Rizal Park, Manila, PO Box 3451 | | |

| Country | Highest peak or point | National Tourist Office | Principal mountaineering clubs and organizations | Principal mountaineering periodicals and their frequency |
|---|---|---|---|---|
| **Poland** | Rysy 2499 | Polskie Towarzystwo Turystyczno-Krajoznawcze, 00–075 Warsaw, Senatorska 11 | Polski Zwiazek Alpinizmu, Ul Sienkiewicza 12/439, 00–010 Warszawa Klub Wysokogorski (address as above) | *Taternik* (KW) (4); *Wierchy* (1) |
| **Portugal** | Pico (Azores) 2351; Malhao da Estrela (mainland) 1991 | Direcção Geral do Turismo, Av António Augusto de Aguiar 86, 3° Lisbon | Club Nacional de Montanhismo, Rua Formosa 303, Porto | *Companheiros* (Club de Campismo de Lisboa) (5) |
| **Romania** | Negoiu 2548 | National Tourist Office – Carpari, Bucharest 1, 7 Blvd Magheru | | |
| **South Africa** | Injasuti 3459 | S African Tourist Corp, 10th floor, Arcadia Centre, 130 Beatrix St, Private Bag X164, Pretoria | Mountain Club of South Africa, PO Box 164, Cape Town | *Journal of the Mountain Club of South Africa* (1) |
| **Spain** | Mt Teide (Canary Islands) 3716 | Min of Trade & Tourism, Avenida del Generalisimo 39, Madrid | Federacion Española de Montanismo, Alberto Aguilera 3pizo 4izq, Madrid 15<br><br>Club Alpino Español, Mayor 6, Madrid 13<br><br>Real Sociedad Española de Alpinismo Peñalara, Avda José Antonio 27, Madrid<br><br>Club Alpino Guadarrama, Avda José Antonio 11, Madrid | *Muntanya* (Cent Excursioista de Catalunya) (6); *Pyrenaica* (Federación Vasca de Montanismo) (1); *Peñalara* (RSE de AP) (4) |
| **Sweden** | Kebnekaise 2117 | Sveriges Turistrad, Hamngatan 27, PO Box 7306, 10385 Stockholm 7<br><br>Svenska Turistföreningen (STF), PO Box 7615, S-103-94, Stockholm | Svenska Fjallklubben, Villagatan 24, 11432 Stockholm<br><br>Svenska Klatterforbundet, PO Box 14036, 70014 Orebro | *Till Fjalls* (SF) (1) |
| **Switzerland** | Mte Rosa (Dufourspitze) 4634 | Swiss National Tourist Office, Talacker 42, 8023 Zurich | Schweizer Alpenklub, Geschäftsstelle SAC, Helvetiaplatz 4, 3005 Berne<br><br>Club Suisse des Femmes Alpinistes, Balderngasse 9, 8001 Zurich | *Die Alpen (Les Alpes)* (SAC) (4+12) |
| **Syria** | Mt Hermon 2814 | Min of Tourism, Abou Firas, El-Hamadani St, Damascus | | |
| **Taiwan (Formosa)** | Yu Shan (Mt Morrison) 3997 | Tourism Bureau, Min of Communications, 9th floor, 280 Chung Hsaio E Rd, Section 4, Taipai | | |

| | | | |
|---|---|---|---|
| **Tanzania** | Kilimanjaro 5894 | Tanzania Nat Tourist Board, Caxton House, Kenyatta Av, PO Box 87, Nairobi, Kenya | Kilimanjaro Mountain Club, PO Box 66, Moshi | |
| **Turkey** | Mt Ararat 5185 | Min of Tourism & Information, Gazi Mustafa Kemal Bulvari 33, Ankara | Turkiye Dagcilik Federasyonu, Beden Terbiyesi Genel Müdürlügü, Ulus Ishani A-Blok, Ankara | |
| **Uganda** | Mt Stanley 5109 | Uganda Tourist Assoc, PO Box 1542, Kampala | Mountain Club of Uganda, PO Box 2927, Kampala | |
| **United States** | Mt McKinley 6193 | US Travel Service, US Dept of Commerce, Washington, DC 20230 | The American Alpine Club, 113 East 90th St, New York, NY 10028 | *The American Alpine Journal* (AAC) (1); *Appalachia* (AMC) (2); *Appalachia Bulletin* (AMC) (6); *Sierra Club Bulletin* (SC) (6); all the following are commercial: *Off Belay* (6); *Climbing* (6); *Ascent* (1); *Backpacker* (6) |
| | | | Appalachian Mountaineering Club, Five Joy Street, Boston, Mass 02108 | |
| | | | Sierra Club, 530 Bush St, San Francisco, Cal 94108 | |
| **USSR** | P Kommunizma 7495 | Intourist, Moscow K9, Prospekt Karla Marxa 16 | Federation d'Alpinisme d'URSS, Skatertnyi pereoulok 4, Moscou 69 | |
| **Venezuela** | P Bolivar 5007 | Corporación de Turismo de Venezuela, Centro Capriles 7, Plaza Venezuela, Apdo 50200, Caracas | | |
| **Yugoslavia** | Triglav 2863 | Turistički Savez Jugoslavije, Mosě Pijade 8/IV, Poštanski fah 595, 11001 Belgrade | Planinarski Savez Jugoslavije, Dobrinjska 10/1, Beograd | *Planinski Vestnik* (12) |
| **International** | | | Union Internationale des Associations d'Alpinisme, 29 rue des Délices, 1211 Geneva 11, Switzerland | *Bulletin* |

# Abbreviations

| | | | | |
|---|---|---|---|---|
| m | metres | | | |
| km | kilometres | | | |
| GWh | Giga-watt hours | | | |
| km/hr | kilometres per hour | | | |
| °C | degree Celsius | | | |
| hr | hour | | | |
| min | minute | | | |

| | | | | |
|---|---|---|---|---|
| Aig | Aiguille | | Mtn | Mountain |
| Bk | Baku | | N | North |
| C | Col | | No | Nevado |
| Ca, Ce | Cima, Cime | | P | Pic, Pik, Piz |
| Co | Cerro | | Pk | Peak |
| c-wy | cableway | | Pt, Pte | Point, Pointe |
| E | East | | R | River |
| fa | first ascent | | Ra | Range |
| G | Gunung, Gebel | | ry | railway |
| Gr | Gross, Grosser | | r-ry | rack railway |
| int | international | | S | South |
| J | Jebel | | Sa | Sierra |
| le | last eruption | | V | Volcano |
| Mt | Mount, Mont | | W | West |
| Mte | Monte | | | |

Note: 'm' is omitted from the heights of features when it follows immediately after the name. Heights below sea level are denoted '−'

**Key to maps:**

- ●     towns, cities, etc.
- ▲     peaks
- ⌒     ridges
- --------     national/state borders
- ............     National Parks
- ⊔     pass/tunnel

# Index

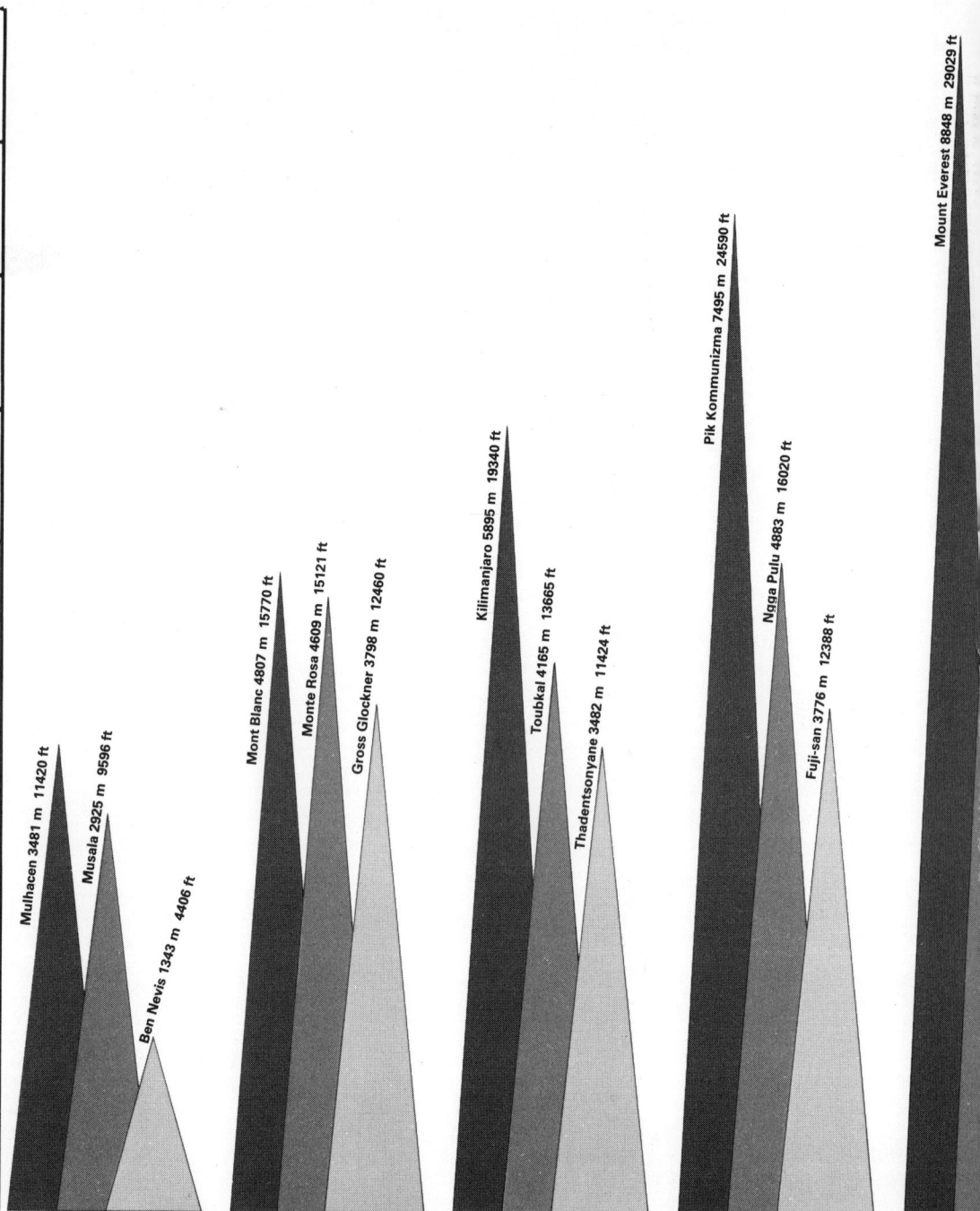

9,000

8,000

7,000

6,000

5,000

4,000

3,000

2,000

1,000

METRES

Mulhacen 3481 m  11420 ft

Musala 2925 m  9596 ft

Ben Nevis 1343 m  4406 ft

Mont Blanc 4807 m  15770 ft

Monte Rosa 4609 m  15121 ft

Gross Glockner 3798 m  12460 ft

Kilimanjaro 5895 m  19340 ft

Toubkal 4165 m  13665 ft

Thadentsonyane 3482 m  11424 ft

Pik Kommunizma 7495 m  24590 ft

Ngga Pulu 4883 m  16020 ft

Fuji-san 3776 m  12388 ft

Mount Everest 8848 m  29029 ft

**EUROPE**

**ALPS**

**AFRICA**

**ASIA**

**HIMALAYA**